Editorial Policy

Editorial Policy

§ 1. Lecture Notes aim to report new developments - quickly, informally, and at a high level. The texts should be reasonably self-contained and rounded off. Thus they may, and often will, present not only results of the author but also related work by other people. Furthermore, the manuscripts should provide sufficient motivation, examples and applications. This clearly distinguishes Lecture Notes manuscripts from journal articles which normally are very concise. Articles intended for a journal but too long to be accepted by most journals, usually do not have this "lecture notes" character. For similar reasons it is unusual for Ph. D. theses to be accepted for the Lecture Notes series.

§ 2. Manuscripts or plans for Lecture Notes volumes should be submitted (preferably in duplicate) either to one of the series editors or to Springer- Verlag, Heidelberg . These proposals are then refereed. A final decision concerning publication can only be made on the basis of the complete manuscript, but a preliminary decision can often be based on partial information: a fairly detailed outline describing the planned contents of each chapter, and an indication of the estimated length, a bibliography, and one or two sample chapters - or a first draft of the manuscript. The editors will try to make the preliminary decision as definite as they can on the basis of the available information.

§ 3. Final manuscripts should preferably be in English. They should contain at least 100 pages of scientific text and should include

- a table of contents;
- an informative introduction, perhaps with some historical remarks: it should be accessible to a reader not particularly familiar with the topic treated;
- a subject index: as a rule this is genuinely helpful for the reader.

Further remarks and relevant addresses at the back of this book.

Lecture Notes in Mathematics 900

Pierre Deligne James S. Milne
Arthur Ogus Kuang-yen Shih

Hodge Cycles, Motives, and Shimura Varieties

Springer-Verlag
Berlin Heidelberg New York
London Paris Tokyo
Hong Kong Barcelona
Budapest

Authors

Pierre Deligne
Institut des Hautes Etudes Scientifiques
91440 Bures-sur-Yvette, France

James S. Milne
Mathematics Department, University of Michigan
Ann Arbor, MI 48109, USA

Arthur Ogus
Mathematics Department, University of California
Berkeley, CA 94720, USA

Kuang-yen Shih
17628 Kornblum Ave.
Torrance, CA 90504, USA

AMS Subject Classifications (1980):

ISBN 3-540-11174-3 Springer-Verlag Berlin Heidelberg New York
ISBN 0-387-11174-3 Springer-Verlag New York Heidelberg Berlin

Printed in Germany

Printing and binding: Beltz Offsetdruck, Hemsbach/Bergstr.
2141/3140-543210

PREFACE

This volume collects six related articles.

The first is the notes (written by J.S. Milne) of a major part of the seminar "Périodes des Intégrales Abéliennes" given by P. Deligne at I.H.E.S., 1978-79.

The second article was written for this volume (by P. Deligne and J.S. Milne) and is largely based on: N Saavedra Rivano, Catégories tannakiennes, Lecture Notes in Math. 265, Springer, Heidelberg 1972.

The third article is a slight expansion of part of: J.S. Milne and Kuang-yen Shih, Shimura varieties: conjugates and the action of complex conjugation 154 pp. (Unpublished manuscript, October 1979).

The fourth article is based on a letter from P. Deligne to R. Langlands, dated 10th April, 1979, and was revised and completed (by Deligne) in July, 1981.

The fifth article is a slight revision of another section of the manuscript of Milne and Shih referred to above.

The sixth article, by A. Ogus, dates from July, 1980.

P. Deligne
J.S. Milne
A. Ogus
Kuang-yen Shih

CONTENTS

General Introduction

Let X be a smooth projective variety over $\mathbb{C}$. Hodge conjectured that certain cohomology classes on X are algebraic. The work of Deligne that is described in the first article of this volume shows that, when X is an abelian variety, the classes considered by Hodge have many of the properties of algebraic classes.

In more detail, let X^{an} be the complex analytic manifold associated with X, and consider the singular cohomology groups $H^n(X^{an},\mathbb{Q})$. The variety X^{an} being of Kähler type (any projective embedding defines a Kähler structure), its cohomology groups $H^n(X^{an},\mathbb{C}) = H^n(X^{an},\mathbb{Q}) \otimes \mathbb{C}$ have a canonical decomposition

$$H^n(X^{an},\mathbb{C}) = \bigoplus_{p+q=n} H^{p,q}, \quad H^{p,q} = H^q(X^{an}, \Omega^p_{X^{an}}).$$

The cohomology class $c\ell(Z) \in H^{2p}(X^{an},\mathbb{C})$ of an algebraic subvariety Z of codimension p in X is rational (it lies in $H^{2p}(X^{an},\mathbb{Q})$) and of bidegree (p,p) (it lies in $H^{p,p}$). The Hodge conjecture states that, conversely, any element of $H^{2p}(X,\mathbb{Q}) \cap H^{p,p}$ is a linear combination over $\mathbb{Q}$ of the classes of algebraic subvarieties. Since the conjecture is unproven, it is convenient to call these rational (p,p)-classes <u>Hodge cycles</u> on X.

Now consider a smooth projective variety X over a field k that is of characteristic zero, algebraically closed, and small enough to be embeddable in $\mathbb{C}$. The algebraic de Rham cohomology groups $H^n_{DR}(X/k)$ have the property that, for any embedding $\sigma\colon k \hookrightarrow \mathbb{C}$, there is a canonical isomorphism $H^n_{DR}(X/k) \otimes_{k,\sigma} \mathbb{C} \xrightarrow{\approx} H^n_{DR}(X^{an}) = H^n(X^{an},\mathbb{C})$. It is natural to say that $t \in H^{2p}_{DR}(X/k)$ is a <u>Hodge cycle on X relative to σ</u> if its image in $H^{2p}(X^{an},\mathbb{C})$ is $(2\pi i)^p$ times a Hodge cycle on $X \otimes_{k,\sigma} \mathbb{C}$. Deligne's results show that, if X is an abelian variety, then an element of $H^{2p}_{DR}(X/k)$ that is a Hodge cycle on X relative to one embedding of k in $\mathbb{C}$ is a Hodge cycle relative to all embeddings; further, for any embedding, $(2\pi i)^p$ times a Hodge cycle in $H^{2p}(X^{an},\mathbb{C})$ always lies in the image of $H^{2p}_{DR}(X/k)$. Thus the notion of a Hodge cycle on an abelian variety is intrinsic to the variety: it is a purely algebraic notion. In the case that $k = \mathbb{C}$ the theorem shows that the image of a Hodge cycle under an automorphism of $\mathbb{C}$ is again a Hodge cycle; equivalently, the notion of a Hodge cycle on an abelian variety X over $\mathbb{C}$ does not depend on the map $X \to \operatorname{spec} \mathbb{C}$. Of course, all of this would be obvious if only one knew the Hodge conjecture.

In fact, in the first article a stronger result is proved in which a Hodge cycle is defined to be an element of $H^n_{DR}(X) \times \prod_{\ell} H^n(X_{et},\mathbb{Q}_\ell)$. As the title of the original seminar suggests, the stronger result has consequences for the algebraicity of the periods of abelian integrals: briefly, Deligne's result allows one to prove all arithmetic properties of abelian periods that would follow from knowing the Hodge conjecture for abelian varieties.

The second article is mainly expository. Since Tannakian categories are used in several articles, we thought it useful to include an account of the essential features of the theory. The exposition largely follows that of Saavedra [1] except at three places: in §3 we point out an error in Saavedra's results concerning a non-neutral Tannakian category; in §4 we eliminate an unnecessary connectedness assumption in the theory of polarized Tannakian categories; and in §6 we discuss motives relative to absolute Hodge cycles rather than algebraic cycles.

A neutralized Tannakian category is a k-linear category $\underline{C}$ with an operation $\otimes: \underline{C} \times \underline{C} \to \underline{C}$ and a functor to finite-dimensional vector spaces over k satisfying certain conditions sufficient to ensure that $\underline{C}$ is equivalent to the category of finite-dimensional representations of an affine group scheme G over k. The importance of this notion is that properties of an abstract category $\underline{C}$ will be faithfully reflected in properties of the associated group G.

The category of polarizable $\mathbb{Q}$-rational Hodge structures is Tannakian. The group associated to its Tannakian subcategory of those Hodge structures of CM-type is called the connected Serre group S^{o}.

It would follow from Grothendieck's standard conjectures that the category of motives, arising from the category of projective smooth varieties over a field k, is Tannakian. If, in the definition of motive, "algebraic cycle" is replaced by "Hodge cycle" and the initial category is taken to consist of abelian varieties over a field k of characteristic zero, then the resulting category of Hodge motives is Tannakian. (Of course, for this to make sense, one needs Deligne's Theorem.) If the initial category is taken to be all abelian varieties over $\mathbb{C}$ of CM-type , then the group associated with the category of motives is again S^o ; if the initial category is taken to be all abelian varieties over $\mathbb{Q}$ that become of CM-type over $\mathbb{C}$, then the associated group is called the Serre group S . The identity component of S is S^o , and S is an extension of $\mathrm{Gal}(\bar{\mathbb{Q}}/\mathbb{Q})$ by S^o . There is a canonical continuous splitting $S(\mathbb{A}^f) \xleftarrow{sp} \mathrm{Gal}(\bar{\mathbb{Q}}/\mathbb{Q})$ over the ring of finite adèles $\mathbb{A}^f$.

The third article is again largely expository: it describes Langlands's construction of his Taniyama group. Langlands's study of the zeta functions of Shimura varieties led him to make a conjecture concerning the conjugates of a Shimura variety (Langlands [1, p 417]). In the belief that this conjecture was too imprecise to be proved by the methods usually applied to Shimura varieties, he then made a second, stronger conjecture (Langlands [2, p 232-33]). This second conjecture is stated in terms of the Taniyama group T which, like the Serre group, is an extension of $\mathrm{Gal}(\bar{\mathbb{Q}}/\mathbb{Q})$ by S^o together with a continuous splitting over $\mathbb{A}^f$. In the following two articles, this conjecture is proved for most Shimura varieties, viz. for those of abelian type (see article V.1 for a definition of this term).

In the first of the two articles, Deligne proves that the Serre group, together with its structure as an extension and its adèlic splitting, is isomorphic to the Taniyama group. This is shown by a uniqueness argument: any two extensions of $Gal(\bar{\mathbb{Q}}/\mathbb{Q})$ by S^o with adèlic splittings are isomorphic provided they possess a certain list of properties. The Taniyama group has these properties by construction while for the Serre group they follow from properties of abelian varieties of potential CM-type (for example, the theorem of Shimura and Taniyama). The significance of this result is that, while the Serre group summarizes a great deal of information about abelian varieties of potential CM-type , the definition of the Taniyama group does not mention them. In particular, the result gives a description (in terms of the Taniyama group) of the action of $Gal(\bar{\mathbb{Q}}/\mathbb{Q})$ on an abelian variety of potential CM-type over $\mathbb{Q}$ and its points of finite order.

In the fifth article it is shown that this last result (called conjecture CM in the article) is equivalent to Langlands's conjecture for the Shimura varieties associated with symplectic groups. Then it is shown that arguments involving symplectic embeddings and connected Shimura varieties suffice to prove the conjecture for all Shimura varieties of abelian type. (We note that when this article was written, Deligne had shown only that the Serre and Taniyama groups were isomorphic as extensions, i.e., not necessarily by an isomorphism preserving the adèlic splittings; thus some parts of the article are now redundant. Also that it appears likely that the methods Kazhdan and Borovoi use to show the existence of canonical models can be combined with the methods of the article to give a proof of Langlands's conjecture for all Shimura varieties.)

The final article returns to questions suggested by the first article. An algebraic cycle on a variety X over a number field K defines a class in the crystalline cohomology of any smooth reduction of X , and this class is acted on in a particularly simple way by the Frobenius endomorphism. Does an absolute Hodge cycle have the same property? If a class $\gamma \in H_{DR}(X/K)$ behaves as an algebraic cycle when embedded in the crystalline cohomology of any smooth reduction of X , is it necessarily absolutely Hodge? Such questions are discussed in the fourth section of the sixth article. The remaining sections are concerned with the following topics: giving a geometric interpretation of the second spectral sequence of de Rham hypercohomology in positive characteristics, relating the filtration on de Rham cohomology in characteristic zero to the filtrations mod p (almost all p) , and the definition of a new invariant ± 1 , the crystalline discriminant, attached to a variety in characteristic p .

For more detailed descriptionsof the contents of the articles, we refer the reader to the individual introductions.

Langlands, R.

[1] Some contemporary problems with origins in the Jugendtraum, Proc. Symp. Pure Math., A.M.S. 28(1976) 401-418.

[2] Automorphic representations, Shimura varieties, and motives. Ein Märchen, Proc. Symp. Pure Math., A.M.S., 33(1979) part 2, 205-246.

Saavedra Rivano, N.

[1] Catégories Tannakiennes, Lecture Notes in Math, 265, Springer, Heidelberg, 1972.

Notations and Conventions.

The ring of finite adèles, $(\varprojlim \mathbb{Z}/m\mathbb{Z}) \otimes \mathbb{Q}$, of $\mathbb{Q}$ is denoted by $\mathbb{A}^f$, and $\mathbb{A}$ denotes the full ring of adèles $\mathbb{R} \times \mathbb{A}^f$. For E a number field, $\mathbb{A}_E^f$ and $\mathbb{A}_E$ denote $E \otimes_{\mathbb{Q}} \mathbb{A}^f$ and $E \otimes \mathbb{A}$. The group of idèles of E is $\mathbb{A}_E^{\times}$, and the idèle class group is $C_E = \mathbb{A}_E^{\times}/E^{\times}$.

The reciprocity isomorphism of class field theory is normalized so that a uniformizing parameter corresponds to the reciprocal of the arithmetic Frobenius element.

Complex conjugation is denoted by ι in articles I, II, III, and V, and by c in IV (it is, of course, also denoted by $z \mapsto \bar{z}$).

The groups referred to in articles II and IV as the connected Serre group S^o and the Serre group S are referred to in articles III and V as the Serre group S and the motive Galois group M respectively.

Item x.y of article V is referred to as (V.x.y).

I HODGE CYCLES ON ABELIAN VARIETIES

P. Deligne (Notes by J.S. Milne)

Introduction

The main result proved in these notes is that any Hodge cycle on an abelian variety (in characteristic zero) is an absolute Hodge cycle -- see §2 for definitions and (2.11) for a precise statement of the result.

The proof is based on the following two principles.

A. Let $t_1,\ldots,t_N$ be absolute Hodge cycles on a projective smooth variety X and let G be the largest algebraic subgroup of $GL(H^*(X,\mathbb{Q})) \times GL(\mathbb{Q}(1))$ fixing the t_i ; any t on X fixed by G is an absolute Hodge cycle (see 3.8).

B. If $(X_s)_{s\in S}$ is an algebraic family of projective smooth varieties with S connected, and t_s is a family of rational cycles (i.e. a global section of ...) such that t_s is an absolute Hodge cycle for one s , then t_s is an absolute Hodge cycle for all s (see 2.12, 2.15).

Using B and the families of abelian varieties parametrized by Shimura varieties, one shows that it suffices to prove the main result for A an abelian variety of CM-type (see §6). Fix a CM-field E, which we can assume to be Galois over $\mathbb{Q}$, and let (A_α) be the family of all abelian varieties, up to E-isogeny, over $\mathbb{C}$ with complex multiplication by E. Principle B is used to construct some absolute Hodge cycles on varieties of the form $\bigoplus_{\alpha=1}^{d} A_\alpha$ -- the principle allows us to replace $\bigoplus A_\alpha$ by an abelian variety of the form $A_0 \otimes_{\mathbb{Q}} E$ (see §4). Let $G \subset GL(\bigoplus H_1(A_\alpha,\mathbb{Q})) \times GL(\mathbb{Q}(1))$ be the subgroup fixing the absolute Hodge cycles just constructed plus some other (obvious) absolute Hodge cycles. It is shown that G fixes every Hodge cycle on an A_α, and Principle A therefore completes the proof (see §5).

On analyzing which properties of absolute Hodge cycles are used in the above proof, one arrives at a slightly stronger result. Call a rational cohomology class c on a projective smooth variety X <u>accessible</u> if it belongs to the smallest family of rational cohomology classes such that:

(a) the cohomology class of any algebraic cycle is accessible;

(b) the pull-back by a map of varieties of an accessible class is accessible;

(c) if $t_1,\ldots,t_N \in H^*(X,\mathbb{Q})$ are accessible, and if a rational class t in some $H^{2p}(X,\mathbb{Q})$ is fixed by the algebraic subgroup G of $\mathrm{Aut}(H^*(X,\mathbb{Q}))$ (automorphisms of $H^*(X,\mathbb{Q})$ as a graded algebra) fixing the t_i, then t is accessible;

(d) Principle B , with "absolute Hodge" replaced by accessible, holds.

Sections 4, 5, 6 of these notes can be interpreted as proving that, when X is an abelian variety, any Hodge cycle (i.e., rational (p,p)-cycle) in $H^{2p}(X,\mathbb{Q})$ is accessible. Sections 2,3 define the notion of an absolute Hodge cycle and show that the family of absolute Hodge cycles satisfies (a), (b), (c), and (d) ; therefore an accessible class is absolutely Hodge. We have the implications:

$$\text{Hodge} \xRightarrow{\text{ab. var.}} \text{accessible} \Rightarrow \text{absolute Hodge} \Rightarrow \text{Hodge} .$$

Only the first implication is restricted to abelian varieties.

The remaining two sections, §1 and §7 , serve respectively to review the different cohomology theories and to give some applications of the main result to the algebraicity of certain products of special values of the Γ-function.

Notations: All algebraic varieties are complete and smooth over fields of characteristic zero unless stated otherwise. (The reader will lose little if he takes all varieties to be projective.) $\mathbb{C}$ denotes an algebraic closure of $\mathbb{R}$ and $i \in \mathbb{C}$ a square root of -1 ; thus i is defined only up to sign. A choice of i determines an orientation of $\mathbb{C}$ as a real manifold we take that for which $1 \wedge i > 0$ -- and hence an orientation of any complex manifold. Complex conjugation on $\mathbb{C}$ is denoted by ι or by $z \mapsto \bar{z}$. Recall that the category of abelian varieties up to isogeny is obtained from the category of abelian varieties by taking the same class of objects but replacing

$\mathrm{Hom}(A,B)$ by $\mathrm{Hom}(A,B) \otimes \mathbb{Q}$. We shall always regard an abelian variety as an object in the category of abelian varieties up to isogeny: thus $\mathrm{Hom}(A,B)$ is a vector space over $\mathbb{Q}$.

If (V_α) is a family of rational representations of an algebraic group G over k and $t_{\alpha,\beta} \in V_\alpha$, then the <u>subgroup of G fixing the</u> $t_{\alpha,\beta}$ is the algebraic subgroup H of G such that, for all k-algebras R, $H(R) = \{g \in G(R) \mid g(t_{\alpha,\beta}\otimes 1) = t_{\alpha,\beta}\otimes 1$, all $\alpha,\beta\}$. Linear duals are denoted by a superscript $\vee$. If X is a variety over a field k and σ is an embedding $\sigma\colon k \hookrightarrow k'$, then σX denotes $X \otimes_{k,\sigma} k'$ $(= X \times_{\mathrm{spec}(k)} \mathrm{spec}(k'))$.

1. <u>Review of cohomology</u>

Let X be a topological manifold and F a sheaf of abelian groups on X. We define

$$H^n(X,F) = H^n(\Gamma(X,F^\bullet))$$

where $F \to F^\bullet$ is any acyclic resolution of F; thus $H^n(X,F)$ is uniquely defined, up to a unique isomorphism.

When F is the constant sheaf defined by a field K, these groups can be identified with singular cohomology groups as follows. Let $S_\bullet(X,K)$ be the complex in which $S_n(X,K)$ is the vector space over K with basis the singular n-simplices in X and the boundary map sends a simplex to the (usual) alternating sum of its faces. Set $S^\bullet(X,K) = \mathrm{Hom}(S_\bullet(X,K),K)$ with the boundary map for which

$$(\alpha,\sigma) \longmapsto \alpha(\sigma): \quad S^{\cdot}(X,K) \otimes S_{\cdot}(X,K) \to K$$

is a morphism of complexes, namely that defined by $(d\alpha)(\sigma) = (-1)^{\deg(\alpha)+1}\alpha(d\sigma)$.

Proposition 1.1. There is a canonical isomorphism $H^n(S^{\cdot}(X,K)) \xrightarrow{\approx} H^n(X,K)$.

Proof: If U is a unit ball, then $H^0(S^{\cdot}(U,K)) = K$ and $H^n(S^{\cdot}(U,K)) = 0$ for $n > 0$. Thus $K \to S^{\cdot}(U,K)$ is a resolution of the group K . Let $\underline{S}^n$ be the sheaf on X associated with the presheaf $V \mapsto S^n(V,K)$. The last remark shows that $K \to \underline{S}^{\cdot}$ is a resolution of the sheaf K . As each $\underline{S}^n$ is fine (Warner [1,5.32]), $H^n(X,K) = H^n(\Gamma(X,\underline{S}^{\cdot}))$. But the obvious map $S^{\cdot}(X,K) \to \Gamma(X,\underline{S}^{\cdot})$ is surjective with an exact complex as kernel (loc. cit.), and so

$$H^n(S^{\cdot}(X,K)) \xrightarrow{\approx} H^n(\Gamma(X,\underline{S}^{\cdot})) = H^n(X,K) .$$

Now assume X is a differentiable manifold. On replacing "singular n-simplex" by "differentiable singular n-simplex" in the above definitions, one obtains complexes $S^{\cdot}_{\infty}(X,K)$ and $S_{\cdot}^{\infty}(X,K)$. The same argument shows there is a canonical isomorphism $H^n_{\infty}(X,K) \overset{df}{=} H^n(S^{\cdot}_{\infty}(X,K)) \xrightarrow{\approx} H^n(X,K)$ (loc. cit.).

Let O_{X^∞} be the sheaf of C^∞ real-valued functions on X , $\Omega^n_{X^\infty}$ the O_{X^∞}-module of C^∞ differential n-forms on X , and $\Omega^{\cdot}_{X^\infty}$ the complex

$$O_{X^\infty} \xrightarrow{d} \Omega^1_{X^\infty} \xrightarrow{d} \Omega^2_{X^\infty} \xrightarrow{d} \cdots$$

The de Rham cohomology groups of X are defined to be

$$H^n_{DR}(X) = H^n(\Gamma(X,\Omega^{\cdot}_{X^\infty})) = \{\text{closed n-forms}\} / \{\text{exact n-forms}\} .$$

If U is the unit ball, Poincaré's lemma shows that $H^0_{DR}(U) = \mathbb{R}$ and $H^n_{DR}(U) = 0$ for $n > 0$. Thus $\mathbb{R} \to \Omega^{\cdot}_{X^\infty}$ is a resolution of the constant sheaf $\mathbb{R}$, and as the sheaves $\Omega^n_{X^\infty}$ are fine (Warner [1,5.28]), we have $H^n(X,\mathbb{R}) = H^n_{DR}(X)$.

For $\omega \in \Gamma(X,\Omega^n_{X^\infty})$ and $\sigma \in S^\infty_n(X,\mathbb{R})$, define $<\omega,\sigma> = (-1)^{\frac{n(n+1)}{2}} \int_\sigma \omega \in \mathbb{R}$. Stokes's theorem states that $\int_\sigma d\omega = \int_{d\sigma} \omega$, and so $<d\omega,\sigma> + (-1)^n<\omega,d\sigma> = 0$. The pairing $<\ ,\ >$ therefore defines a map of complexes $f\colon\ \Gamma(X,\Omega^{\cdot}_{X^\infty}) \to S^{\cdot}_\infty(X,\mathbb{R})$.

<u>Theorem</u> 1.2 (de Rham): The map $H^n_{DR}(X) \to H^n_\infty(X,\mathbb{R})$ defined by f is an isomorphism for all n.

<u>Proof</u>: The map is inverse to the map $H^n_\infty(X,\mathbb{R}) \xrightarrow{\approx} H^n(X,\mathbb{R}) = H^n_{DR}(X)$ defined in the previous two paragraphs (Warner [1,5.36]). (Our signs differ from the usual because the standard sign conventions $\int_\sigma d\omega = \int_{d\sigma}\omega$, $\int_{X\times Y} pr_1^*\omega \wedge pr_2^*\eta = \int_X \omega \int_Y \eta$ etc. violate the standard sign conventions for complexes.)

A number $\int_\sigma \omega$, $\sigma \in H_n(X,\mathbb{Q})$, is called a <u>period</u> of ω. The map in (1.2) identifies $H^n(X,\mathbb{Q})$ with the space of classes of closed forms whose periods are all rational. Theorem 1.2 can be restated as follows: a closed differential form is exact if all its periods are zero; there exists a closed differential form having arbitrarily assigned periods on an independent set of cycles.

<u>Remark</u> 1.3 (Singer-Thorpe [1,6.2]). If X is compact then it has a smooth triangulation T. Define $S.(X,T,K)$ and $S^{\cdot}(X,T,K)$ as before, but using only simplices in T. Then the map $\Gamma(X,\Omega^{\cdot}_{X^{\infty}}) \to S^{\cdot}(X,T,K)$, defined by the same formulas as f above, induces isomorphisms $H^n_{DR}(X) \xrightarrow{\approx} H^n(S^{\cdot}(X,T,K))$.

Next assume that X is a complex manifold, and write $\Omega^{\cdot}_{X^{an}}$ for the complex

$$O_{X^{an}} \xrightarrow{d} \Omega^1_{X^{an}} \xrightarrow{d} \Omega^2_{X^{an}} \xrightarrow{d} \cdots$$

in which $\Omega^n_{X^{an}}$ is the sheaf of holomorphic differential n-forms. (Thus locally a section of $\Omega^n_{X^{an}}$ is of the form $\omega = \sum \alpha_{i_1 \ldots i_n} dz_{i_1} \wedge \cdots \wedge dz_{i_n}$ with $\alpha_{i_1 \ldots i_n}$ a holomorphic function and the z_i local coordinates.) The complex form of Poincaré's lemma shows that $\mathbb{C} \to \Omega^{\cdot}_{X^{an}}$ is a resolution of the constant sheaf $\mathbb{C}$, and so there is a canonical isomorphism $H^n(X,\mathbb{C}) \xrightarrow{\approx} \mathbb{H}^n(X,\Omega^{\cdot}_{X^{an}})$ (hypercohomology).

If X is a compact Kähler manifold, the spectral sequence

$$E_1^{p,q} = H^q(\Omega^p_{X^{an}}) \Longrightarrow \mathbb{H}^{p+q}(\Omega^{\cdot}_{X^{an}})$$

degenerates, and so provides a canonical splitting $H^n(X,\mathbb{C}) = \bigoplus_{p+q=n} H^q(X,\Omega^p_{X^{an}})$ (the Hodge decomposition); moreover $H^{p,q} \overset{df}{=\!=} H^q(X,\Omega^p_{X^{an}})$ is the complex conjugate of $H^{q,p}$ relative to the real structure $H^n(X,\mathbb{R}) \otimes \mathbb{C} \xrightarrow{\approx} H^n(X,\mathbb{C})$ (Weil [2]).

The decomposition has the following explicit description: the complex $\Omega^{\cdot}_{X^{\infty}} \otimes \mathbb{C}$ of sheaves of complex-valued differential forms on the underlying differentiable manifold is an acyclic resolution of $\mathbb{C}$, and so $H^n(X,\mathbb{C}) = H^n(\Gamma(X,\Omega^{\cdot}_{X^{\infty}} \otimes \mathbb{C}))$; Hodge theory shows that each element of the second group is represented by a unique harmonic n-form, and the decomposition corresponds to the decomposition of harmonic n-forms into sums of harmonic (p,q)-forms, $p + q = n$.

Finally, let X be an algebraic variety over a field k . If $k = \mathbb{C}$ then X defines a compact complex manifold X^{an} , and there are therefore groups $H^n(X^{an},\mathbb{Q})$, depending on the map $X \to \mathrm{spec}(\mathbb{C})$, that we shall write $H^n_B(X)$ (here B abbreviates Betti). There exist canonical Hodge decompositions: $H^n_B(X) = \bigoplus_{p+q=n} H^{p,q}(X)$, $\overline{H^{p,q}} = H^{q,p}$. If X is projective, then the choice of a projective embedding determines a Kähler structure on X^{an} , and hence a Hodge decomposition (which is independent of the choice of the embedding because it is determined by the Hodge filtration, and the Hodge filtration depends only on X ; see 1.4). In the general case we refer to Deligne [1,5.3,5.5] for the existence of the decompositions.

For an arbitrary k and an embedding $\sigma: k \hookrightarrow \mathbb{C}$ we write $H^n_\sigma(X)$ for $H^n_B(\sigma X)$ and $H^{p,q}_\sigma(X)$ for $H^{p,q}(\sigma X)$. As ι defines a homeomorphism $\sigma X^{an} \to \iota\sigma X^{an}$, it induces an isomorphism $H^n_{\iota\sigma}(X) \xrightarrow{\approx} H^n_\sigma(X)$.

Let $\Omega^{\cdot}_{X/k}$ be the complex in which $\Omega^n_{X/k}$ is the sheaf of algebraic differential n-forms, and define the (algebraic) de Rham cohomology group $H^n_{DR}(X/k)$ to be $\mathbb{H}^n(X_{Zar},\Omega^\cdot_{X/k})$ (hypercohomology relative to the Zariski topology). For any map $\sigma: k \hookrightarrow k'$ there is a canonical isomorphism $H^n_{DR}(X/k) \otimes_{k,\sigma} k' \xrightarrow{\approx} H^n_{DR}(X \otimes_k k'/k')$. The spectral sequence

$$E_1^{p,q} = H^q(X_{Zar},\Omega^p_{X/k}) \Longrightarrow \mathbb{H}^{p+q}(X_{Zar},\Omega^\cdot_{X/k})$$

defines a filtration (the Hodge filtration) $F^pH^n_{DR}(X)$ on $H^n_{DR}(X)$ which is stable under base change.

Theorem 1.4. If $k = \mathbb{C}$ the obvious maps $X^{an} \to X_{Zar}, \Omega^\cdot_{X^{an}} \leftarrow \Omega^\cdot_X$, induce an isomorphism $H^n_{DR}(X) \xrightarrow{\approx} H^n_{DR}(X^{an}) = H^n(X^{an},\mathbb{C})$ under which $F^pH^n_{DR}(X)$ corresponds to $F^pH^n(X^{an},\mathbb{C}) \stackrel{df}{=} \bigoplus_{p' \geq p} H^{p',q'}$.

Proof: The initial terms of the spectral sequences

$$E_1^{p,q} = H^q(X_{Zar},\Omega^p_X) \Longrightarrow H^{p+q}_{DR}(X)$$

$$E_1^{p,q} = H^q(X^{an},\Omega^p_{X^{an}}) \Longrightarrow H^{p+q}_{DR}(X^{an})$$

are isomorphic. (See Serre [1] for the projective case, and Grothendieck [2] for the general case.) The theorem follows from this because, by definition of the Hodge decomposition, the filtration of $H^n_{DR}(X^{an})$ defined by the above spectral sequence is equal to the filtration of $H^n(X^{an},\mathbb{C})$ defined in the statement of the theorem.

It follows from the theorem and the discussion preceding it that any embedding $\sigma: k \hookrightarrow \mathbb{C}$ defines an isomorphism $H^n_{DR}(X) \otimes_{k,\sigma} \mathbb{C} \xrightarrow{\approx} H^n_\sigma(X) \otimes_{\mathbb{Q}} \mathbb{C}$ and, in particular, a k-structure on

$H^n_\sigma(X) \otimes_{\mathbb{Q}} \mathbb{C}$. When $k = \mathbb{Q}$, this structure should be distinguished from the $\mathbb{Q}$-structure defined by $H^n_\sigma(X)$: the two are related by the periods (see below).

When k is algebraically closed we write $H^n(X, \mathbb{A}^f)$, or $H^n_{et}(X)$, for $H^n(X_{et},\hat{\mathbb{Z}}) \otimes_{\mathbb{Z}} \mathbb{Q}$, where $H^n(X_{et},\hat{\mathbb{Z}}) = \varprojlim_m H^n(X_{et},\mathbb{Z}/m\mathbb{Z})$ (étale cohomology). If X is connected, $H^0(X,\mathbb{A}^f) = \mathbb{A}^f$, the ring of finite adèles for $\mathbb{Q}$, which justifies the first notation. By definition, $H^n_{et}(X)$ depends only on X (and not on the map $X \to \operatorname{spec} k$). The map $H^n_{et}(X) \to H^n_{et}(X \otimes k')$ defined by an inclusion of algebraically closed fields $k \hookrightarrow k'$ is an isomorphism (special case of the proper base change theorem, Artin et.al.[1, XII]). The comparison theorem (ibid. XI) shows that, when $k = \mathbb{C}$, there is a canonical isomorphism $H^n_B(X) \otimes \mathbb{A}^f \xrightarrow{\approx} H^n_{et}(X)$. It follows that $H^n_B(X) \otimes \mathbb{A}^f$ is independent of the map $X \to \operatorname{spec} \mathbb{C}$, and that, over any (algebraically closed) field, $H^n_{et}(X)$ is a free $\mathbb{A}^f$-module.

$H^n(X,\mathbb{A}^f)$ can also be described as the restricted product of the spaces $H^n(X,\mathbb{Q}_\ell)$, ℓ prime, with respect to the subspaces $H^n(X,\mathbb{Z}_\ell)$.

Next we define the notion of a "Tate twist" in each of our three cohomology theories. For this we shall define objects $\mathbb{Q}(1)$, and set $H^n(X)(m) = H^n(X) \otimes \mathbb{Q}(1)^{\otimes m}$. We want $\mathbb{Q}(1)$ to be $H_2(\mathbb{P}^1)$ (realization of the Tate motive in the cohomology theory) but to avoid the possibility of introducing sign ambiguities, we shall define it directly:

$\mathbb{Q}_B(1) = 2\pi i\,\mathbb{Q}$

$\mathbb{Q}_{et}(1) = \mathbb{A}^f(1) \overset{df}{=\!=} (\varprojlim_m \mu_m(k)) \otimes_{\mathbb{Z}} \mathbb{Q}, \mu_m(k) = \{\zeta \in k \mid \zeta^m = 1\}$

$\mathbb{Q}_{DR}(1) = k$,

and so

$H^n_B(X)(m) = H^n_B(X) \otimes_{\mathbb{Q}} (2\pi i)^m\mathbb{Q} = H^n(X^{an}, (2\pi i)^m\mathbb{Q}) \qquad (k = \mathbb{C})$

$H^n_{et}(X)(m) = H^n_{et}(X) \otimes_{\mathbb{A}^f} (\mathbb{A}^f(1))^{\otimes m} = (\varprojlim_r H^n(X_{et}, \mu_r(k)^{\otimes m})) \otimes_{\mathbb{Z}} \mathbb{Q}$

(k alg. cl.)

$H^n_{DR}(X)(m) = H^n_{DR}(X)$.

These definitions extend in an obvious way to negative m ; for example we set $\mathbb{Q}_{et}(-1) = \mathrm{Hom}_{\mathbb{A}^f}(\mathbb{A}^f(1), \mathbb{A}^f)$ and define $H^n_{et}(X)(-m) = H^n_{et}(X) \otimes \mathbb{Q}_{et}(-1)^{\otimes m}$.

There are canonical isomorphisms

$\mathbb{Q}_B(1) \otimes_{\mathbb{Q}} \mathbb{A}^f \xrightarrow{\approx} \mathbb{Q}_{et}(1) \qquad (k \text{ alg. cl.}, k \subset \mathbb{C})$

$\mathbb{Q}_B(1) \otimes_{\mathbb{Q}} \mathbb{C} \xrightarrow{\approx} \mathbb{Q}_{DR}(1) \otimes_k \mathbb{C} \qquad (k \subset \mathbb{C})$

and hence canonical isomorphisms (the <u>comparison</u> isomorphisms)

$H^n_B(X)(m) \otimes \mathbb{A}^f \xrightarrow{\approx} H^n_{et}(X)(m) \qquad (k \text{ alg. cl.}, k \subset \mathbb{C})$

$H^n_B(X)(m) \otimes \mathbb{C} \xrightarrow{\approx} H^n_{DR}(X)(m) \otimes_k \mathbb{C} \qquad (k \subset \mathbb{C})$.

To define the first, we note that exp defines an isomorphism $2\pi i\,\mathbb{Z}/m2\pi i\,\mathbb{Z} \xrightarrow{\approx} \mu_m(k)$; after passing to the limit over m, and tensoring with $\mathbb{Q}$, we obtain the required isomorphism $2\pi i\,\mathbb{A}^f \xrightarrow{\approx} \mathbb{A}^f(1)$. The second isomorphism is induced by the inclusions $2\pi i\,\mathbb{Q} \hookrightarrow \mathbb{C} \hookrightarrow k$. Although the Tate twist for de Rham cohomology is trivial, it should not be ignored. For example,

$$\begin{array}{ccccl} H^n_B(X) \otimes \mathbb{C} & \xrightarrow{\approx} & H^n_B(X)(m) \otimes \mathbb{C} & & (1 \mapsto (2\pi i)^m;\ \text{defined up to sign}) \\ \approx \downarrow \text{canon.} & & \approx \downarrow \text{canon.} & & \\ H^n_{DR}(X) & \xrightarrow{=} & H^n_{DR}(X)(m) & & (k = \mathbb{C}) \end{array}$$

fails to commute by a factor of $(2\pi i)^m$.

In each cohomology theory there is a canonical way of associating a class $cl(Z)$ in $H^{2p}(X)(p)$ with an algebraic cycle Z on X of pure codimension p . Since our cohomology groups are without torsion, we can do this using Chern classes (Grothendieck [1]). Starting with a functorial homomorphism $c_1 : Pic(X) \to H^2(X)(1)$, one use the splitting principle to define the Chern polynomial $c_t(E) = \Sigma\, c_p(E)t^p$, $c_p(E) \in H^{2p}(X)(p)$, of a vector bundle E on X. The map $E \longmapsto c_t(E)$ is additive, and therefore factors through the Grothendieck group of the category of vector bundles on X . But, as X is smooth, this group is the same as the Grothendieck group of the category of coherent O_X-modules, and we can therefore define

$$cl(Z) = \frac{1}{(p-1)!}\, c_p(O_Z)$$

(loc. cit 4.3).

In defining c_1 for the Betti and étale theories, we begin with the maps

$$\mathrm{Pic}(X) \longrightarrow H^2(X^{an}, 2\pi i\,\mathbb{Z})$$

$$\mathrm{Pic}(X) \longrightarrow H^2(X_{et}, \mu_m(k))$$

arising (as boundary maps) from the sequences

$$0 \longrightarrow 2\pi i\,\mathbb{Z} \longrightarrow O_{X^{an}} \xrightarrow{\exp} O^{\times}_{X^{an}} \longrightarrow 0$$

$$0 \longrightarrow \mu_m \longrightarrow O^{\times}_X \xrightarrow{m} O^{\times}_X \longrightarrow 0 .$$

For the de Rham theory, we note that the dlog map , $f \longmapsto \frac{df}{f}$, defines a map of complexes

$$\begin{array}{ccccccc} 0 & \longrightarrow & O^{\times}_S & \longrightarrow & 0 & \longrightarrow & \cdots \\ \downarrow & & \downarrow \mathrm{dlog} & & \downarrow & & \\ O_X & \xrightarrow{d} & \Omega^1_X & \xrightarrow{d} & \Omega^2_X & \xrightarrow{d} & \cdots \end{array}$$

and hence a map

$$c_1 : \mathrm{Pic}(X) = H^1(X, O^{\times}_X) = \mathbb{H}^2(X, 0 \to O^{\times}_X \to \cdots) \longrightarrow \mathbb{H}^2(X, \Omega^{\cdot}_X)$$

$$= H^2_{DR}(X) = H^2_{DR}(X)(1) .$$

It can be checked that the three maps c_1 are compatible with the comparison isomorphisms and it follows formally that the maps cl are also compatible. (At least, it does once one has checked that Gysin maps and multiplicative structures are compatible with the comparison isomorphisms.)

When $k = \mathbb{C}$, there is a direct way of defining a class $cl(Z) \in H_{2d-2p}(X(\mathbb{C}), \mathbb{Q})$ (singular homology; $d = \dim(X)$, $p = \mathrm{codim}(Z)$): the choice of an i determines an orientation

X and of the smooth part of Z, and there is therefore a topologically defined class $cl(Z) \in H_{2d-2p}(X(\mathbb{C}),\mathbb{Q})$. This class has the property that for $[\omega] \in H^{2d-2p}(X^{\infty},\mathbb{R}) = H^{2d-2p}(\Gamma(X,\Omega^{\cdot}_{X^{\infty}}))$ represented by the closed form ω,

$$< cl(Z), [\omega] > = \int_Z \omega .$$

By Poincaré duality, $cl(Z)$ corresponds to a class $cl_{top}(Z) \in H^{2p}_B(X)$, whose image in $H^{2p}_B(X)(p)$ under the map induced by $1 \longmapsto 2\pi i: \mathbb{Q} \longrightarrow \mathbb{Q}(1)$ is known to be $cl_B(Z)$. The above formula becomes

$$\int_X cl_{top}(Z) \cup [\omega] = \int_Z \omega .$$

There are trace maps $(d = \dim X)$

$$Tr_B: H^{2d}_B(X)(d) \xrightarrow{\approx} \mathbb{Q}$$

$$Tr_{et}: H^{2d}_{et}(X)(d) \xrightarrow{\approx} \mathbb{A}^f$$

$$Tr_{DR}: H^{2d}_{DR}(X)(d) \xrightarrow{\approx} k$$

that are determined by the requirement $Tr(cl(point)) = 1$; they are compatible with the comparison isomorphisms. When $k = \mathbb{C}$, Tr_B and Tr_{DR} are equal to the composites

$$\begin{array}{llllllll} Tr_B: & H^{2d}_B(X)(d) & \xrightarrow{\approx} & H^{2d}_B(X) & \longrightarrow & H^{2d}(\Gamma(\Omega^{\cdot}_{X^{\infty}})) & \longrightarrow & \mathbb{C} \\ & 1 & \longmapsto & 2\pi i & & [\omega] & \longmapsto & \int_X \omega \end{array}$$

$$\begin{array}{llllll} Tr_{DR}: & H^{2d}_{DR}(X)(d) = H^{2d}_{DR}(X) & \xrightarrow{\approx} & H^{2d}(\Gamma(\Omega^{\cdot}_{X^{\infty}})) & \longrightarrow & \mathbb{C} \\ & & & [\omega] & \longmapsto & \frac{1}{(2\pi i)^d}\int_X \omega \end{array}$$

where we have chosen an i and used it to orientate X. (Note that the composite maps are obviously independent of the choice of i.) The formulas in the last paragraph show that

$$\mathrm{Tr}_{DR}(\mathrm{cl}_{DR}(Z) \cup [\omega] = \frac{1}{(2\pi i)^{\dim Z}} \int_Z \omega \ .$$

A definition of Tr_{et} can be found in (Milne [1, VI.11]).

We now deduce some consequences concerning periods.

<u>Proposition</u> 1.5. Let X be a variety over an algebraically closed field $k \subset \mathbb{C}$ and let Z be an algebraic cycle on $X_{\mathbb{C}}$ of dimension r. For any C^∞ differential r-form ω on $X_{\mathbb{C}}$ whose class $[\omega]$ in $H^{2r}_{DR}(X_{\mathbb{C}})$ lies in $H^{2r}_{DR}(X)$,

$$\int_Z \omega \in (2\pi i)^r k \ .$$

<u>Proof</u>: We first note that Z is algebraically equivalent to a cycle Z_0 defined over k. In proving this we can assume Z to be prime. There exists a smooth (not necessarily complete) variety T over k, a subvariety $\tilde{Z} \subset X \times T$ that is flat over T, and a point spec $\mathbb{C} \to T$ such that $Z = \tilde{Z} \times_T \operatorname{spec} \mathbb{C}$ in $X \times_T \operatorname{spec} \mathbb{C} = X_{\mathbb{C}}$. We can therefore take Z_0 to be $\tilde{Z} \times_T \operatorname{spec} k \subset X \times_T \operatorname{spec} k = X$ for any point spec $k \to T$. From this it follows that $\mathrm{cl}_{DR}(Z) = \mathrm{cl}_{DR}(Z_0) \in H^{2r}_{DR}(X)(r)$ and $\mathrm{Tr}_{DR}(\mathrm{cl}_{DR}(Z) \cup [\omega]) \in k$. But we saw above that $\int_Z \omega = (2\pi i)^r \mathrm{Tr}_{DR}(\mathrm{cl}_{DR}(Z) \cup [\omega])$.

We next derive a classical relation between the periods of an elliptic curve. For a complete smooth curve X and an open affine subset U, the map

$$H^1_{DR}(X) \longrightarrow H^1_{DR}(U) = \Gamma(U,\Omega^1_X)/d\Gamma(U,O_X)$$

$$= \frac{\{\text{meromorphic differentials, holomorphic on } U\}}{\{\text{exact differentials}\}}$$

is injective with image the set of classes represented by forms whose residues are all zero (such forms are said to be of the second kind). When $k = \mathbb{C}$, $Tr_{DR}([\alpha] \cup [\beta])$, where α and β are differential 1-forms of the second kind, can be computed as follows. Let Σ be the finite set of points where α or β has a pole. For z a local parameter at $P \in \Sigma$, α can be written $\alpha = (\sum_{-\infty \ll i}^{\infty} a_i z^i)dz$, with $a_{-1} = 0$. There therefore exists a meromorphic function a defined near P such that $da = \alpha$. We write $\int\alpha$ for any such function; it is defined up to a constant. As $Res_P\beta = 0$, $Res_P(\int\alpha)\beta$ is well-defined, and one proves that

$$Tr_{DR}([\alpha] \cup [\beta]) = \sum_{P \in \Sigma} Res_P(\textstyle\int\alpha)\beta .$$

Now let X be the elliptic curve

$$y^2 z = 4x^3 - g_2 xz^2 - g_3 z^3$$

There is a lattice Λ in $\mathbb{C}$ and corresponding Weierstrass function $\wp(z)$ such that $z \mapsto (\wp(z), \wp'(z), 1)$ defines an isomorphism $\mathbb{C}/\Lambda \xrightarrow{\approx} X(\mathbb{C})$. Let γ_1 and γ_2 be generators of Λ such that the bases $\{\gamma_1, \gamma_2\}$ and $\{1, i\}$ of $\mathbb{C}$ have the same orientation. We can regard γ_1 and γ_2 as elements of $H_1(X, \mathbb{Z})$, and then $\gamma_1 \cdot \gamma_2 = 1$. The differentials

$\omega = \frac{dx}{y}$ and $\eta = \frac{xdx}{y}$ on X pull back to dz and $\wp(z)dz$ respectively on $\mathbb{C}$; the first is therefore holomorphic and the second has a single pole at $\infty = (0,1,0)$ on X with residue zero (because $0 \in \mathbb{C}$ maps to $\infty \in X$ and $\wp(z) = \frac{1}{z^2} + a_2 z^2 + \ldots$). We find that

$$\mathrm{Tr}_{DR}([\omega] \cup [\eta]) = \mathrm{Res}_0(\textstyle\int dz)\wp(z)dz = \mathrm{Res}_0(z\,\wp(z)dz) = 1 .$$

Let $\int_{\gamma_i} \frac{dx}{y} = \int_{\gamma_i} \frac{dx}{\sqrt{4x^3-g_2x-g_3}} = \omega_i \qquad (i=1,2)$

and $\int_{\gamma_i} \frac{xdx}{y} = \int_{\gamma_i} \frac{xdx}{\sqrt{4x^3-g_2x-g_3}} = \eta_i \qquad (i=1,2)$

be the periods of ω and η. Under the map $H^1_{DR}(X) \to H^2(X,\mathbb{C})$, ω maps to $\omega_1\gamma_1' + \omega_2\gamma_2'$ and η maps to $\eta_1\gamma_1' + \eta_2\gamma_2'$, where $\{\gamma_1',\gamma_2'\}$ is the basis dual to $\{\gamma_1, \gamma_2\}$. Thus

$$\begin{aligned} 1 = \mathrm{Tr}_{DR}([\omega] \cup [\eta]) &= \mathrm{Tr}_B((\omega_1\gamma_1' + \omega_2\gamma_2') \cup (\eta_1\gamma_1' + \eta_2\gamma_2')) \\ &= (\omega_1\eta_2 - \omega_2\eta_1)\mathrm{Tr}_B(\gamma_1' \cup \gamma_2') \\ &= \frac{1}{2\pi i}(\omega_1\eta_2 - \omega_2\eta_1) . \end{aligned}$$

Hence $\omega_1\eta_2 - \omega_2\eta_1 = 2\pi i$. This is the Legendre relation.

The next proposition shows how the existence of algebraic cycles can force algebraic relations between the periods of abelian integrals. Let X be an abelian variety over a subfield k of $\mathbb{C}$. Recall that $H^r(X) = \Lambda^r(H^1(X)$ and $H^1(X \times X \times \ldots) = H^1(X) \oplus H^1(X) \oplus \ldots$ (any cohomology theory). Let $\nu \in \mathbb{G}_m(\mathbb{Q})$ act on $\mathbb{Q}_B(1)$ as ν^{-1}; there is then a natural action of $GL(H_B^1(X)) \times \mathbb{G}_m$ on $H_B^r(X^n)(m)$ for any r, n, and m. We define G to be the subgroup of $GL(H_B^1(X)) \times \mathbb{G}_m$ fixing all tensors of the form $cl_B(Z)$, Z an algebraic cycle on some X^n. (See Notations for a precise description of what this means.)

Consider the canonical isomorphisms

$$H_{DR}^1(X) \otimes_k \mathbb{C} \xrightarrow{\approx} H^1(X^{an}, \mathbb{C}) \xleftarrow{\approx} H_B^1(X) \otimes_{\mathbb{Q}} \mathbb{C} .$$

The <u>periods</u> p_{ij} of X are defined by the equations

$$\alpha_i = \sum p_{ji} a_j$$

where $\{\alpha_i\}$ and $\{a_i\}$ are bases for $H_{DR}^1(X)$ and $H_B^1(X)$ over k and $\mathbb{Q}$ respectively. The field $k(p_{ij})$ generated over k by the p_{ij} is independent of the bases chosen.

<u>Proposition 1.6.</u> With the above definitions

$$\text{tr.deg}_k\, k(p_{ij}) \leq \dim(G) .$$

<u>Proof:</u> We can replace k by its algebraic closure in $\mathbb{C}$, and hence assume that each algebraic cycle on $X_{\mathbb{C}}$ is equivalent to an algebraic cycle on X (see the proof of 1.5). Define P to be the functor of k-algebras such that an element of $P(A)$ is an isomorphism $p\colon H_B^1 \otimes_{\mathbb{Q}} A \xrightarrow{\approx} H_{DR}^1 \otimes_k A$ mapping $cl_B(Z) \otimes 1$ to $cl_{DR}(Z) \otimes 1$ for all algebraic cycles Z on a power of X. When $A = \mathbb{C}$, the comparison isomorphism is such a p, and so $P(\mathbb{C})$ is not empty. It is easily seen that P is represented by an algebraic variety that becomes a G_k-torsor under the

the obvious action. The bases $\{\alpha_i\}$ and $\{a_i\}$ can be used to identify the points of P with matrices. The matrix (p_{ij}) is a point of P with coordinates in $\mathbb{C}$, and so the proposition is a consequence of the following well-known lemma.

Lemma 1.7. Let $\mathbb{A}^N$ be an affine space over k, and let $z \in \mathbb{A}^N(\mathbb{C})$; the transcendence degree of $k(z_1,\ldots,z_N)$ over k is the dimension of the Zariski closure of $\{z\}$.

Remark 1.8. If X is an elliptic curve then $\dim G$ is 2 or 4 according as X has complex multiplication or not. Chudnovsky has shown that $\text{tr deg}_k\, k(p_{ij}) = \dim (G)$ when X is an elliptic curve with complex multiplication. Does equality hold for all abelian varieties?

One of the main purposes of the seminar was to show (1.5) and (1.6) make sense, and remain true. if "algebraic cycle" is replaced by "Hodge cycle" (in the case the X is an abelian variety).

2. Absolute Hodge cycles; principle B.

Let k be an algebraically closed field of finite transcendence degree over $\mathbb{Q}$, and let X be a variety over k. Write $H^n_{\mathbb{A}}(X)(m) = H^n_{DR}(X)(m) \times H^n_{et}(X)(m)$; it is a free $k \times \mathbb{A}^f$-module. Corresponding to an embedding $\sigma\colon k \hookrightarrow \mathbb{C}$ there are canonical isomorphisms

$$\sigma^*_{DR}\colon\ H^n_{DR}(X)(m) \otimes_{k,\sigma} \mathbb{C} \xrightarrow{\approx} H^n_{DR}(\sigma X)(m)$$

$$\sigma^*_{et}\colon\ H^n_{et}(X)(m) \xrightarrow{\approx} H^n_{et}(\sigma X)(m)$$

whose product we write σ^* . The diagonal embedding $H^n_\sigma(X)(m) \hookrightarrow H^n_{DR}(\sigma X)(m) \times H^n_{et}(\sigma X)(m)$ induces an isomorphism $H^n_\sigma(X)(m) \otimes (\mathbb{C} \times \mathbb{A}^f) \xrightarrow{\approx} H^n_{DR}(\sigma X)(m) \times H^n_{et}(\sigma X)(m)$ (product of the comparison isomorphisms, §1). An element $t \in H^{2p}_{\mathbb{A}}(X)(p)$ is a <u>Hodge cycle relative to σ</u>, if

(a) t is rational relative to σ , i.e., $\sigma^*(t)$ lies in the rational subspace $H^{2p}_\sigma(X)(p)$ of $H^{2p}_{DR}(\sigma X)(p) \times H^{2p}_{et}(\sigma X)(p)$;

(b) the first component of t lies in $F^0H^{2p}_{DR}(X)(p) \stackrel{df}{=} F^pH^{2p}_{DR}(X)$.

Under the assumption (a), condition (b) is equivalent to requiring that the image of t in $H^{2p}_{DR}(X)(p)$ is of bidegree (0,0). If t is a Hodge cycle relative to every embedding $\sigma: k \hookrightarrow \mathbb{C}$ then it is called an <u>absolute Hodge cycle</u>.

<u>Example</u> 2.1 (a) For any algebraic cycle Z on X , $t = (cl_{DR}(Z), cl_{et}(Z))$ is an absolute Hodge cycle. (The Hodge conjecture asserts there are no others.) Indeed, for any $\sigma: k \hookrightarrow \mathbb{C}$, $\sigma^*(t) = cl_B(Z)$, and is therefore rational, and it is well-known that $cl_{DR}(\sigma Z)$ is of bidegree (p,p) in $H^{2p}_{DR}(\sigma X)$.

(b) Let X be a variety of dimension d , and consider the diagonal $\Delta \subset X \times X$. Corresponding to the decomposition

$$H^{2d}(X \times X)(d) = \bigoplus_{i=0}^{2d} H^{2d-i}(X) \otimes H^i(X)$$

(Künneth formula)

we have $cl(\Delta) = \sum_{i=0}^{2d} \pi^i$. The π^i are absolute Hodge cycles.

(c) Suppose that X is given with a projective embedding, and let $\gamma \in H^2_{DR}(X)(1) \times H^2_{et}(X)(1)$ be the class of a hyperplane section. The hard Lefschetz theorem shows that

$$H^{2p}(X)(p) \longrightarrow H^{2d-2p}(X)(d-p),\ x \longmapsto \gamma^{d-2p}.x$$

is an isomorphism. The class x is an absolute Hodge cycle if and only if $\gamma^{d-2p}.x$ is an absolute Hodge cycle.

(d) Loosely speaking, any cycle that is constructed from a set of absolute Hodge cycles by a canonical rational process will again be an absolute Hodge cycle.

<u>Open Question</u> 2.2. Does there exist a cycle rational relative to every σ but which is not absolutely Hodge?

More generally, consider a family $(X_\alpha)_{\alpha\in A}$ of varieties over a field k (as above). Choose $(m(\alpha)) \in \mathbb{N}^{(A)}$, $(n(\alpha)) \in \mathbb{N}^{(A)}$, and $m \in \mathbb{Z}$, and write

$$T_{DR} = \otimes_\alpha H^{m(\alpha)}_{DR}(X_\alpha) \otimes \otimes_\alpha H^{n(\alpha)}_{DR}(X_\alpha)^\vee(m)$$

$$T_{et} = \otimes_\alpha H^{m(\alpha)}_{et}(X_\alpha) \otimes \otimes_\alpha H^{n(\alpha)}_{et}(X_\alpha)^\vee(m)$$

$$T_{\mathbb{A}} = T_{DR} \times T_{et}$$

$$T_\sigma = \otimes_\alpha H^{m(\alpha)}_\sigma(X_\alpha) \otimes \otimes_\alpha H^{n(\alpha)}_\sigma(X_\alpha)^\vee(m),\quad \sigma\colon k \hookrightarrow \mathbb{C}.$$

Then we say that $t \in T_{\mathbb{A}}$ is <u>rational relative to σ</u> if its image in $T_{\mathbb{A}} \otimes_{k \times \mathbb{A}^f,(\sigma,1)} \mathbb{C} \times \mathbb{A}^f$ is in T_σ, that it is a <u>Hodge cycle relative to σ</u> if it is rational relative to σ and its first component lies in F^o, and that it is an <u>absolute Hodge cycle</u> if it is a Hodge cycle relative to every σ.

Note that, for there to exist Hodge cycles in $T_{\mathbb{A}}$ it is necessary that $\Sigma\, m(\alpha) - \Sigma\, n(\alpha) = 2m$.

<u>Example</u> 2.3. Cup-product defines a map $T_{\mathbb{A}}^{m,n}(p) \times T_{\mathbb{A}}^{m',n'}(p') \longrightarrow T_{\mathbb{A}}^{m+m',n+n'}(p+p')$, and hence an element of $T_{\mathbb{A}}^{\vee} \otimes T_{\mathbb{A}}^{\vee} \otimes T_{\mathbb{A}}$; this element is an absolute Hodge cycle.

<u>Open Question</u> 2.4. Let $t \in H_{DR}^{2p}(X)(p)$ and suppose that $t \in F^o H_{DR}^{2p}(X)(p)$ and that, for all $\sigma\colon k \hookrightarrow \mathbb{C}$, $\sigma_{DR}^*(t) \in H_\sigma^{2p}(X)(p)$. Do these conditions imply that t is the first component of an absolute Hodge cycle?

In order to develop the theory of absolute Hodge cycles, we shall need to use the Gauss-Manin connection (Katz-Oda [1], Katz [1], Deligne [2]). Let k_0 be a field of characteristic zero and S a smooth k_0-scheme (or the spectrum of a finitely generated field over k_0). A <u>k_0-connection</u> on a coherent O_S-module $\mathcal{E}$ is a homomorphism of sheaves of abelian group

$$\nabla\colon \mathcal{E} \longrightarrow \Omega^1_{S/k_0} \otimes_{O_S} \mathcal{E}$$

such that

$$\nabla(fe) = f\nabla(e) + df \otimes e$$

for sections f of O_S and e of $\mathcal{E}$. The kernel of ∇, $\mathcal{E}^{\nabla}$, is the sheaf of horizontal sections of $(\mathcal{E},\nabla)$. Such a ∇ can be extended to a homomorphism of abelian sheaves,

$$\nabla_n: \Omega^n_{S/k_0} \otimes_{O_S} \mathcal{E} \longrightarrow \Omega^{n+1}_{S/k_0} \otimes_{O_S} \mathcal{E}$$

$$\omega \otimes e \longmapsto d\omega \otimes e + (-1)^n \omega \wedge \nabla(e) ,$$

and ∇ is said to be integrable if $\nabla_1 \circ \nabla = 0$. Moreover ∇ gives rise to an O_S-linear map

$$D \longmapsto \nabla_D: \mathrm{Der}(S/k_0) \rightarrow \mathrm{End}_{k_0}(\mathcal{E})$$

$$\nabla_D = (\mathcal{E} \xrightarrow{\nabla} \Omega^1_{S/k_0} \otimes_{O_S} \mathcal{E} \xrightarrow{D \otimes 1} O_S \otimes_{O_S} \mathcal{E} = \mathcal{E}) .$$

Note that $\nabla_D(fe) = D(f)e + f\nabla_D(e)$. One checks that ∇ is integrable if and only if $D \longmapsto \nabla_D$ is a Lie algebra homomorphism.

Now consider a proper smooth morphism $\pi: X \rightarrow S$, and write $\underline{H}^n_{DR}(X/S)$ for $\mathbb{R}^n \pi_*(\Omega^{\cdot}_{X/S})$. This is a locally free sheaf of O_S-modules and has a canonical connection ∇, the Gauss-Manin connection, which is integrable. It therefore defines a Lie algebra homomorphism $\mathrm{Der}(S/k_0) \longrightarrow \mathrm{End}_{k_0}(\underline{H}^n_{DR}(X/S))$. If $k_0 \hookrightarrow k_0'$ is an inclusion of fields and $X'/S' = (X/S) \otimes_{k_0} k_0'$, then the Gauss-Manin connection on $H^n_{DR}(X'/S')$ is $\nabla \otimes 1$. In the case that $k_0 = \mathbb{C}$, the relative form of Serre's GAGA theorem [1] shows that $\underline{H}^n_{DR}(X/S)^{an} = \underline{H}^n_{DR}(X^{an}/S^{an})$

and ∇ gives rise to a connection ∇^{an} on $\underline{H}^n_{DR}(X^{an}/S^{an})$. The relative Poincaré lemma shows that $(R^n\pi_* \mathbb{C}) \otimes O_{S^{an}} \xrightarrow{\approx} \underline{H}^n_{DR}(X^{an}/S^{an})$, and it is known that ∇^{an} is the unique connection such that

$$R^n \pi_*(\mathbb{C}) \xrightarrow{\approx} \underline{H}^n_{DR}(X^{an}/S^{an})^{\nabla^{an}} .$$

<u>Proposition</u> 2.5. Let $k_0 \subset \mathbb{C}$ have finite transcendence degree over $\mathbb{Q}$, let k be a field which is finitely generated over k_0, let X be a variety over k, and let ∇ be the Gauss-Manin connection on $H^n_{DR}(X)$ relative to $X \to \text{spec } k \to \text{spec } k_0$. If $t \in H^n_{DR}(X)$ is rational relative to all embeddings of k into $\mathbb{C}$ then $\nabla t = 0$.

<u>Proof</u>: Choose a regular k_0-algebra A of finite type and a smooth projective map π: $X_A \to \text{spec } A$ whose generic fibre is $X \to \text{spec } k$ and which is such that t extends to an element of $\Gamma(\text{spec } A, \underline{H}^n_{DR}(X/\text{spec } A))$. After a base change relative to $k_0 \hookrightarrow \mathbb{C}$, we obtain maps

$$X_S \longrightarrow S \longrightarrow \text{spec } \mathbb{C} \text{ , } S = \text{spec } A_{\mathbb{C}} \text{ ,}$$

and a global section $t' = t \otimes 1$ of $\underline{H}^n_{DR}(X_S^{an}/S^{an})$. We have to show that $(\nabla \otimes 1)\, t' = 0$ or, equivalently, that t' is a global section of $\underline{H}^n(X_S^{an},\mathbb{C}) \stackrel{df}{=} R^n\pi_*^{an}\, \mathbb{C}$.

An embedding σ: $k \hookrightarrow \mathbb{C}$ gives rise to an injection $A \hookrightarrow \mathbb{C}$ (i.e. a generic point of spec A in the sense of Weil) and hence a point s of S. The hypotheses show that, at each of these points, $t(s) \in H^n(X_s^{an},\mathbb{Q}) \subset H^n_{DR}(X_s^{an})$. Locally on S, $\underline{H}^n_{DR}(X_S^{an}/S^{an})$ will be the sheaf of holomorphic sections of the trivial bundle, $S \times \mathbb{C}^m$, and $\underline{H}^n(X^{an},\mathbb{C})$ the sheaf of locally constant sections. Thus, locally, t' is a function $S \to S \times \mathbb{C}^m$, $s \longmapsto (t_1(s),\ldots,t_m(s))$. Each $t_i(s)$ is a holomorphic function which, by hypothesis, takes real (even rational) values on a dense subset of S. It is therefore constant.

Remark 2.6. In the situation of (2.5), assume that $t \in H^n_{DR}(X)$ is rational relative to one σ and horizontal for ∇. An argument similar to the above then shows that t is rational relative to all embeddings that agree with σ on k_0.

Corollary 2.7. Let $k_0 \subset k$ be algebraically closed fields of finite transcendence degree over $\mathbb{Q}$, and let X be a variety over k_0. If $t \in H^n_{DR}(X_k)$ is rational relative to all $\sigma\colon k \hookrightarrow \mathbb{C}$ then it is defined over k_0, i.e. it is in the image of $H^n_{DR}(X) \longrightarrow H^n_{DR}(X_k)$.

Proof: Let k' be a subfield of k which is finitely generated over k_0 and such that $t \in H^n_{DR}(X \otimes_{k_0} k')$. The hypothesis shows that $\nabla t = 0$, where ∇ is the Gauss-Manin connection for $X_{k'} \to \operatorname{spec} k' \to \operatorname{spec} k_0$. Thus, for any $D \in \operatorname{Der}(k'/k_0)$, $\nabla_D(t) = 0$. But $X_{k'}$ arises from a variety over k_0, and so the action of $\operatorname{Der}(k'/k_0)$ on $H^n_{DR}(X_{k'}) = H^n_{DR}(X) \otimes_{k_0} k'$ is through k': $\nabla_D = 1 \otimes D$. Thus the corollary follows from the next well-known lemma.

<u>Lemma</u> 2.8. Let $k_0 \subset k'$ be as above, and let $V = V_0 \otimes_{k_0} k'$ where V_0 is a vector space over k_0. If $t \in V$ is fixed (i.e. killed) by all derivations of k/k_0, then $t \in V_0$.

Let $C^p_{AH}(X)$ denote the subset of $H^{2p}_{\mathbb{A}}(X)(p)$ of absolute Hodge cycles; it is a finite-dimensional vector space over $\mathbb{Q}$.

<u>Proposition</u> 2.9 (a) Let X be a variety over an algebraically closed field k_0, let k be an algebraically closed field containing k_0, and assume that k_0 and k have finite transcendence degree over $\mathbb{Q}$. Then the canonical map

$$H^{2p}_{\mathbb{A}}(X)(p) \longrightarrow H^{2p}_{\mathbb{A}}(X_k)(p)$$

induces an isomorphism

$$C^p_{AH}(X) \xrightarrow{\approx} C^p_{AH}(X_k) .$$

(b) Let k be an algebraically closed field of finite transcendence degree over $\mathbb{Q}$, and let X_0 be a variety defined over a subfield k_0 of k whose algebraic closure is k; write $X = X_0 \otimes_{k_0} k$. Then $\mathrm{Gal}(k/k_0)$ acts on $C^p_{AH}(X)$ through a finite quotient.

<u>Proof</u> (a) The map is injective, and a cycle on X is absolutely Hodge if and only if it is absolutely Hodge on X_k, and so the only non-obvious step is to show that an absolute Hodge cycle t on X_k arises from a cycle on X. But (2.8) shows that t_{DR} arises from an element of $H^{2p}_{DR}(X)(p)$, and

$H^{2p}_{et}(X)(p) \to H^{2p}_{et}(X_k)(p)$ is an isomorphism.

(b) It is obvious that the action of $Gal(k/k_0)$ on $H^{2p}_{DR}(X)(p) \times H^{2p}_{et}(X)(p)$ stabilizes $C^p_{AH}(X)$. We give three proofs that it factors through a finite quotient.

(i) Note that $C^p_{AH}(X) \to H^{2p}_{DR}(X)$ is injective. Clearly $H^{2p}_{DR}(X) = \bigcup H^{2p}_{DR}(X_0 \otimes k_i)$, where the k_i run through the finite extensions of k_0 contained in k, and hence all elements of a basis for $C^p_{AH}(X)$ lie in $H^{2p}_{DR}(X_0 \otimes k_i)$ for some i.

(ii) Note that $C^p_{AH}(X) \to H^{2p}(X_{et}, \mathbb{Q}_\ell)(p)$ is injective for any ℓ. The subgroup H of $Gal(k/k_0)$ fixing $C^p_{AH}(X)$ is closed. Thus $Gal(k/k_0)/H$ is a profinite group, which is countable since it is a subgroup of $GL_m(\mathbb{Q})$ for some m. It follows that it is finite.

(iii) A polarization on X gives a positive definite form on $C_{AH}(X)$, which is stable under $Gal(k/k_0)$. This shows that the action factors through a finite quotient.

<u>Remark</u> 2.10 (a) All of the above is still valid if we work with a family of varieties (X_α) rather than a single X.
(b) Proposition (2.9) would remain true if we had defined an absolute Hodge cycle to be an element t of $F^0H^{2p}_{DR}(X)(p)$ such that, for all $\sigma\colon k \hookrightarrow \mathbb{C}$, $\sigma^*_{DR}(t) \in H^{2p}_\sigma(X)$.

Proposition (2.9) allows us to define the notion of an absolute Hodge cycle on any (complete smooth) variety X over a field k (of characteristic zero). If k is algebraically closed then we choose an algebraically closed subfield k_0 that is of finite transcendence degree over $\mathbb{Q}$ and such that

X has a model X_0 over k_0 ; then $t \in H^{2p}_{\mathbb{A}}(X)(p)$ is an <u>absolute Hodge cycle</u> if it lies in the subspace $H^{2p}_{\mathbb{A}}(X_0(p))$ and is an absolute Hodge cycle there. The proposition shows that this definition is independent of the choice of k_0 and X_0 . (This definition is forced on us if we want (2.9a) to hold without restriction on the transcendence degrees of k and k_0 .) If k is not algebraically closed we choose an algebraic closure $\bar{k}$ of k and define an <u>absolute Hodge cycle</u> on X to be an absolute Hodge cycle on $X \otimes_k \bar{k}$ that is fixed by $\mathrm{Gal}(\bar{k}/k)$.

One can show that if k is algebraically closed and of cardinality not greater than that of $\mathbb{C}$, then $t \in H^{2p}_{DR}(X)(p) \times H^{2p}_{et}(X)(p)$ is an absolute Hodge cycle if it is rational relative to all embeddings $\sigma: k \hookrightarrow \mathbb{C}$ and $t_{dR} \in F^0H^{2p}_{DR}(X)(p)$. If $k = \mathbb{C}$ then the first condition has to be checked only for isomorphisms σ . (Provided the axiom of choice is assumed!) When $k = \mathbb{C}$ we define a <u>Hodge cycle</u> to be a cycle that is Hodge relative to $\sigma = \mathrm{id}: \mathbb{C} \hookrightarrow \mathbb{C}$.

<u>Main Theorem</u> 2.11. If X is an abelian variety over an algebraically closed field k , and t is a Hodge cycle on X relative to one embedding $\sigma: k \hookrightarrow \mathbb{C}$, then it is an absolute Hodge cycle.

The proof will occupy most of the rest of these notes. We begin with a result concerning families of varieties parametrized by smooth (not necessarily complete) algebraic varieties over $\mathbb{C}$. Let S be such a parameter variety and let $\pi: X \to S$ be a smooth proper map. We write $\underline{H}^n_{et}(X)(r)$ for $\varprojlim_m (R^n\pi_*^{et}\mu_m^{\otimes r}) \otimes_{\mathbb{Z}} \mathbb{Q}$.

<u>Theorem</u> 2.12 (Principle B). Let t be a global section of $\underline{H}_{DR}^{2p}(X/S)(p) \times \underline{H}_{et}^{2p}(X)(p)$ such that $\nabla t_{DR} = 0$ and $(t_{DR})_s \in F^o H_{DR}^{2p}(X_s)(p)$ for all $s \in S$. If $t_s \in H_{\mathbb{A}}^{2p}(X_s)(p)$ is an absolute Hodge cycle for one s, it is an absolute Hodge cycle for all s.

<u>Proof</u>: Suppose that t_s is an absolute Hodge cycle for $s = s_1$, and let s_2 be a second point of S. We have to show that t_{s_2} is rational relative to an isomorphism $\sigma: \mathbb{C} \xrightarrow{\approx} \mathbb{C}$. On applying σ we obtain a map $\sigma X \to \sigma S$ and a global section σt of $\underline{H}_{DR}^{2p}(\sigma X/\sigma S)(p) \times \underline{H}_{et}^{2p}(\sigma X)(p)$. We know that $\sigma(t)_{\sigma(s_1)}$ is rational and have to show $\sigma(t)_{\sigma(s_2)}$ is rational. Clearly σ only translates the problem, and so we can omit it.

First consider the component t_{DR} of t. By assumption $\nabla t_{DR} = 0$, and so t_{DR} is a global section of $\underline{H}^{2p}(X^{an},\mathbb{C})$. Since it is rational at one point, it must be rational at every point.

Next consider t_{et}. As $\underline{H}_B^{2p}(X) \stackrel{df}{=} R^{2p}\pi_*^{an}\mathbb{Q}$ and $\underline{H}_{et}^{2p}(X)$ are local systems (i.e. are locally constant), for any point s of S there are isomorphisms

$$\Gamma(S,\underline{H}_B^{2p}(X)(p)) \xrightarrow{\approx} H_B^{2p}(X_s)(p)^{\pi_1(S,s)} \quad \text{and}$$

$$\Gamma(S,\underline{H}_{et}^{2p}(X)(p)) \xrightarrow{\approx} H_{et}^{2p}(X_s)(p)^{\pi_1(S,s)} .$$

Consider,

$$\begin{array}{ccccc}
\Gamma(S,\underline{H}_B^{2p}(X)(p)) & \hookrightarrow & \Gamma(S,\underline{H}_B^{2p}(X)(p)) \otimes \mathbb{A}^f & = & \Gamma(S,\underline{H}_{et}^{2p}(X)(p)) \\
\downarrow \approx & & \downarrow \approx & & \\
H_B^{2p}(X_s)(p)^{\pi_1} & \hookrightarrow & H_B^{2p}(X_s)(p)^{\pi_1} \otimes \mathbb{A}^f & = & H_{et}^{2p}(X_s)(p)^{\pi_1} \\
\cap\downarrow & & \cap\downarrow & & \\
H_B^{2p}(X_s)(p) & \hookrightarrow & H_B^{2p}(X_s)(p) \otimes \mathbb{A}^f & = & H_{et}^{2p}(X_s)(p) .
\end{array}$$

We are given $t_{et} \in \Gamma(S,\underline{H}_{et}^{2p}(X)(p))$ and are told that its image in $H_{et}^{2p}(X_s)(p)$ is in $H_B^{2p}(X_s)(p)$ if $s = s_1$. The next easy lemma shows that t_{et} lies in $\Gamma(S,\underline{H}_B^{2p}(X)(p))$, and therefore is in $H_B^{2p}(X_s)(p)$ for all s.

<u>Lemma</u> 2.13. Let $V \hookrightarrow W$ be an inclusion of vector spaces, and let Z be a third vector space. Then $V \otimes Z \hookrightarrow W \otimes Z$, and $(V \otimes Z) \cap W = V$.

<u>Remark</u> 2.14. The assumption in the theorem that $(t_{dR})_s \in F^0H_{dR}^{2p}(X_s)(p)$ for all s is unnecessary; it is implied by the condition that $\nabla t_{DR} = 0$ (Deligne [4, 4.1.2, Théorème de la partie fixe]).

We shall also need a slight generalization of (2.12).

<u>Theorem</u> 2.15. Let π: $X \to S$ be as in (2.12), and let V be a local subsystem of $R^{2p}\pi_*\mathbb{Q}(p)$ such that V_s consists of (0,0)-cycles for all s and of absolute Hodge cycles for at least one s. Then V_s consists of absolute Hodge cycles for all s.

Proof. If V is constant, so that every element of V_s extends to a global section, then this follows from (2.12). The following argument reduces the general case to that case.

At each point $s \in S$, $R^{2p}\pi_*\mathbb{Q}(p)_s$ has a Hodge structure. Moreover $R^{2p}\pi_*\mathbb{Q}(p)$ has a polarization, i.e., there is a form $\psi\colon R^{2p}\pi_*\mathbb{Q}(p) \times R^{2p}\pi_*\mathbb{Q}(p) \to \mathbb{Q}$ which at each point s defines a polarization on the Hodge structure $R^{2p}\pi_*\mathbb{Q}(p)_s$. On $R^{2p}\pi_*\mathbb{Q}(p) \cap (R^{2p}\pi_*\mathbb{C}(p))^{0,0}$ the form is symmetric, bilinear, rational, and positive definite. Since the action of $\pi_1(S,s_0)$ preserves the form, the image of $\pi_1(S,s_0)$ in $\mathrm{Aut}(V_{s_0})$ is finite. Thus, after passing to a finite covering we can assume that V is constant.

Remark 2.16. Both (2.12) and (2.15) generalize, in an obvious way, to families $\pi_\alpha\colon X_\alpha \to S$.

3. Mumford-Tate groups; principle A.

Let G be a reductive algebraic group over a field k of characteristic zero, and let $(V_\alpha)_{\alpha\in A}$ be a faithful family of finite-dimensional representations over k of G, so that the map $G \hookrightarrow \prod GL(V_\alpha)$ is injective. For any $m \in \mathbb{N}^{(A)}$, $n \in \mathbb{N}^{(A)}$ we can form $T^{m,n} = \otimes V_\alpha^{\otimes m(\alpha)} \otimes V_\alpha^{\vee\otimes n(\alpha)}$, which is again a representation of G. For any subgroup H of G we write H' for the subgroup of G fixing all tensors, occurring in some $T^{m,n}$, that are fixed by H. Clearly $H \subset H'$, and we shall need criteria guaranteeing their equality.

Proposition 3.1. The notations are as above.

(a) Any finite-dimensional representation of G is contained in a direct sum of representations $T^{m,n}$.

(b) (Chevalley's theorem). Any subgroup H of G is the stabilizer of a line D in some finite-dimensional representation of G.

(c) If H is reductive, or if $X_k(G) \to X_k(H)$ is surjective (or has finite cokernel), then $H = H'$. (Here $X_k(G)$ denotes $\mathrm{Hom}_k(G,\mathbb{G}_m)$).

Proof. (a) Let W be a representation of G, and let W_0 be the trivial representation (meaning $gw = w$, all $g \in G, w \in W$) with the same underlying vector space as W. Then $G \times W \to W$ defines a map $W \to W_0 \otimes k[G]$ which is G-equivariant (Waterhouse [1, 3.5]). Since $W_0 \otimes k[G] \approx k[G]^{\dim W}$, it suffices to prove (a) for the regular representation. There is a finite sum $V = \oplus V_\alpha$ such $G \to GL(V)$ is injective (because G is Noetherian). The map $GL(V) \to End(V) \times End(V^\vee)$ identifies $GL(V)$ (and hence G) with a closed subvariety of $End(V) \times End(V^\vee)$ (loc. cit.). There is therefore a surjection $Sym(End(V)) \times Sym(End(V^\vee)) \twoheadrightarrow k[G]$, where Sym denotes a symmetric algebra, and (a) now follows from the fact that representations of reductive groups are semisimple (see II.2).

(b) Let I be the idea of functions on G which are zero on H. Then, in the regular representation of G on $k[G]$, H is the stabilizer of I. Choose a finite-dimensional subspace W of $k[G]$ that is G-stable and contains a generating set for the ideal I . Then H is the stabilizer of the subspace $I \cap W$ of W , and of $\Lambda^d(I \cap W)$ in $\Lambda^d W$, where d is the dimension of $I \cap W$ (Borel [1, 5.1]).

(c) According to (b), H is the stabilizer of a line D in some representation V of G and it follows from (a) that V can be taken to be a direct sum of $T^{m,n}$ s .

Assume that H is reductive. Then $V = V' \oplus D$ for some H-stable V' and $\check{V} = \check{V}' \oplus \check{D}$. Thus H is the group fixing a generator of $D \otimes \check{D}$ in $V \otimes \check{V}$.

Assume that $X_k(G) \to X_k(H)$ is surjective, i.e. that any character of H extends to a character of G . The one-dimensional representation of H on D can be regarded as the restriction to H of a representation of G . Now H is the group fixing a generator of $D \otimes \check{D}$ in $V \otimes \check{D}$.

Remark 3.2 (a) It is clearly necessary to have some condition on H in order to have $H' = H$. For example, let B be a Borel subgroup of the reductive group G and let $v \in V$ be fixed by B . Then $g \longmapsto gv$ defines a map of algebraic varieties $G/B \to V$ which is constant because G/B is complete. Thus v is fixed by G , and $B' = G$.

However, the above argument shows the following: let H' be the group fixing all tensors occurring in subquotients of $T^{m,n}$s that are fixed by H ; then $H = H'$.

(b) In fact, in all our applications of (3.1c), H will be the Mumford-Tate group of a polarizable Hodge structure, and hence will be reductive. However the Mumford-Tate groups of mixed Hodge structure (even polarizable) need not be reductive, but may satisfy the second condition of (3.1c) (with $G = GL$) .

(c) The Theorem of Haboush (Demazure [1]) can be used to show that the second from of (3.1c) holds when k has characteristic p .

Let V be a finite-dimensional vector space over $\mathbb{Q}$. A $\mathbb{Q}$-rational Hodge structure of weight n on V is a decomposition $V_{\mathbb{C}} = \bigoplus_{p+q=n} V^{p,q}$ such that $\overline{V^{p,q}} = V^{q,p}$. Such a structure determines a map $\mu\colon \mathbb{G}_m \to GL(V_{\mathbb{C}})$ such that $\mu(\lambda)v^{p,q} = \lambda^{-p}v^{p,q}$, $v^{p,q} \in V^{p,q}$. The complex conjugate $\overline{\mu}$ of μ satisfies $\overline{\mu}(\lambda)v^{p,q} = \overline{\lambda}^{-q}v^{p,q}$. Since μ and $\overline{\mu}$ commute, their product determines a map of real algebraic groups $h\colon \mathbb{C}^{\times} \to GL(V_{\mathbb{R}})$, $h(\lambda)v^{p,q} = \lambda^{-p}\overline{\lambda}^{-q}v^{p,q}$. Conversely, a homomorphism $h\colon \mathbb{C}^{\times} \to GL(V_{\mathbb{R}})$ such that $\mathbb{R}^{\times} \hookrightarrow \mathbb{C}^{\times} \longrightarrow GL(V_{\mathbb{R}})$ is $\lambda \longmapsto \lambda^{-n}.\mathrm{id}$ defines a Hodge structure of weight n on V .

We write $F^pV = \bigoplus_{p'\geq p} V^{p',q'}$, so that $\ldots \supset F^pV \supset F^{p+1}V \supset \ldots$ is a decreasing filtration on V .

Let $\mathbb{Q}(1)$ denote the vector space $\mathbb{Q}$ with the unique Hodge structure such that $\mathbb{Q}(1)_{\mathbb{C}} = \mathbb{Q}(1)^{-1,-1}$; it has weight -2 and $h(\lambda)1 = \lambda\overline{\lambda}1$. For any integer m, $\mathbb{Q}(1)^{\otimes m} \overset{\mathrm{df}}{=} \mathbb{Q}(m)$ $\mathbb{Q}(m)^{-m,-m}$ has weight $-2m$. (Strictly speaking, we should define $\mathbb{Q}(1) = 2\pi i\,\mathbb{Q}$ ....)

Remark 3.3. The notation $h(\lambda)\, v^{p,q} = \lambda^{-p}\overline{\lambda}^{-q}v^{p,q}$ is the negative of that used in Deligne [2] and Saavedra [1]. It is perhaps justified by the following. Let A be an abelian variety over $\mathbb{C}$. The exact sequences,

$$0 \longrightarrow \mathrm{Lie}(A^\vee)^\vee \longrightarrow H_1(A,\mathbb{C}) \longrightarrow \mathrm{Lie}(A) \longrightarrow 0$$

and

$$\begin{array}{ccccccccc} 0 & \longrightarrow & F^1H^1(A,\mathbb{C}) & \longrightarrow & H^1(A,\mathbb{C}) & \longrightarrow & F^1/F^2 & \longrightarrow & 0 \\ & & \| & & & & \| & & \\ & & H^{1,0} = H^0(\Omega^1) & & & & H^{0,1} = H^1(\mathcal{O}_X) & & \end{array}$$

are canonically dual. Since $H^1(A,\mathbb{C})$ has a natural Hodge structure of weight 1 with (1,0)-component $H^0(\Omega^1)$, $H_1(A,\mathbb{C})$ has a natural Hodge structure of weight -1 with (-1,0)-component $\mathrm{Lie}(A)$. Thus $h(\lambda)$ acts as λ on $\mathrm{Lie}(A)$, the tangent space to A at zero.

Let V be a vector space over $\mathbb{Q}$ with Hodge structure h of weight n. For $m_1, m_2 \in \mathbb{N}$ and $m_3 \in \mathbb{Z}$, $T = V^{\otimes m_1} \otimes V^{\vee\otimes m_2} \otimes \mathbb{Q}(1)^{\otimes m_3}$ has a Hodge structure of weight $(m_1-m_2)n - 2m_3$. An element of $T_{\mathbb{C}}$ is said to be rational of bidegree (p,q) if it lies in $T \cap T^{p,q}$. We let $\nu \in \mathbb{G}_m$ act on $\mathbb{Q}(1)$ as ν^{-1}; there is then a canonical action of $GL(V) \times \mathbb{G}_m$ on T. The Mumford-Tate group G of (V,h) is the subgroup of $GL(V) \times \mathbb{G}_m$ fixing all rational tensors of bidegree (0,0) belonging to any T. Thus projection on the first factor identifies $G(\mathbb{Q})$ with the set of $g \in GL(V)$ for which there exists a $\nu(g) \in \mathbb{Q}^\times$ with the property that $gt = \nu(g)^p t$ for any $t \in V^{\otimes m_1} \otimes V^{\vee\otimes m_2}$ of bidegree (p,p).

Proposition 3.4. The group G is the smallest algebraic subgroup of $GL(V) \times \mathbb{G}_m$ defined over $\mathbb{Q}$ for which $\mu(\mathbb{G}_m) \subset G_{\mathbb{C}}$.

Proof: Let $cl(\mu)$ be the intersection of all $\mathbb{Q}$-rational algebraic subgroups of $GL(V) \times \mathbb{G}_m$ which, over $\mathbb{C}$, contain $\mu(\mathbb{G}_m)$. For any $t \in T$, t is of type $(0,0)$ if and only if it is fixed by $\mu(\mathbb{G}_m)$ or, equivalently, it is fixed by $cl(\mu)$. Thus $G = cl(\mu)'$ in the notation of (3.1) and the next lemma completes the proof.

Lemma 3.5. Any $\mathbb{Q}$-rational character of $cl(\mu)$ extends to a $\mathbb{Q}$-rational character of $GL(V) \times \mathbb{G}_m$.

Proof: Let $\chi: cl(\mu) \to GL(W)$ be a representation of dimension one defined over $\mathbb{Q}$, i.e. a $\mathbb{Q}$-rational character. The restriction of the representation to $\mathbb{G}_m$ is isomorphic to $\mathbb{Q}(n)$ for some n. After tensoring W with $\mathbb{Q}(-n)$, we can assume that $\chi \circ \mu = 1$, i.e. $\mu(\mathbb{G}_m)$ acts trivially. But then $cl(\mu)$ must act trivially, and the trivial character extends to the trivial character.

Proposition 3.6. If V is polarizable then G is reductive.

Proof: Choose an i and write $C = h(i)$. (C is often called the Weil operator.) For $v^{p,q} \in V^{p,q}$, $Cv^{p,q} = i^{-p+q}\, v^{p,q}$, and so C^2 acts as $(-1)^n$ on V ($n = p+q$ is the weight of (V,h)).

We choose a polarization ψ for V. Recall that ψ is a morphism $\psi: V \otimes V \to \mathbb{Q}(-n)$ of Hodge structures such that the real-valued form $\psi(x,Cy)$ on $V_{\mathbb{R}}$ is symmetric and positive definite. Under the canonical isomorphism $Hom(V \otimes V,\mathbb{Q}(-n)) \xrightarrow{\approx} V^{\vee} \otimes V^{\vee}(-n)$, ψ corresponds to a tensor of bidegree $(0,0)$ (because it is a morphism of Hodge structures) and therefore it is fixed by G: $\psi(g_1v, g_1v') = g_2^n\psi(v,v')$ for $(g_1,g_2) \in G \subset GL(V) \times \mathbb{G}_m$ and $v,v' \in V$.

Recall that if H is a real algebraic group and σ is an involution of $H_{\mathbb{C}}$, then the real-form of H defined by σ is a real algebraic group H_σ together with an isomorphism $H_{\mathbb{C}} \xrightarrow{\approx} (H_\sigma)_{\mathbb{C}}$ under which complex conjugation on $H(\mathbb{C})$ corresponds to $\sigma \circ$ (complex conjugation) on $H_\sigma(\mathbb{C})$. We are going to use the following criterion: a connected algebraic group H over $\mathbb{R}$ is reductive if it has a compact real-form H_σ. To prove the criterion it suffices to show that H_σ is reductive. On any finite-dimensional representation of V of H there is an H_σ-invariant positive-definite symmetric form, namely $\langle u,v \rangle_0 = \int_{H_\sigma} \langle hu,hv \rangle \, dh$ where $\langle\ ,\ \rangle$ is any positive-definite symmetric form on V. If W is an H_σ-stable subspace of V, then its orthogonal complement is also H_σ-stable. Thus every finite-dimensional representation of H_σ is semisimple, and this implies H_σ is reductive (see [II.2]).

We shall apply the criterion to the special Mumford-Tate group of (V,h), $G^0 \overset{df}{=} \mathrm{Ker}(G \to \mathbb{G}_m)$. Let G^1 be the smallest $\mathbb{Q}$-rational subgroup of $GL(V) \times \mathbb{G}_m$ such that $G^1_{\mathbb{R}}$ contains $h(U^1)$, where $U^1(\mathbb{R}) = \{z \in \mathbb{C}^\times \mid z\bar{z} = 1\}$. Then $G^1 \subset G$, and in fact $G^1 \subset G^0$. Since $G^1_{\mathbb{R}} \cdot h(\mathbb{C}^\times) = G_{\mathbb{R}}$ and $h(U^1) = \mathrm{Ker}(h(\mathbb{C}^\times) \to \mathbb{G}_m)$, it follows that $G^0 = G^1$, and therefore G^0 is connected.

Since $C = h(i)$ acts as $i\bar{i} = 1$ on $\mathbb{Q}(1)$, $C \in G^0(\mathbb{R})$. Its square C^2 acts as $(-1)^n$ on V and therefore lies in the centre of $G^0(\mathbb{R})$. The inner automorphism ad C of $G_{\mathbb{R}}$ defined by C is therefore an involution. For $u,v \in V_{\mathbb{C}}$ and $g \in G^0(\mathbb{C})$ we have

$$\psi(u,C\bar{v}) = \psi(gu,gC\bar{v}) = \psi(gu,CC^{-1}gC\bar{v}) = \psi(gu,C\overline{g^*v})$$

where $g^* = C^{-1}\bar{g}C = (\mathrm{ad}\ C)(\bar{g})$. Thus the positive definite form $\phi(u,v) \overset{\mathrm{df}}{=} \psi(u,Cv)$ on $V_{\mathbb{R}}$ is invariant under the real-form of G^0 defined by ad C, and so the real-form is compact.

Example 3.7. (Abelian variety of CM-type). Let F be a finite product of totally real number fields F_i, and E a product of fields, each of which is a quadratic imaginary extension of exactly one of the fields F_i. Let $S = \mathrm{Hom}(E,\mathbb{C}) = \mathrm{Hom}(E,\bar{\mathbb{Q}}) = \mathrm{spec}\ E_{\mathbb{C}}$. $\mathrm{Gal}(\bar{\mathbb{Q}}/\mathbb{Q})$ acts on S and for any $\sigma \in S, \iota\sigma = \sigma\iota_{E/F}$, where $\iota_{E/F}$ is the canonical involution of E with fixed algebra F. A CM-type for E is a subset $\Sigma \subset S$ such that $S = \Sigma \cup \iota\Sigma$ (disjoint union). Correspondingly we define A to be $\mathbb{C}^{\Sigma}/\Sigma(O_E)$ where O_E, the ring of integers in E, is embedded in $\mathbb{C}^{\Sigma}$ by $u \longrightarrow (\sigma u)_{\sigma\in\Sigma}$. Obviously E acts on A; moreover $H_1(A,\mathbb{Q}) = E$, and

$$H_1(A) \otimes \mathbb{C} = E \otimes \mathbb{C} \xrightarrow{\approx} \mathbb{C}^S = \mathbb{C}^{\Sigma} \oplus \mathbb{C}^{\iota\Sigma}, \quad u \otimes 1 \longrightarrow (\sigma u)_{\sigma\in S}$$

with $\mathbb{C}^{\Sigma}$ the $(-1,0)$-component of $H_1(A) \otimes \mathbb{C}$ and $\mathbb{C}^{\iota\Sigma}$ the $(0,-1)$-component. Thus $\mu(\lambda)$ acts as λ on $\mathbb{C}^{\Sigma}$ and 1 on $\mathbb{C}^{\iota\Sigma}$.

Let G be the Mumford-Tate group of $H_1(A)$. The actions of $\mu(\mathbb{C}^\times)$ and $E^\times$ on $H_1(A) \otimes \mathbb{C}$ commute. As $E^\times$ is its own commutant in $GL(H_1(A))$ this means that $\mu(\mathbb{C}^\times) \subset (E \otimes \mathbb{C})^\times$ and $G = cl(\mu) \subset E^\times$. In particular G is a torus, and can be described by its cocharacter group $Y(G) \overset{df}{=} Hom_{\bar{\mathbb{Q}}}(\mathbb{G}_m, G)$.

Clearly $Y(G) \subset Y(E^\times) \times Y(\mathbb{G}_m) = \mathbb{Z}^S \times \mathbb{Z}$. Note that $\mu \in Y(G)$ is equal to $\sum_{s\in\Sigma} e_s + e_0$, where $\{e_s\} \subset \mathbb{Z}^S$ is the basis dual to $S = \{s\} \subset X(E^\times)$ and e_0 is the element 1 of the last copy of $\mathbb{Z}$. The following are obvious:

(a) $(\mathbb{Z}^S \times \mathbb{Z})/Y(G)$ is torsion-free.

(b) $\mu \in Y(G)$.

(c) $Y(G)$ is stable under $Gal(\bar{\mathbb{Q}}/\mathbb{Q})$; thus $Y(G)$ is the Gal-module generated by μ.

(d) Since $\mu + \iota\mu = 1$ on S, $Y(G) \subset \{y \in \mathbb{Z}^S \times \mathbb{Z} \mid y = \sum n_s e_s + n_0 e_0, n_s + n_{\iota s} = cnst\}$.

This means $G(\mathbb{Q}) \subset Ker(N_{E/F}\colon E^\times \to F^\times/\mathbb{Q}^\times) \times \mathbb{Q}^\times$.

<u>Theorem</u> 3.8 (Principle A) Let $(X_\alpha)_\alpha$ be a family of varieties over $\mathbb{C}$ and consider spaces T obtained by tensoring spaces of the form $H_B^{n_\alpha}(X_\alpha)$, $H_B^{n_\alpha}(X_\alpha)^\vee$, and $\mathbb{Q}(1)$. Let $t_i \in T_i$, $i = 1,\ldots,N$, (T_i of the above type) be absolute Hodge cycles and let G be the subgroup of $\prod_{\alpha,n_\alpha} GL(H_B^{n_\alpha}(X_\alpha)) \times \mathbb{G}_m$ fixing the t_i. If t belongs to some T and is fixed by G, then it is an absolute Hodge cycle.

<u>Proof</u>: We remove the identification of the ground field k with $\mathbb{C}$. Let $\sigma\colon k \xrightarrow{\approx} \mathbb{C}$ be the isomorphism implicit in

statement of the theorem and let $\tau\colon k \xrightarrow{\approx} \mathbb{C}$ be a second isomorphism. We can assume that the t_i and t all belong to the same space T. The canonical inclusions of cohomology groups

$$H_\sigma(X_\alpha) \hookrightarrow H_\sigma(X_\alpha) \otimes (\mathbb{C} \times \mathbb{A}^f) \hookleftarrow H_\tau(X_\alpha)$$

induce maps

$$T_\sigma \hookrightarrow T \otimes (\mathbb{C} \times \mathbb{A}^f) \hookleftarrow T_\tau \,.$$

We shall regard these maps as inclusions. Thus $\{t_1,\dots,t_N,t\} \subset T_\sigma \subset T \otimes (\mathbb{C} \times \mathbb{A}^f)$ and $\{t_1,\dots,t_N\} \subset T_\tau$. To show that t is rational we have to show that $t \in T_\tau$.

Let P be the functor of $\mathbb{Q}$-algebras such that

$$P(R) = \{p\colon\ H_\sigma \otimes R \xrightarrow{\approx} H_\tau \otimes R \mid p \text{ maps } t_i \,(\text{in } T_\sigma) \text{ to } t_i\,(\text{in } T_\tau),\ i = 1,\dots,N\} \,.$$

The existence of the canonical inclusions mentioned above shows that $P(\mathbb{C} \times \mathbb{A}^f)$ is non-empty, and it is easily checked that P is a G-torsor.

<u>Lemma</u> 3.9 Let P be a $\mathbb{Q}$-rational G-torsor of maps $H^\alpha_\sigma \xrightarrow{\approx} H^\alpha_\tau$ where $(H^\alpha_\sigma)_\alpha$ and $(H^\alpha_\tau)_\alpha$ are families of $\mathbb{Q}$-rational representations of G. Let T_σ and T_τ be like spaces of tensors constructed out of H_σ and H_τ respectively. Then P defines a map $T^G_\sigma \to T_\tau$.

Proof: Locally for the étale topology on spec($\mathbb{Q}$), points of P define maps $T_\sigma \xrightarrow{\approx} T_\tau$. The restriction to T_σ^G of such a map is independent of the point. Thus, by étale descent theory, they define a map of vector spaces $T_\sigma^G \longrightarrow T_\tau$.

On applying the lemma (and its proof) in the above situation we obtain a map $T_\sigma^G \longrightarrow T_\tau$ such that

$$\begin{array}{ccc} T_\sigma^G & \longrightarrow & T_\tau \\ \downarrow & & \downarrow \\ T_\sigma & \longrightarrow & T \otimes (\mathbb{C} \times \mathbb{A}^f) \end{array}$$

commutes. This means that $T_\sigma^G \subset T_\tau$, and therefore $t \in T_\tau$.

It remains to show that the first component t_{DR} of t, lying in $T \otimes \mathbb{C} = T_{DR}$, is in $F^0 T_{DR}$. But in general, if s is rational and $s \in T_{DR}$, where T_{DR} is constructed out of spaces $H_{DR}^{n_\alpha}(X_\alpha)$, $H_{DR}^{n_\alpha}(X_\alpha)^\vee$, $\mathbb{Q}(1)$, then $s \in F^0 T_{DR}$ is equivalent to s being fixed by $\mu(\mathbb{C}^\times)$. Thus $(t_i)_{DR} \in F^0$, $i = 1,\ldots,N$, implies $G \supset \mu(\mathbb{C}^\times)$, which implies $t_{DR} \in F^0$.

4. Construction of some absolute Hodge cycles

Recall that a number field E is a CM-field if, for any embedding $E \hookrightarrow \mathbb{C}$, complex conjugation induces a nontrivial automorphism $e \longmapsto \bar{e}$ of E that is independent of the embedding. The fixed field of the automorphism is then a totally real field F over which E has degree two.

A bi-additive form

$$\phi: V \times V \longrightarrow E$$

on a vector space V over such a field E is Hermitian if $\phi(ev,w) = e\phi(v,w)$ and $\phi(v,w) = \overline{\phi(w,v)}$ for $v,w \in V$, $e \in E$. For any embedding $\tau: F \hookrightarrow \mathbb{R}$ we obtain a Hermitian form ϕ_τ in the usual sense on the vector space $V_\tau = V \otimes_{F,\tau} \mathbb{R}$, and we let a_τ and b_τ be the dimensions of maximal subspaces of V_τ on which ϕ_τ is positive definite and negative definite respectively. If $d = \dim V$ then ϕ defines a Hermitian form on $\Lambda^d V$ that, relative to some basis vector, is of the form $(x,y) \longmapsto fx\bar{y}$. The element f is in F, and is independent of the choice of the basis vector up to multiplication by an element of $N_{E/F}E^\times$. It is called the discriminant of ϕ. Let $\{v_1,\ldots,v_d\}$ be an orthogonal basis for ϕ and let $\phi(v_i,v_i) = c_i$; then a_τ is the number of i for which $\tau c_i > 0$, b_τ the number of i for which $\tau c_i < 0$, and $f = \Pi c_i \pmod{N_{E/F}E^\times}$. Note that $\operatorname{sign}(\tau f) = (-1)^{b_\tau}$.

Proposition 4.1. Suppose given integers (a_τ, b_τ) for each τ, and an element $f \in F^\times/N_{E/F}E^\times$, such that $a_\tau + b_\tau = d$ all τ and $\operatorname{sign}(\tau f) = (-1)^{b_\tau}$. Then there exists a non-degenerate Hermitian form ϕ on a vector space V of dimension d with invariants (a_τ,b_τ) and f; moreover (V,ϕ) is unique up to isomorphism.

Proof: This result is due to Landherr [1]. Today one prefers to regard it as a consequence of the Hasse principle for simply-connected semisimple algebraic groups and the classification of Hermitian forms over local fields.

Corollary 4.2. Assume that the Hermitian space (V,ϕ) is non-degenerate and let $d = \dim(V)$. The following are equivalent:

(a) $a_\tau = b_\tau$ for all τ , and $\mathrm{disc}(f) = (-1)^{d/2}$;

(b) there is a totally isotropic subspace of V of dimension $d/2$.

Proof: Let W be a totally isotropic subspace of V of dimension $d/2$. The map $v \longmapsto \phi(-,v)\colon V \to W^{\vee}$ induces an anti-linear isomorphism $V/W \xrightarrow{\approx} W$. Thus a basis $v_1,\ldots,v_{d/2}$ of W can be extended to a basis $\{v_i\}$ of V such that

$$\phi(v_i,v_{d/2+i}) = 1,\ 1 \le i \le d/2 ,$$

$$\phi(v_i,v_j) = 0,\quad j \neq i \pm d/2 .$$

It is now easily checked that (V,ϕ) satisfies (a). Conversely (E^d,ϕ), where

$$\phi((a_i),(b_i)) = \sum_{1\le i\le d/2} a_i\bar{b}_{d/2+i} + \sum_{d/2<i\le d} a_{d/2+i}\bar{b}_i ,$$

is, up to isomorphism, the only Hermitian space satisfying (a), and it also satisfies (b).

A Hermitian form satisfying the equivalent conditions of the corollary will be said to be split (because then $\mathrm{Aut}_E(V,\phi)$ is an E-split algebraic group).

We shall need the following (trivial) lemma.

Lemma 4.3. Let k be a field, let k' be an étale k-algebra (i.e., a finite product of separable finite field extensions of k), and let V be a free k'-module of finite rank.

(a) For any k'-linear map $f: V \to k'$, define $Tr_{k'/k} f$ to be the k-linear map $v \longmapsto Tr_{k'/k}(f(v)): V \to k$; then $f \longmapsto Tr_{k'/k} f: Hom_{k'}(V,k') \to Hom_k(V,k)$ is an isomorphism.

(b) $\bigwedge^n_{k'} V$ is, in a natural way, a direct summand of $\bigwedge^n_k V$.

<u>Proof</u>: (a) Since the pairing $Tr_{k'/k}: k' \times k' \to k$ is non-degenerate, it is obvious that $f \longmapsto Tr_{k'/k} f$ is injective, and the two spaces have the same dimension over k.

(b) There are obvious maps $\bigwedge^n_{k'} V \longrightarrow \bigwedge^n_k V$

and $\bigwedge^n_{k'} \check{V} \longrightarrow \bigwedge^n_k \check{V}$

where $\check{V}$ is the k'-linear dual of V. But $(\check{\bigwedge V}) = (\bigwedge V)^{\vee}$, and so the second map gives rise to a map $\bigwedge^n_k V \longrightarrow \bigwedge^n_{k'} V$, which is inverse to the first. (More elegantly, descent theory shows that it suffices to prove the proposition with $k' = k^S$, $S = Hom_k(k', \bar{k})$. Then $V = \bigoplus_{s \in S} V_s$ and the map in (a) sends $f = (f_s)$ to Σf_s, which is obviously an isomorphism. For (b), note that

$$\bigwedge^n_k V = \bigoplus_{\Sigma n_s = n} (\bigotimes_{s \in S} \bigwedge^{n_s} V_s) \supset \bigoplus_{s \in S} \bigwedge^n V_s = \bigwedge^n_{k'} V.)$$

Let A be an abelian variety over $\mathbb{C}$, E a CM-field, and $\nu: E \longrightarrow End(A)$ a homomorphism (so, in particular, $\nu(1) = id$). Let d be the dimension of $H_1(A,\mathbb{Q})$ over E, so that $d[E:\mathbb{Q}] = 2\dim(A)$. When $H_1(A,\mathbb{R})$ is identified with the tangent space to A at zero it acquires a complex structure; we denote by J the $\mathbb{R}$-linear endomorphism "multiplication by i" of $H_1(A,\mathbb{R})$. If $h: \mathbb{C}^\times \to GL(H^1(A,\mathbb{R})) = GL(H_1(A,\mathbb{R}))$ is the homomorphism determined by the Hodge structure on $H^1(A,\mathbb{R})$ then $h(i) = J$.

Corresponding to the decomposition

$$e \otimes z \longmapsto (\ldots,\sigma(e)z,\ldots): E \otimes_{\mathbb{Q}} \mathbb{C} \xrightarrow{\approx} \bigoplus_{\sigma \in S} \mathbb{C},\ S = \mathrm{Hom}(E,\mathbb{C})$$

there is a decomposition

$$H^1_B(A) \otimes \mathbb{C} \xrightarrow{\approx} \bigoplus_{\sigma \in S} H^1_{B,\sigma} \qquad \text{(E-linear isomorphism)}$$

such that $e \in E$ acts on the complex vector space $H_{B,\sigma}$ as $\sigma(e)$. Each $H^1_{B,\sigma}$ has dimension d , and (as E respects the Hodge structure on $H^1_B(A)$) acquires a Hodge structure,

$$H^1_{B,\sigma} = H^{1,0}_{B,\sigma} \oplus H^{0,1}_{B,\sigma} .$$

Let $a_\sigma = \dim H^{1,0}_{B,\sigma}$ and $b_\sigma = \dim H^{0,1}_{B,\sigma}$; thus $a_\sigma + b_\sigma = d$.

<u>Proposition</u> 4.4: The subspace $\bigwedge^d_E H^1_B(A)$ of $H^d(A,\mathbb{Q})$ is purely of bidegree $(\frac{d}{2},\frac{d}{2})$ if and only if $a_\sigma = \frac{d}{2} = b_\sigma$.

<u>Proof</u>: Note that $H^d(A,\mathbb{Q}) = \bigwedge^d_{\mathbb{Q}} H^1(A,\mathbb{Q})$, and so (4.3) canonically identifies $\bigwedge^d_E H^1_B(A)$ with a subspace of $H^d_B(A)$.

As in the last line of the proof of (4.3) we have

$$(\Lambda_E^d H_B^1) \otimes \mathbb{C} = \Lambda_{E\otimes\mathbb{C}}^d H_B^1 \otimes \mathbb{C} = \bigoplus_{\sigma\in S} \Lambda^d H_{B,\sigma}^1 = \bigoplus_{\sigma\in S} \Lambda^d (H_{B,\sigma}^{1,0} \oplus H_{B,\sigma}^{0,1})$$

$$= \bigoplus_{\sigma\in S} \Lambda^{a_\sigma} H_{B,\sigma}^{1,0} \otimes \Lambda^{b_\sigma} H_{B,\sigma}^{0,1} ,$$

and $\Lambda^{a_\sigma} H_{B,\sigma}^{1,0}$ and $\Lambda^{b_\sigma} H_{B,\sigma}^{0,1}$ are purely of bidegree $(a_\sigma,0)$ and $(0,b_\sigma)$ respectively.

Thus, in this case, $(\Lambda_E^d H_B^1(A))(\frac{d}{2})$ consists of Hodge cycles, and we would like to show that it consists of absolute Hodge cycles. In one special case this is easy.

<u>Lemma</u> 4.5. Let A_0 be an abelian variety of dimension $\frac{d}{2}$ and let $A = A_0 \otimes_{\mathbb{Q}} E$. Then $\Lambda_E^d H^1(A,\mathbb{Q})(\frac{d}{2}) \subset H^d(A,\mathbb{Q})(\frac{d}{2})$ consists of absolute Hodge cycles.

<u>Proof</u>: There is a commutative diagram

$$\begin{array}{ccc} H_B^d(A_0)(\frac{d}{2}) \otimes_{\mathbb{Q}} E & \longrightarrow & H_{\mathbb{A}}^d(A_0)(\frac{d}{2}) \otimes_{\mathbb{Q}} E \\ \downarrow \approx & & \downarrow \approx \\ (\Lambda_E^d H_B^1(A_0 \otimes_{\mathbb{Q}} E))(\frac{d}{2}) & \longrightarrow & (\Lambda_{E\otimes\mathbb{A}}^d H_{\mathbb{A}}^1(A_0 \otimes E))(\frac{d}{2}) \subset H_{\mathbb{A}}^d(A_0 \otimes E)(\frac{d}{2}) \end{array}$$

in which the vertical maps are induced by $H^1(A_0) \otimes E \xrightarrow{\approx} H^1(A_0 \otimes E)$. From this, and similar diagrams corresponding to isomorphisms $\sigma\colon \mathbb{C} \xrightarrow{\approx} \mathbb{C}$, one sees that $H_{\mathbb{A}}^d(A_0)(\frac{d}{2}) \otimes E \hookrightarrow H_{\mathbb{A}}^d(A_0 \otimes E)(\frac{d}{2})$ induces an inclusion $C_{AH}^d(A_0) \otimes E \hookrightarrow C_{AH}^d(A_0 \otimes E)$. But $C_{AH}^d(A_0) = H_B^d(A_0)(\frac{d}{2})$ since $H_B^d(A_0)(\frac{d}{2})$ is a one-dimensional space generated by the class of any point on A_0 .

In order to prove the general result we need to consider families of abelian varieties (ultimately, we wish to apply (2.15)), and for this we need to consider polarized abelian varieties. A polarization θ on A is determined by a Riemann form, i.e. a $\mathbb{Q}$-bilinear alternating form ψ on $H_1(A,\mathbb{Q})$ such that the form $(z,w) \longmapsto \psi(z,Jw)$ on $H_1(A,\mathbb{R})$ is symmetric and definite; two Riemann forms ψ and ψ' on $H_1(A,\mathbb{Q})$ correspond to the same polarizaton if and only if there is an $a \in \mathbb{Q}^\times$ such that $\psi' = a\psi$. We shall consider only triples (A,θ,ν) in which the Rosati involution defined by θ induces complex conjugation on E. (The Rosati involution $e \longmapsto {}^te$: $\mathrm{End}(A) \to \mathrm{End}(A)$ is determined by the condition $\psi(ev,w) = \psi(v,{}^tew)$, $v,w \in H_1(A,\mathbb{Q})$.)

<u>Lemma</u> 4.6. Let $f \in E^\times$ be such that $\bar{f} = -f$, and let ψ be a Riemann form for A. There exists a unique E-Hermitian form ϕ on $H_1(A,\mathbb{Q})$ such that $\psi(x,y) = \mathrm{Tr}_{E/\mathbb{Q}}(f\phi(x,y))$.

<u>Proof</u>: We first need:

<u>Sublemma</u> 4.7. Let V and W be finite-dimensional vector spaces over E, and let $\psi: V \times W \to \mathbb{Q}$ be a $\mathbb{Q}$-bilinear form such that $\psi(ev,w) = \psi(v,ew)$. Then there exists a unique E-bilinear form ϕ such that $\psi(v,w) = \mathrm{Tr}_{E/\mathbb{Q}}\phi(v,w)$.

<u>Proof</u>: ψ defines a $\mathbb{Q}$-linear map $V \otimes_E W \to \mathbb{Q}$, i.e. an element of $(V \otimes_E W)^\vee$. But $\mathrm{Tr}_{E/\mathbb{Q}}$ identifies the $\mathbb{Q}$-linear dual of $V \otimes_E W$ with the E-linear dual, and ψ with a ϕ.

To prove (4.6), we take V to be $H_1(A,\mathbb{Q})$ and W to be $H_1(A,\mathbb{Q})$ with E acting through complex conjugation, and apply (4.7). This shows that $\psi(x,y) = Tr_{E/\mathbb{Q}}\phi_1(x,y)$ with ϕ_1 sesquilinear. Let $\phi = f^{-1}\phi_1$, so that $\psi(x,y) = Tr_{E/\mathbb{Q}}(f\phi(x,y))$. Since ϕ is sesquilinear it remains to show that $\phi(x,y) = \overline{\phi(y,x)}$. As $\psi(x,y) = -\psi(y,x)$ for all $x,y \in H_1(A,\mathbb{Q})$, $Tr(f\phi(x,y)) = - Tr(f\phi(y,x)) = Tr(\bar{f}\phi(y,x))$. On replacing x by ex with $e \in E$, we find that $Tr(fe\phi(x,y)) = Tr(\overline{fe}\phi(y,x))$. On the other hand $Tr(fe\phi(x,y)) = Tr(\overline{fe}\,\overline{\phi(x,y)})$ and, as $\overline{fe}$ is an arbitrary element of E, the non-degeneracy of the trace implies $\overline{\phi(x,y)} = \phi(y,x)$. The uniqueness of ϕ is obvious from (4.7).

<u>Theorem</u> 4.8. Let A be an abelian variety over $\mathbb{C}$, and let $\nu\colon E \longrightarrow End(A)$ be a homomorphism, where E is a CM-field. Assume there exists a polarization θ for A such that:

(a) the Rosati involution of θ induces complex conjugation on E;

(b) there exists a split E-Hermitian form ϕ on $H_1(A,\mathbb{Q})$ and on $f \in E^\times$, with $\bar{f} = - f$, such that $\psi(x,y) \stackrel{df}{=} Tr_{E/\mathbb{Q}}(f\phi(x,y))$ is a Riemann form for θ.

Then the subspace $(\Lambda_E^d H^1(A,\mathbb{Q}))(\frac{d}{2}) \subset H^d(A,\mathbb{Q})(\frac{d}{2})$, where $d = \dim_E H_1(A,\mathbb{Q})$, consists of absolute Hodge cycles.

<u>Proof</u>: In the course of the proof we shall see that (b) implies that A satisfies the equivalent statements of (4.4). Thus the theorem will follow from (2.15), (4.4), and (4.5) once we have show there exists a connected smooth (not necessarily complete) variety S over $\mathbb{C}$ and an abelian scheme Y over S

together with an action ν of E on Y/S such that:

(a) for all $s \in S$, (Y_s,ν_s) satisfies the equivalent statements in (4.4);

(b) for some $s_0 \in S$, $Y_{s_0} = A_0 \otimes_{\mathbb{Q}} E$, with $e \in E$ acting as $id \otimes e$;

(c) for some $s_1 \in S$, $(Y_s,\nu_{s_1}) = (A,\nu)$.

We shall first construct an analytic family of abelian varieties satisfying these conditions, and then pass to the quotient by a discrete group to obtain an algebraic family.

Let $H = H_1(A,\mathbb{Q})$, regarded as an E-space, and choose a θ, ϕ, f, and ψ as in the statement of the theorem. We choose i such that $\psi(x,h(i)y)$ is positive definite.

Consider the set of all quadruples (A_1,θ_1,ν_1,k_1) in which A_1 is an abelian variety over $\mathbb{C}$, ν_1 is an action of E on A_1, θ_1 is a polarization of A, and k_1 is an E-linear isomorphism $H_1(A_1,\mathbb{Q}) \xrightarrow{\approx} H$ carrying a Riemann form for θ_1 into $c\psi$ for some $c \in \mathbb{Q}^\times$. From such a quadruple we obtain a complex structure on $H(\mathbb{R})$ (corresponding via k_1 to the complex structure on $H_1(A_1,\mathbb{R}) = \mathrm{Lie}(A_1)$) such that:

(a) the action of E commutes with the complex structure;

(b) ψ is a Riemann form relative to the complex structure.

Conversely, a complex structure on $H \otimes \mathbb{R}$ satisfying (a) and (b) determines a quadruple (A_1,θ_1,ν_1,k_1) with $H_1(A_1,\mathbb{Q}) = H$

(as an E-module), $\mathrm{Lie}(A_1) = H \otimes \mathbb{R}$ (provided with the given complex structure), θ_1 the polarization with Riemann form ψ, and k_1 the identity map. Moreover two quadruples $(A_1, \theta_1, \nu_1, k_1)$ and $(A_2, \theta_2, \nu_2, k_2)$ are isomorphic if and only if they define the same complex structure on H. Let X be the set of complex structures on H satisfying (a) and (b). Our first task will be to turn X into an analytic manifold in such a way that the family of abelian varieties that it parametrizes becomes an analytic family.

A point of X is determined by an $\mathbb{R}$-linear map $J\colon H \otimes \mathbb{R} \to H \otimes \mathbb{R}$, $J^2 = -1$, such that

(a') J is E-linear, and

(b') $\psi(x, Jy)$ is symmetric and definite.

Note that $\psi(x, Jy)$ is symmetric if and only if $\psi(Jx, Jy) = \psi(x, y)$. Fix an isomorphism

$$E \otimes_{\mathbb{Q}} \mathbb{R} \xrightarrow{\approx} \bigoplus_{\tau \in T} \mathbb{C} \qquad (T = \mathrm{Hom}(F, \mathbb{R}),\ F = \text{real subfield of } E)$$

such that $f \otimes 1 \longmapsto (if_\tau)$ with $f_\tau \in \mathbb{R}$, $f_\tau > 0$. Corresponding to this isomorphism there is a decomposition

$$H \otimes_{\mathbb{Q}} \mathbb{R} \xrightarrow{\approx} \bigoplus_{\tau \in T} H_\tau$$

in which each H_τ is a complex vector space. Condition (a') implies $J = \oplus J_\tau$, where J_τ is a $\mathbb{C}$-linear isomorphism

$H_\tau \xrightarrow{\approx} H_\tau$ such that $J_\tau^2 = -1$. Let

$$H_\tau = H_\tau^+ \oplus H_\tau^-$$

where H_τ^+ and H_τ^- are the eigenspaces of J_τ with eigenvalues $+i$ and $-i$ respectively. The compatibility of ψ and ν implies

$$(H,\psi) \otimes \mathbb{R} \xrightarrow{\approx} \bigoplus_{\tau \in T} (H_\tau, \psi_\tau)$$

with ψ_τ an $\mathbb{R}$-bilinear alternating from on H_τ such that $\psi_\tau(zx,y) = \psi_\tau(x,\bar{z}y)$ for $z \in \mathbb{C}$. The condition $\psi(Jx,Jy) = \psi(x,y)$ implies that H_τ^+ is the orthogonal complement of H_τ^- relative to ψ_τ: $H_\tau = H_\tau^+ \perp H_\tau^-$. We also have

$$(H,\phi) \otimes \mathbb{R} \xrightarrow{\approx} \bigoplus_{\tau \in T} (H_\tau, \phi_\tau)$$

and $\psi_\tau(x,y) = \mathrm{Tr}_{\mathbb{C}/\mathbb{R}}(if_\tau \phi_\tau(x,y))$. As $\psi(x,y) = \sum_\tau \mathrm{Tr}_{\mathbb{C}/\mathbb{R}}(if_\tau \phi_\tau(x_\tau,y_\tau))$, we find

$$\begin{aligned}
\psi(x,Jx) > 0, \text{ all } x &\iff \mathrm{Tr}_{\mathbb{C}/\mathbb{R}}(if_\tau \phi_\tau(x_\tau,Jx_\tau)) > 0, \\
&\qquad \text{all } x_\tau, \tau, \\
&\iff \mathrm{Tr}_{\mathbb{C}/\mathbb{R}}(i\phi_\tau(x_\tau,Jx_\tau)) > 0 \text{ all } x_\tau, \tau, \\
&\iff \begin{cases} \phi_\tau \text{ is positive definite on } H_\tau^+, \text{ and} \\ \phi_\tau \text{ is negative definite on } H_\tau^- . \end{cases}
\end{aligned}$$

This shows, in particular, that $H_\tau^+ = H_\tau^{-1,0}$ and $H_\tau^- = H_\tau^{0,-1}$ each have dimension $d/2$ (cf. 4.4). Let X^+ and X^- be the sets of $J \in X$ for which $\psi(x,Jy)$ is positive definite and negative definite respectively. Then X is a disjoint union $X = X^+ \cup X^-$. As J is determined by its $+i$ eigenspace we see that X^+ can be identified with

$$\{(V_\tau)_{\tau\in T} | V_\tau \text{ a maximal subspace of } H_\tau \text{ such that } \phi_\tau > 0 \text{ on } V_\tau\} .$$

This is an open connected complex submanifold of a product of Grassman manifolds

$$X^+ \subset \prod_{\tau\in T} \mathrm{Grass}_{d/2}(V_\tau) .$$

Moreover, there is an analytic structure on $X^+ \times V(\mathbb{R})$ such that $X^+ \times V(\mathbb{R}) \to X^+$ is analytic and the inverse image of $J \in X^+$ is $V(\mathbb{R})$ with the complex structure provided by J . On dividing $V(\mathbb{R})$ by an O_E-stable lattice $V(\mathbb{Z})$ in V , we obtain the sought analytic family B of abelian varieties.

Note that A is a member of the family. We shall next show that there is also an abelian variety of the form $A_0 \otimes E$ in the family. To do this we only have to show that there exists a quadruple $(A_1, \theta_1, \nu_1, k_1)$ of the type discussed above with $A_1 = A_0 \otimes E$. Let A_0 be any abelian variety of dimension $d/2$ and define $\nu_1 \colon E \to \mathrm{End}(A_0 \otimes E)$ so that $e \in E$ acts on $H_1(A_0 \otimes E) = H_1(A_0) \otimes E$ through its action on E . A Riemann form ψ_0 on A_0 extends in an obvious way to a Riemann form ψ_1 on A_1 that is compatible with the action of E. We define θ_1 to be the corresponding polarization, and let ϕ_1 be the Hermitian form on $H_1(A_0 \otimes E, \mathbb{Q})$ such that $\psi_1 = Tr_{E/\mathbb{Q}}(f\phi_1)$ (see 4.6). If $I_0 \subset H_1(A_0, \mathbb{Q})$ is a totally isotropic subspace of $H_1(A_0, \mathbb{Q})$ of (maximum) dimension $d/2$ then $I_0 \otimes E$ is a totally isotropic subspace of dimensiona $d/2$ over E, which (by 4.2) shows that the Hermitian space $(H_1(A_0 \otimes E, \mathbb{Q}), \phi_1)$ is split. There is therefore an E-linear isomorphism $k_1: (H_1(A_0 \otimes E, \mathbb{Q}), \phi_1) \xrightarrow{\approx} (H, \phi)$, which carries $\psi_1 = Tr_{E/\mathbb{Q}}(f\phi_1)$ to $\psi = Tr_{E/\mathbb{Q}}(f\phi)$ This completes this part of the proof.

Let n be an integer ≥ 3, and let Γ be the set of O_E-isomorphisms $g: V(\mathbb{Z}) \to V(\mathbb{Z})$ preserving ψ and such that $(g-1)V(\mathbb{Z}) \subset nV(\mathbb{Z})$. Then Γ acts on X^+ by $J \longmapsto g \circ J \circ g^{-1}$ and (compatibly) on B. On forming the quotients, we obtain a map $\Gamma \backslash B \to \Gamma \backslash X^+$ which is an algebraic family of abelian varieties. In fact $\Gamma \backslash X^+$ is the moduli variety for quadruples $(A_1, \theta_1, \nu_1, k_1)$ in which A_1, θ_1 and ν_1 are essentially as before, but now k_1 is a level n structure $k_1: A_n(\mathbb{C}) = H_1(A, \mathbb{Z}/n\mathbb{Z}) \xrightarrow{\approx} V(\mathbb{Z})/nV(\mathbb{Z})$; the map $X^+ \to \Gamma \backslash X^+$ can be interpreted as "regard k_1 modulo n". To prove these facts, one can use the theorem of Baily-Borel [1] to show that $\Gamma \backslash X^+$ is algebraic and a theorem of Borel [2] to show that $\Gamma \backslash B$ is algebraic — see §6 where we discuss a similar question in greater detail.

Remark 4.9. With the notations of the theorem, let G be the $\mathbb{Q}$-rational algebraic group such that

$G(\mathbb{Q}) = \{g \in GL_E(H) \mid \exists \nu(g) \in \mathbb{Q}^\times$ such that $\psi(gx,gy) = \nu(g)\psi(x,y), \forall x,y \in H\}$. The homomorphism $h: \mathbb{C}^\times \to GL(H \otimes \mathbb{R})$ defined by the Hodge structure on $H_1(A,\mathbb{Q})$ factors through $G_{\mathbb{R}}$, and X can be identified with the $G(\mathbb{R})$-conjugacy class of homomorphisms $\mathbb{C}^\times \to G_{\mathbb{R}}$ containing h. Let K be the compact open subgroup of $G(\mathbb{A}^f)$ of g such that $(g-1)V(\hat{\mathbb{Z}}) \subset nV(\hat{\mathbb{Z}})$. Then $\Gamma\backslash X^+$ is a connected component of the Shimura variety $Sh_K(G,X)$. The general theory shows that $Sh_K(G,X)$ is a fine moduli scheme (see Deligne [3,§4] or V.2 below) and so, from this point of view, the only part of the above proof that is not immediate is that the connected component of $Sh_K(G,X)$ containing A also contains a variety $A_0 \otimes E$.

Remark 4.10. It is easy to construct algebraic cycles on $A_0 \otimes E$; any $\mathbb{Q}$-linear map $\lambda: E \to \mathbb{Q}$ defines a map $A_0 \otimes E \to A_0 \otimes \mathbb{Q} = A_0$, and we can take $cl(\lambda)$ = image of the class of a point in $H^d(A_0) \to H^d(A_0 \otimes E)$. More generally we have $\mathrm{Sym}^*(\mathrm{Hom}_{\mathbb{Q}\text{-linear}}(E,\mathbb{Q})) \to \{\text{algebraic cycles on } A_0 \otimes E\}$. If $E = \mathbb{Q}^r$, this gives the obvious cycles.

Remark 4.11. The argument in the proof of (4.8) is similar to, and was suggested by, an argument of B. Gross [1].

5. Completion of the proof for abelian varieties of CM-type.

The Mumford-Tate, or Hodge, group of an abelian variety A over $\mathbb{C}$ is defined to be the Mumford-Tate group of the rational Hodge structure $H_1(A,\mathbb{Q})$: it is therefore the subgroup of $GL(H_1(A,\mathbb{Q})) \times \mathbb{G}_m$ fixing all Hodge cycles (see §3). In the language of the next article, the category of rational Hodge structures is Tannakian with an obvious fibre functor, and the Mumford-Tate group of A is the group associated with the subcategory generated by $H_1(A,\mathbb{Q})$ and $\mathbb{Q}(1)$.

An abelian variety A is said to be of <u>CM-type</u> if its Mumford-Tate group is commutative. Since any abelian variety A is a product $A = \Pi A_\alpha$ of simple abelian varieties (up to isogeny) and A is of CM-type if and only if each A_α is of CM-type (the Mumford-Tate group of A is contained in the product of those of the A_α), in understanding this concept we can assume A is simple.

<u>Proposition</u> 5.1. A simple abelian variety A over $\mathbb{C}$ is of CM-type if and only if $E = \text{End } A$ is a commutative field over which $H_1(A,\mathbb{Q})$ has dimension 1. Then E is a CM-field, and the Rosati involution on $E = \text{End}(A)$ defined by any polarization of A is complex conjugation.

<u>Proof:</u> Let A be simple and of CM-type, and let μ: $\mathbb{G}_m \to GL(H_1(A,\mathbb{C}))$ be defined by the Hodge structure on $H_1(A,\mathbb{C})$ (see §3). As A is simple, $E = \text{End}(A)$ is a field (possibly noncommutative) of degree $\leq \dim H_1(A,\mathbb{Q})$ over $\mathbb{Q}$. As for any abelian variety, $\text{End}(A)$ is the subalgebra of $\text{End}(H_1(A,\mathbb{Q}))$ of elements commuting with the Hodge structure or, equivalently that commute with $\mu(\mathbb{G}_m)$ in $GL(H_1(A,\mathbb{C}))$.

If G is the Mumford-Tate group of A then $G_{\mathbb{C}}$ is generated by the groups $\{\sigma\mu(\mathbb{G}_m) | \sigma \in \mathrm{Aut}(\mathbb{C})\}$ (see 3.4). Therefore E is the commutant of G in $\mathrm{End}(H_1(A,\mathbb{Q}))$. By assumption G is a torus, and so $H_1(A,\mathbb{C}) = \bigoplus_{\chi \in X(G)} H_\chi$. The commutant of G therefore contains étalc commutative algebras of rank $\dim H_1(A,\mathbb{Q})$ over $\mathbb{Q}$. It follows that E is a commutative field of degree $\dim H_1(A,\mathbb{Q})$ over $\mathbb{Q}$ (and that it is generated, as a $\mathbb{Q}$-algebra, by $G(\mathbb{Q})$; in particular, $h(i) \in E \otimes \mathbb{R}$).

Let ψ be a Riemann form corresponding to some polarization on A. The Rosati involution $e \longmapsto e^*$ on $\mathrm{End}(A) = E$ is determined by the condition $\psi(x,ey) = \psi(e^*x,y)$, $x,y \in H_1(A,\mathbb{Q})$. It follows from $\psi(x,y) = \psi(h(i)x,h(i)y)$ that $h(i)^* = h(i)^{-1} (= -h(i))$. The Rosati involution therefore is non-trivial on E, and E has degree 2 over its fixed field F. We can write $E = F[\sqrt{\alpha}]$, $\alpha \in F$, $\sqrt{\alpha}^* = -\sqrt{\alpha}$; α is uniquely determined up to multiplication by a square in F. If E is identified with $H_1(A,\mathbb{Q})$ through the choice of an appropriate basis vector, then $\psi(x,y) = \mathrm{Tr}_{E/\mathbb{Q}}\, \alpha x y^*$, $x,y \in E$ (cf. 4.6). The positivity condition on ψ implies $\mathrm{Tr}_{E\otimes\mathbb{R}/\mathbb{R}}(\alpha h(i)^{-1} x x^*) > 0$, $x \neq 0$, $x \in E \otimes_{\mathbb{Q}} \mathbb{R}$. In particular, $\mathrm{Tr}_{F\otimes\mathbb{R}/\mathbb{R}}(f x^2) > 0$, $x \neq 0$, $x \in F \otimes \mathbb{R}$, $f = \alpha/h(i)$ which implies that F is totally real. Moreover, for every embedding $\sigma\colon F \hookrightarrow \mathbb{R}$ we must have $\sigma(\alpha) < 0$, for otherwise $E \otimes_{F,\sigma} \mathbb{R} = \mathbb{R} \times \mathbb{R}$ with $(r_1,r_2)^* = (r_2,r_1)$, and the positivity condition is impossible. Thus $\sigma(\alpha) < 0$, and $*$ is complex conjugation relative to any embedding of E in $\mathbb{C}$.

For the converse we only have to observe that $\mu(\mathbb{G}_m)$ commutes with $E \otimes \mathbb{R}$ in $\mathrm{End}(H_1(A,\mathbb{R}))$, and so if $H_1(A,\mathbb{Q})$ is of dimension 1 over E then $\mu(\mathbb{G}_m) \subset (E \otimes \mathbb{R})^{\times}$ and $G \subset E^{\times}$.

Let (A_α) be a finite family of abelian varieties over $\mathbb{C}$ of CM-type. We shall show that every element of a space

$$T_{\mathbb{A}} = (\otimes_\alpha H^1_{\mathbb{A}}(X_\alpha)^{\otimes m_\alpha}) \otimes (\otimes_\alpha H^1_{\mathbb{A}}(X_\alpha)^{\vee \otimes n_\alpha})(m)$$

that is a Hodge cycle (relative to $\mathbb{C} \xrightarrow{\mathrm{id}} \mathbb{C}$) is an absolute Hodge cycle. According to (3.8) (Principle A) to do this it suffices to show that the following two subgroups of $GL(\Pi H_1(A_\alpha,\mathbb{Q})) \times \mathbb{G}_m$ are equal:

G^H = group fixing all Hodge cycles;

G^{AH} = group fixing all absolute Hodge cycles.

Obviously $G^H \subset G^{AH}$.

After breaking up each A_α into its simple factors, we can assume A_α itself is simple. Let E_α be the CM-field $\mathrm{End}(A_\alpha)$ and let E be the smallest Galois extension of $\mathbb{Q}$ containing all E_α; it is again a CM-field. Let $B_\alpha = A_\alpha \otimes_{E_\alpha} E$. It suffices to prove the theorem for the family (B_α) (because the Tannakian category generated by the $H_1(B_\alpha)$ and $\mathbb{Q}(1)$ contains every $H_1(A_\alpha)$; cf. the next article).

In fact we consider an even larger family. Fix E, a CM-field Galois over $\mathbb{Q}$, and consider the family (A_α) of all abelian varieties with complex multiplication by E

(so $H_1(A_\alpha)$ has dimension 1 over E) up to E-isogeny. This family is indexed by $\mathcal{S}$, the set of CM-types for E. Thus, if $S = \mathrm{Hom}(E,\mathbb{C})$ then each element of $\mathcal{S}$ is a set $\Phi \subset S$ such that $S = \Phi \cup \iota\Phi$ (disjoint union). We often identify Φ with the characteristic function of Φ, i.e. we write

$$\Phi(s) = 1 , \; s \in \Phi$$
$$\Phi(s) = 0 , \; s \notin \Phi .$$

With each Φ we associate the isogeny class of abelian varieties containing the abelian variety $\mathbb{C}^\Phi/\Phi(O_E)$ where O_E = ring of integers in E and $\Phi(O_E) = \{(\sigma e)_{\sigma\in\Phi} \in \mathbb{C}^\Phi | \sigma \in O_E\}$.

With this new family we have to show that $G^H = G^{AH}$. We begin by determining G^H (cf. 3.7). The Hodge structure on each $H_1(A_\Phi,\mathbb{Q})$ is compatible with the action of E. This implies that

$$G^H \subset \prod_{\Phi\in\mathcal{S}} GL(H_1(A_\Phi)) \times \mathbb{G}_m$$

commutes with $\prod_{\Phi\in\mathcal{S}} E^\times$. It is therefore contained in $\prod E^\times \times \mathbb{G}_m$. In particular G^H is a torus, and can be described by its group of cocharacters $Y(G^H) \overset{\mathrm{df}}{=} \mathrm{Hom}_{\overline{\mathbb{Q}}}(\mathbb{G}_m,G^H)$ or its group of characters $X(G^H)$. Note that $Y(G^H) \subset Y(\prod_{\Phi\in\mathcal{S}} E^\times \times \mathbb{G}_m) = \mathbb{Z}^{S\times\mathcal{S}} \times \mathbb{Z}$. There is a canonical basis for $X(E^\times)$, namely S, and therefore a canonical basis for $X(\prod_{\Phi\in\mathcal{S}} E^\times \times \mathbb{G}_m)$ which we denote $((x_{s,\Phi}),x_0)$. We denote the dual basis for $Y(\Pi E^{\times} \times \mathbb{G}_m)$ by $(y_{s,\phi}, y_0)$. The element $\mu \in Y(G^H)$ equals $\sum_{s,\phi} \phi(s) y_{s,\phi} + y_0$ (see 3.7). As $G^H_{\mathbb{C}}$ is generated by $\{\sigma\mu(\mathbb{G}_m) | \sigma \in \operatorname{Aut} \mathbb{C}\}$, $Y(G^H)$ is the $\operatorname{Gal}(\overline{\mathbb{Q}}/\mathbb{Q})$-submodule of $Y(\Pi E^{\times} \times \mathbb{G}_m)$ generated by μ. ($\operatorname{Gal}(\overline{\mathbb{Q}}/\mathbb{Q})$ acts on S by $\sigma s = s \circ \sigma^{-1}$; it acts on $Y(\Pi E^{\times} \times \mathbb{G}_m) = \mathbb{Z}^{S \times \mathcal{S}} \times \mathbb{Z}$ through its action on S: $\sigma y_{s,\phi} = y_{\sigma s,\phi}$; these actions factor through $\operatorname{Gal}(E/\mathbb{Q})$).

To begin the computation of G^{AH}, we make a list of tensors that we know to be absolute Hodge cycles on the A_α.

(a) The endomorphisms $E \subset \operatorname{End}(A_\phi)$ for each ϕ. (More precisely we mean the classes $cl_{\mathbb{A}}(\Gamma_e) \in H_{\mathbb{A}}(A_\phi) \otimes H_{\mathbb{A}}(A_\phi)$, Γ_e = graph of e, $e \in E$.)

(b) Let $(A_\phi, \nu: E \hookrightarrow \operatorname{End}(A_\phi))$ correspond to $\phi \in \mathcal{S}$, and let $\sigma \in \operatorname{Gal}(E/\mathbb{Q})$. Define $\sigma\phi = \{\sigma s | s \in \phi\}$. There is an isomorphism $A_\phi \to A_{\sigma\phi}$ induced by

$$\begin{array}{ccc} \mathbb{C}^\phi & \longrightarrow & \mathbb{C}^\phi \\ \downarrow & & \downarrow \\ \mathbb{C}^\phi/\phi(O_E) & \longrightarrow & \mathbb{C}^{\sigma\phi}/\sigma\phi(O_E) \end{array} \qquad (\ldots, z(\tau), \ldots) \mapsto (\ldots, z(\sigma\tau), \ldots)$$

whose graph is an absolute Hodge cycle. (Alternatively, we could have used the fact that $(A_\phi, \sigma\nu: E \to \operatorname{End}(A_\phi))$, where $\sigma\nu = \nu \circ \sigma^{-1}$, is of type $\sigma\phi$ to show that A_ϕ and $A_{\sigma\phi}$ are isomorphic.)

(c) Let $(\phi_i)_{1 \le i \le d}$ be a family of elements of $\mathcal{S}$ and let $A = \bigoplus_{i=1}^{d} A_i$ where $A_i = A_{\phi_i}$. Then E acts on A and

$H_1(A,\mathbb{Q}) = \bigoplus_{i=1}^{d} H_1(A_i,\mathbb{Q})$ has dimension d over E. Under the assumption that $\sum_i \Phi_i$ = constant (so that $\sum_i \Phi_i(s) = d/2$, all $s \in S$) we shall apply (4.8) to construct absolute Hodge cycles on A.

For each i, there is an E-linear isomorphism

$$H_1(A_i,\mathbb{Q}) \otimes_{\mathbb{Q}} \mathbb{C} \xrightarrow{\approx} \bigoplus_{s \in S} H_1(A_i)_s$$

such that $s \in E$ acts on $H_1(A_i)_s$ as $s(e)$. From the definitions one sees that

$$\begin{aligned} H_1(A_i)_s &= H_1(A_i)_s^{-1,0}, \quad & s \in \Phi_i \\ &= H_1(A_i)_s^{0,-1} & s \notin \Phi_i . \end{aligned}$$

Thus, with the notations of (4.4),

$$\begin{aligned} a_s &= \sum_i \Phi_i(s) \\ b_s &= \sum_i (1-\Phi_i(s)) = \sum_i \Phi_i(\iota s) = a_{\iota s} . \end{aligned}$$

The assumption that $\sum \Phi_i$ = constant therefore implies $a_s = b_s = d/2$, all s.

For each i, choose a polarization θ_i for A_i whose Rosati involution stablizes E, and let ψ_i be the corresponding Riemann form. For any totally positive elements f_i in F (the maximal totally real subfield of E) $\theta = \oplus f_i\theta_i$ is a polarization for A. Choose $v_i \neq 0$, $v_i \in H_1(A_i,\mathbb{Q})$; then $\{v_i\}$ is a basis for $H_1(A_i,\mathbb{Q})$ over E. There exists a $\zeta_i \in E^{\times}$ such that $\bar{\zeta}_i = -\zeta_i$ and $\psi_i(xv_i,yv_i) = Tr_{E/\mathbb{Q}}(\zeta_i x\bar{y})$

for all $x,y \in E$. Thus ϕ_i, where $\phi_i(xv_i, yv_i) = (\zeta_i/\zeta_1)x\bar{y}$, is an E-Hermitian form on $H_1(A_i,\mathbb{Q})$ such that $\psi(v,w) = \mathrm{Tr}_{E/\mathbb{Q}}(\zeta_1\phi_i(v,w))$. The E-Hermitian form on $H_1(A,\mathbb{Q})$

$$\phi(\sum x_i v_i, \sum y_i v_i) = \sum_i f_i \phi_i(x_i v_i, y_i v_i)$$

is such that $\psi(v,w) = \mathrm{Tr}_{E/\mathbb{Q}}(\zeta_1\phi(v,w))$ is the Riemann form of θ. The discriminant of ϕ is $\prod_i f_i(\zeta_i/\zeta_1)$. On the other hand, if $s \in S$ restricts to τ on F, then $\mathrm{sign}(\tau\mathrm{disc}(\phi)) = (-1)^{b_s} = (-1)^{d/2}$. Thus $\mathrm{disc}\,\phi = f(-1)^{d/2}$ for some totally positive element f of F. After replacing one f_i with f_i/f, we have that $\mathrm{disc}(f) = (-1)^{d/2}$, and that ϕ is split. Hence (4.8) applies.

In summary: let $A = \bigoplus_{i=1}^{d} A_{\phi_i}$ be such that $\sum_i \phi_i = \text{constant}$; then $(\Lambda_E^d H^1(A,\mathbb{Q}))(d/2) \subset H^d(A,\mathbb{Q})(d/2)$ consists of absolute Hodge cycles.

Since G^{AH} fixes the absolute Hodge cycles of type (a), $G^{AH} \subset \prod_{\Phi} E^\times \times \mathbb{G}_m$. It is therefore a torus, and we have an inclusion

$$Y(G^{AH}) \subset Y(\Pi E^\times \times \mathbb{G}_m) = \mathbb{Z}^{S \times \mathcal{S}} \times \mathbb{Z}$$

and a surjection,

$$X(\Pi E^\times \times \mathbb{G}_m) = \mathbb{Z}^{S \times \mathcal{S}} \times \mathbb{Z} \twoheadrightarrow X(G^{AH}).$$

Let W be a space of absolute Hodge cycles. Under the action of the torus $\Pi E^\times \times \mathbb{G}_m$, $W \otimes \mathbb{C} \approx \bigoplus W_\chi$ where the sum is over $\chi \in X(\Pi E^\times \times \mathbb{G}_m)$ and the torus acts on W_χ through χ.

Since G^{AH} fixes the elements of W, the χ for which $W_\chi \neq 0$ map to zero in $X(G^{AH})$.

On applying this remark with W equal to the space of absolute Hodge cycles described in (b), we find that $x_{s,\phi} - x_{\sigma s,\sigma\phi}$ maps to zero in $X(G^{AH})$, all $\sigma \in \mathrm{Gal}(E/\mathbb{Q})$, $s \in S$, and $\phi \in \mathcal{S}$. As $\mathrm{Gal}(E/\mathbb{Q})$ acts simply transitively on S, this implies that, for a fixed $s_0 \in S$, $X(G^{AH})$ is generated by the image of $\{x_{s_0,\phi}, x_0 \mid \phi \in \mathcal{S}\}$.

Let $d(\phi) \geq 0$ be integers such that $\Sigma\, d(\phi)\phi = d/2$ (constant function on S) where $d = \Sigma d(\phi)$. Then (c) shows that

$$W = \otimes_E H_1(A_\phi,\mathbb{Q})^{\otimes_E d(\phi)}(-d/2) = \bigwedge_E^d H_1(\oplus A_\phi^{d(\phi)},\mathbb{Q})(-d/2)$$
$$\subset H_d(\oplus A_\phi^{d(\phi)},\mathbb{Q})(-d/2)\ .$$

consists of absolute Hodge cycles. The remark then shows that $\sum d(\phi)x_{s,\phi} - d/2$ maps to zero in $X(G^{AH})$ for all s.

Let $X = X(\Pi E^\times \times \mathbb{G}_m)/\sum \mathbb{Z}\,(x_{\sigma s,\sigma\phi} - x_{s,\phi})$, and regard $\{x_{s_0,\phi}, x_0 \mid \phi \in \mathcal{S}\}$ as a basis for X. We know that $X(\Pi E^\times \times \mathbb{G}_m) \twoheadrightarrow X(G^{AH})$ factors through X, and that therefore $Y \supset Y(G^{AH})$ $(\supset Y(G^H))$ where Y is the submodule of $Y(\Pi E^\times \times \mathbb{G}_m)$ dual to X.

<u>Lemma</u> 5.2. The submodule $Y(G^H)^\perp$ of X orthogonal to $Y(G^H)$ is equal to $\{\sum d(\phi)x_{s_0,\phi} - \frac{d}{2}x_0 \mid \sum d(\phi)\phi = \frac{d}{2}, \sum d(\phi) = d\}$; it is generated by elements $\sum d(\phi)x_{s_0,\phi} - (d/2)x_0$, $\sum d(\phi)\phi = d/2$, $d(\phi) \geq 0$ all ϕ.

Proof: As $Y(G^H)$ is the $Gal(E/\mathbb{Q})$-submodule of Y generated by μ, we see that $x = \sum d(\Phi)\, x_{s_0,\Phi} - d/2\, x_0 \in Y(G^H)^{\perp}$ if and only if $\langle \sigma\mu, x \rangle = 0$ all $\sigma \in Gal(E/\mathbb{Q})$. But $\mu = \sum \Phi(s) y_{s,\Phi} + y_0$ and $\sigma\mu = \sum \Phi(s) y_{\sigma s,\Phi} + x_0$, and so $\langle \sigma\mu, x \rangle = \sum d(\Phi)\Phi(\sigma^{-1}s_0) - d/2$. The first assertion is now obvious.

As $\Phi + \iota\Phi = 1$, $x_{s_0,\Phi} + x_{s_0,\iota\Phi} - x_0 \in Y(G^H)^{\perp}$ and has positive coefficients $d(\Phi)$. By adding enough elements of this form to an arbitrary element $x \in Y(G^H)^{\perp}$ we obtain an element with coefficients $d(\Phi) \geq 0$, which completes the proof of the lemma.

The lemma shows that $Y(G^H)^{\perp} \subset \mathrm{Ker}(X \twoheadrightarrow X(G^{AH})) = Y(G^{AH})^{\perp}$. Hence $Y(G^H) \subset Y(G^{AH})$ and it follows that $G^H = G^{AH}$; the proof is complete.

6. Completion of the proof; consequences.

Let A be an abelian variety over $\mathbb{C}$ and let t_α, $\alpha \in I$, be Hodge cycles on A (relative to $\mathbb{C} \xrightarrow{id} \mathbb{C}$). To prove the Main Theorem 2.11 we have to show the t_α are absolute Hodge cycles. Since we know the result for abelian varieties of CM-type, (2.15) shows that it remains only to prove the following proposition.

Proposition 6.1. There exists a connected, smooth (not necessarily complete) algebraic variety S over $\mathbb{C}$ and an abelian scheme $\pi: Y \to S$ such that

(a) for some $s_0 \in S$, $Y_{s_0} = A$;

(b) for some $s_1 \in S$, Y_{s_1} is of CM-type;

(c) the t_α extend to elements that are rational and of bidegree (0,0) everywhere in the family.

The last condition means the following. Suppose that t_α belongs to the tensor space $T_\alpha = H_B^1(A)^{\otimes m(\alpha)} \otimes \ldots$; then there is a section t of $R^1\pi_*\mathbb{Q}^{\otimes m(\alpha)}\otimes\ldots$ over the universal covering $\tilde{S}$ of S (equivalently, over a finite covering of S) such that for $\tilde{s}_0$ mapping to s_0, $t_{\tilde{s}_0} = t_\alpha$, and for all $\tilde{s} \in \tilde{S}$, $t_{\tilde{s}} \in H_B^1(Y_{\tilde{s}})^{\otimes m(\alpha)} \otimes \ldots$ is a Hodge cycle.

We sketch a proof of (6.1). (See also V.2). The parameter variety S will be a Shimura variety and (b) will hold for a dense set of points s_1.

We can assume that one of the t_α is a polarization θ for A. Let $H = H_1(A,\mathbb{Q})$ and let G be the subgroup of $GL(H) \times \mathbb{G}_m$ fixing the t_α. The Hodge structure on H defines a homomorphism $h_0\colon \mathbb{C}^\times \to G(\mathbb{R})$. Let $G^0 = \operatorname{Ker}(G \to \mathbb{G}_m)$; then $\operatorname{ad}(h_0(i))$ is a Cartan involution on $G^0_{\mathbb{C}}$ because the real form of $G^0_{\mathbb{C}}$ corresponding to it fixes the positive definite form $\psi(x,h(i)y)$ on $H \otimes \mathbb{R}$ where ψ is a Riemann form for θ. In particular, G is reductive (see 3.6).

Let $X = \{h\colon \mathbb{C}^\times \to G(\mathbb{R}) \mid h$ conjugate to h_0 under $G(\mathbb{R})\}$. Each $h \in X$ defines a Hodge structure on H of type $\{(-1,0),(0,-1)\}$ relative to which each t_α is of bidegree

(0,0). Let $F^{o}(h) = H^{0,-1} \subset H \otimes \mathbb{C}$. Since $G(\mathbb{R})/K_\infty \xrightarrow{\approx} X$, where K_∞ is the centralizer of h_0 , there is an obvious real differentiable structure on X , and the tangent space to X at h_0 , $tgt_{h_0}(X) \approx Lie(G_{\mathbb{R}})/Lie(K_\infty)$. In fact X is a Hermitian symmetric domain. The Grassmanian, $Grass_d(H\otimes\mathbb{C}) \stackrel{df}{=} \{W \subset H \otimes \mathbb{C} \mid W \text{ of dimension } d\}$, $d = \dim(A)$, is a complex analytic manifold (even an algebraic variety). The map ϕ: $X \to Grass_d(H \otimes \mathbb{C})$, $h \longmapsto F^{o}(h)$, is a real differentiable map, and is injective (because the Hodge filtration determines the Hodge decomposition). The map on tangent spaces factors into

$$
\begin{array}{ccc}
tgt_{h_0}(X) = Lie(G_{\mathbb{R}})/Lie(K_\infty) & \hookrightarrow & End(H\otimes\mathbb{C})/F^{o}End(H\otimes\mathbb{C}) = tgt_{\phi(h_0)}(Grass) \\
\downarrow \approx & \nearrow & \\
Lie(G_{\mathbb{C}})/F^{o}(Lie(G_{\mathbb{C}})), & &
\end{array}
$$

the maps being induced by $G(\mathbb{R}) \hookrightarrow G(\mathbb{C}) \hookrightarrow GL(H\otimes\mathbb{C})$. (The filtrations on $Lie(G_{\mathbb{C}})$ and $End(H\otimes\mathbb{C})$ are those corresponding to the Hodge structures defined by h_0). Thus $d\phi$ identifies $tgt_{h_0}(X)$ with a complex subspace of $tgt_{\phi(h_0)}(Grass)$, and so X is an almost-complex submanifold of $Grass_d(H\otimes\mathbb{C})$. It follows that it is a complex manifold (see Deligne [6,1.1] for more details). (There is an alternative, more group-theoretic description of the complex structure; see Knapp [1 , 2.4, 2.5]).

To each point h of X we can associate a complex torus $F^{o}(h)\backslash H\otimes\mathbb{C}/H(\mathbb{Z})$, where $H(\mathbb{Z})$ is some fixed lattice in H . For example, to h_0 is associated $F^{o}(h_0)\backslash H\otimes\mathbb{C}/H(\mathbb{Z}) = tgt_0(A)/H(\mathbb{Z})$, which is an abelian variety representing A .

From the definition of the complex structure on X it is clear that these tori form an analytic family B over X.

Let $\Gamma = \{g \in G(\mathbb{Q}) \mid (g-1)H(\mathbb{Z}) \subset nH(\mathbb{Z})\}$ some fixed integer n. For a suitably large $n \geq 3$, Γ will act freely on X, and so $\Gamma \backslash X$ will again be a complex manifold. The theorem of Baily and Borel [1] shows that $S = \Gamma \backslash X$ is an algebraic variety.

Γ acts compatibly on B, and on forming the quotients we obtain a complex analytic map $\pi\colon Y \to S$ with $Y = \Gamma \backslash B$. For $s \in S$, Y_s is a polarized complex torus (hence an abelian variety) with level n structure (induced by $H_1(B_h, \mathbb{Z}) \xrightarrow{\approx} H(\mathbb{Z})$ where h maps to s). The solution M_n of the moduli problem for polarized abelian varieties with level n-structure in the category of algebraic varieties is also a

solution in the category of complex analytic manifolds. There is therefore an analytic map $\psi\colon S \longrightarrow M_n$ such that Y is the pull-back of the universal family on M_n. A theorem of Borel [2,3.10] shows that ψ is automatically algebraic, from which it follows that Y/S is an algebraic family.

For some connected component S^o of S, $\pi^{-1}(S^o) \to S^o$ will satisfying (a) and (c) of the proposition. To prove (b) we shall show that, for some $h \in X$ close to h_0, B_h is of CM-type (cf. Deligne [3,5.2]).

Recall (§5) that an abelian variety is of CM-type if and only if its Mumford-Tate group is a torus. From this it follows that B_h, $h \in X$, is of CM-type if and only if h factors through a subtorus of G defined over $\mathbb{Q}$.

Let T be a maximal torus, defined over $\mathbb{R}$, of the algebraic group K_∞. (See Borel-Springer [1] for a proof that T exists.) Since $h_0(\mathbb{C}^\times)$ is contained in the centre of K_∞, $h_0(\mathbb{C}^\times) \subset T(\mathbb{R})$. If T' is any torus in $G_{\mathbb{R}}$ containing T then T' will centralize h_0 and so $T' \subset K_\infty$; T is therefore maximal in $G_{\mathbb{R}}$. For a general (regular) element λ of Lie(T), T is the centralizer of λ. Choose a $\lambda' \in \mathrm{Lie}(G)$ that is close to λ in $\mathrm{Lie}(G_{\mathbb{R}})$ and let T' be the centralizer of λ' in G. Then T' is a maximal torus of G that is defined over $\mathbb{Q}$ and $T' = gTg^{-1}$ where g is an element of $G(\mathbb{R})$ that is close to 1. Thus $h = \mathrm{ad}(g)\circ h_0$ is close to h_0 and B_h is of CM-type.

This completes the proof of the main theorem. We end this section by giving two immediate consequences.

Let X be a variety over a field k and let $\gamma \in H^{2p}(X_{et},\mathbb{Q}_\ell)(p)$, $\ell \neq \mathrm{char}(k)$; then Tate's conjecture asserts that γ is algebraic if and only if there exists a subfield k_0 of k finitely generated over the prime field, a model X_0 of X over k_0, and a $\gamma_0 \in H^{2p}(X_0\otimes\bar{k}_0,\mathbb{Q}_\ell)(p)$ mapping to γ that is fixed by $\mathrm{Gal}(\bar{k}_0/k_0)$. (Only the last condition is not automatic.)

Corollary 6.2. Let A be an abelian variety over $\mathbb{C}$. If Tate's conjecture is true for A then so also is the Hodge conjecture.

Proof: We first remark that, for any variety X over $\mathbb{C}$, Tate's conjecture implies that all absolute Hodge cycles on X are algebraic. For (2.9) shows that there exists a subfield k_0 of $\mathbb{C}$ finitely generated over $\mathbb{Q}$ and a model X_0 of

X over k_0 such that $\mathrm{Gal}(\bar{k}_0/k_0)$ acts trivially on $C^p_{AH}(X_0 \otimes \bar{k}_0)$. If we let $C^p_{alg}(X_0 \otimes \bar{k}_0)$ be the $\mathbb{Q}$-linear subspace of $C^p_{AH}(X_0 \otimes \bar{k}_0)$ of algebraic cycles, then Tate's conjecture shows that the images of C^p_{AH} and C^p_{alg} in $H^{2p}(X_{et}, \mathbb{Q}_\ell)(p)$ generate the same $\mathbb{Q}_\ell$-linear subspaces. Thus $C^p_{alg} \otimes \mathbb{Q}_\ell = C^p_{AH} \otimes \mathbb{Q}_\ell$, and $C^p_{alg} = C^p_{AH}$.

Now let A be an abelian variety over $\mathbb{C}$ and let $t \in H^{2p}(A,\mathbb{C})$ be rational of bidegree (p,p). If $t_0 \in H^{2p}(A,\mathbb{Q})$ maps to t, then the image t' of t_0 in $H^{2p}_{\mathbb{A}}(A)(p)$ is a Hodge cycle relative to $\mathbb{C} \xrightarrow{id} \mathbb{C}$. The main theorem shows that t' is an absolute Hodge cycle, and the remark shows that it is algebraic.

<u>Remark</u> 6.3. The above result was first proved independently by Piatetski-Shapiro [1] and Deligne (unpublished) by an argument similar to that which concluded the proof of the main theorem. ((6.2) is easy to prove for varieties of CM-type; in fact, Pohlmann [1] shows that the two conjectures are equivalent in that case.) We mention also that Borovoĭ [1] shows that, for an abelian variety X over a field k, the $\mathbb{Q}_\ell$-subspace of $H^{2p}(X_{et}, \mathbb{Q}_\ell)(p)$ generated by cycles that are Hodge relative to an embedding $\sigma\colon k \hookrightarrow \mathbb{C}$ is independent of the embedding.

<u>Corollary</u> 6.4. Let A be an abelian variety over $\mathbb{C}$ and let G^H be the Mumford-Tate group of A. Then $\dim(G^H) \geq \mathrm{tr.deg}_k k(p_{ij})$ where the p_{ij} are the periods of A.

<u>Proof</u>: Same as that of (1.6).

7. Algebraicity of values of the Γ-function

The following result generalizes (1.5)

Proposition 7.1. Let $\bar{k}$ be an algebraically closed subfield of $\mathbb{C}$, and let V be a variety of dimension n over $\bar{k}$. If $\sigma \in H_{2r}^{B}(V)$ maps to an absolute Hodge cycle γ under

$$H_{2r}^{B}(V) \xrightarrow{(2\pi i)^{-r}} H_{2r}^{B}(V)(-r) \xrightarrow{\approx} H_{B}^{2n-2r}(V)(n-r) \hookrightarrow H_{\mathbb{A}}^{2n-2r}(V_{\mathbb{C}})(n-r)$$

then, for any C^{∞} differential r-form ω on $V_{\mathbb{C}}$ whose class $[\omega]$ in $H_{DR}^{2r}(V/\mathbb{C})$ lies in $H_{DR}^{2r}(V/\bar{k})$,

$$\int_{\sigma} \omega \in (2\pi i)^{r}\bar{k}.$$

Proof: Proposition 2.9 shows that γ arises from an absolute Hodge cycle γ_{o} on $V/\bar{k}$. Let $(\gamma_{o})_{DR}$ be the component of γ in $H_{DR}^{2n-2r}(V/\bar{k})$. Then, as in the proof of (1.5),

$$\int_{\sigma} \omega = (2\pi i)^{r}\, \mathrm{Tr}_{DR}((\gamma_{o})_{DR} \cup [\omega]) \in (2\pi i)^{r}\, H_{DR}^{2n}(V/\bar{k}) = (2\pi i)^{r}\bar{k}.$$

In the most important case of the proposition, k will be the algebraic closure $\bar{\mathbb{Q}}$ of $\mathbb{Q}$ in $\mathbb{C}$, and it will then be important to know not only that the period

$$P(\sigma,\omega) \overset{df}{=} (2\pi i)^{-r} \int_{\sigma} \omega$$

is algebraic, but also which field it lies in. We begin by describing a general procedure for finding this field and then illustrate it by an example in which V is a Fermat hyperspace and the period is a product of values of the Γ-function.

Let V now be a variety over a number field $k \subset \mathbb{C}$ and let S be a finite abelian group acting on V over k.

If $\alpha: S \to \mathbb{C}^{\times}$ is a character of S taking values in $k^{\times}$ and H is a k vector space on which S acts k-linearly, then we write

$$H_{\alpha} = \{v \in H \mid sv = \alpha(s)v, \quad \text{all} \quad s \in S\} .$$

Assume that all Hodge cycles on $V_{\mathbb{C}}$ are absolutely Hodge and that $H^{2r}(V(\mathbb{C}),\mathbb{C})_{\alpha}$ has dimension 1 and is of bidegree (r,r). Then $(C^{r}_{AH}(\bar{V}) \otimes k)_{\alpha}$, where $\bar{V} = V \otimes_{k} \bar{\mathbb{Q}}$, has dimension one over k. The actions of S and $\mathrm{Gal}(\bar{\mathbb{Q}}/k)$ on $H^{2r}_{DR}(\bar{V}/\bar{\mathbb{Q}}) = H^{2r}_{DR}(V/k) \otimes_{k} \bar{\mathbb{Q}}$ commute because the latter acts through its action on $\bar{\mathbb{Q}}$; they therefore also commute on $C^{r}_{AH}(\bar{V}) \otimes k$, which embeds into $H^{2r}_{DR}(\bar{V}/\bar{\mathbb{Q}})$. It follows that $\mathrm{Gal}(\bar{\mathbb{Q}}/k)$ stabilizes $(C^{r}_{AH}(\bar{V}) \otimes k)_{\alpha}$ and, as this has dimension 1, there is a character $\chi: \mathrm{Gal}(\bar{\mathbb{Q}}/k) \to k^{\times}$ such that

$$\tau\gamma = \chi(\tau)^{-1}\gamma, \quad \tau \in \mathrm{Gal}(\bar{\mathbb{Q}}/k), \quad \gamma \in (C^{r}_{AH}(\bar{V}) \otimes k)_{\alpha}.$$

Proposition 7.2. With the above assumptions, let $\sigma \in H^{B}_{2r}(V)$ and let ω be a C^{∞} differential $2r$-form on $V(\mathbb{C})$ whose class $[\omega]$ in $H^{2r}_{DR}(V/\mathbb{C})$ lies in $H^{2r}_{DR}(V/k)_{\alpha}$; then $P(\sigma,\omega)$ lies in an abelian algebraic extension of k, and

$$\tau(P(\sigma,\omega)) = \chi(\tau)\, P(\sigma,\omega), \quad \text{all} \quad \tau \in \mathrm{Gal}(\bar{\mathbb{Q}}/k) .$$

Proof: Regard $[\omega] \in H^{2r}_{DR}(V/\mathbb{C})_{\alpha} = (C^{r}_{AH}(\bar{V}) \otimes \mathbb{C})_{\alpha}$; then $[\omega] = z\gamma$ for some $z \in \mathbb{C}$, $\gamma \in (C^{r}_{AH}(\bar{V}) \otimes k)_{\alpha}$. Moreover $P(\sigma,\omega) \overset{df}{=} (\frac{1}{2\pi i})^{r} \int_{\sigma} \omega = z\gamma(\sigma \otimes (2\pi i)^{-r}) \in zk$, where we are regarding γ as an element of $H^{2r}_{B}(V)(r) \otimes k = H^{B}_{2r}(V)(-r)^{\vee} \otimes k$. Thus $P(\sigma,\omega)^{-1}[\omega] \in (C^{r}_{AH}(\bar{V}) \otimes k)$. As $[\omega] \in H^{2r}_{DR}(\bar{V}/\bar{\mathbb{Q}}) = C^{r}_{AH}(\bar{V}) \otimes \bar{\mathbb{Q}}$, this shows that $P(\sigma,\omega) \in \bar{\mathbb{Q}}$. Moreover, $\tau(P(\sigma,\omega)^{-1}[\omega]) = \chi(\tau)^{-1}(P(\sigma,\omega)^{-1}[\omega])$. On using that $\tau[\omega] = [\omega]$, we deduce that $\tau(P(\sigma,\omega)) = \chi(\tau)\ P(\sigma,\omega)$.

Remark 7.3 (a) Because $C^r_{AH}(\bar{V})$ injects into $H^{2r}(\bar{V}_{et},\mathbb{Q}_\ell)(r)$, χ can be calculated from the action of $\mathrm{Gal}(\bar{\mathbb{Q}}/k)$ on $H^{2r}(\bar{V}_{et},\mathbb{Q}_\ell)_\alpha(r)$.

(b) The argument in the proof of the proposition shows that $\sigma \otimes (2\pi i)^{-r} \in H^B_{2r}(V)(-r)$ and $P(\sigma,\omega)^{-1}[\omega] \in H^{2r}_{DR}(\bar{V}/\bar{\mathbb{Q}})$ are different manifestations of the same absolute Hodge cycle.

The Fermat hypersurface

We shall apply (7.2) to the Fermat hypersurface

$$V:\ X_0^d + X_1^d + \ldots + X_{n+1}^d = 0$$

of degree d and dimension n , which will be regarded as a variety over $k \overset{\mathrm{df}}{=} \mathbb{Q}(e^{2\pi i/d})$. As above we write $\bar{V} = V \otimes_k \bar{\mathbb{Q}}$, and we shall often drop the subscript on $V_{\mathbb{C}}$.

It is known that the motive of V is contained in the category of motives generated by motives of abelian varieties (see (II 6.26)), and therefore (2.11) shows that every Hodge cycle on V is absolutely Hodge (cf. (II 6.27)) .

Let μ_d be the group of dth roots of 1 in $\mathbb{C}$, and let $S = \bigoplus_{i=0}^{n+1} \mu_d/(\text{diagonal})$. Then S acts on V/k according to the formula:

$$(\zeta_0 : \ldots)(x_0 : \ldots) = (\zeta_0 x_0 : \ldots)\ , \quad \text{all } (x_0 : \ldots) \in V(\mathbb{C})\ .$$

The character group of S will be identified with

$$X(S) = \{\underline{a} \in (\mathbb{Z}/d\,\mathbb{Z})^{n+2} \mid \underline{a} = (a_o, \dots, a_{n+1}) ,\ \Sigma a_i = o\} ;$$

$\underline{a} \in X(S)$ corresponds to the character

$$\underline{\zeta} = (\zeta_o : \dots) \longmapsto \underline{\zeta}^{\underline{a}} \overset{df}{=} \Pi\ \zeta_i^{a_i} .$$

For $a \in \mathbb{Z}/d\,\mathbb{Z}$ we let $\langle a\rangle$ denote the representative of a in $\mathbb{Z}$ with $1 \leq \langle a\rangle \leq d$, and for $\underline{a} \in X(S)$ we let $\langle \underline{a}\rangle = d^{-1}\ \Sigma \langle a_i\rangle \in \mathbb{N}$.

If $H(V)$ is a cohomology group on which there is a natural action of k , we have a decomposition

$$H(V) = \oplus H(V)_{\underline{a}} ,\ H(V)_{\underline{a}} = \{v \mid \underline{\zeta}v = \underline{\zeta}^{\underline{a}} v ,\ \underline{\zeta} \in S\}$$

Let $(\mathbb{Z}/d\,\mathbb{Z})^{\times}$ act on $X(S)$ in the obvious way, $u(a_o, \dots) = (ua_o, \dots)$, and let $[\underline{a}]$ be the orbit of $\underline{a}$. The irreducible representations of S over $\mathbb{Q}$ (and hence the idempotents of $\mathbb{Q}[S]$) are classified by these orbits, and so $\mathbb{Q}[S] = \Pi\ \mathbb{Q}[\underline{a}]$ where $\mathbb{Q}[\underline{a}]$ is a field whose degree over $\mathbb{Q}$ is equal to the order of $[\underline{a}]$. The map $\underline{\zeta} \longmapsto \underline{\zeta}^{\underline{a}} : S \to \mathbb{C}$ induces an embedding $\mathbb{Q}[\underline{a}] \hookrightarrow k$. Any cohomology group decomposes as $H(V) = \oplus H(V)_{[\underline{a}]}$ where $H(V)_{[\underline{a}]} \otimes \mathbb{C} = \oplus_{\underline{a}' \in [\underline{a}]} (H(V) \otimes \mathbb{C})_{\underline{a}'}$.

<u>Calculation of the cohomology</u>

<u>Proposition</u> 7.4 The dimension of $H^n(V,\mathbb{C})_{\underline{a}}$ is 1 if no $a_i = o$ or if all $a_i = 0$; otherwise $H^n(V,\mathbb{C})_{\underline{a}} = 0$.

<u>Proof</u>: The map

$$(x_o : x_1 : \dots) \longmapsto (x_o^d : x_1^d : \dots) : \mathbb{P}^{n+1} \longrightarrow \mathbb{P}^{n+1}$$

defines a finite surjective map $\pi: V \to P^n$ where $P^n(\approx \mathbb{P}^n)$ is the hyperplane $\Sigma X_i = 0$. There is an action of S on $\pi_*\mathbb{C}$ which induces a decomposition $\pi_*\mathbb{C} \approx \oplus(\pi_*\mathbb{C})_{\underline{a}}$, and $H^r(V,\mathbb{C}) \xrightarrow{\approx} H^r(P^n,\pi_*\mathbb{C})$, being compatible with the actions of S, gives rise to isomorphisms $H^r(V,\mathbb{C})_{\underline{a}} \xrightarrow{\approx} H^r(P^n,(\pi_*\mathbb{C})_{\underline{a}})$. The sheaf $(\pi_*\mathbb{C})_{\underline{a}}$ is locally constant of dimension 1 except over the hyperplanes $H_i: X_i = o$ corresponding to i for which $a_i \neq o$, where it is ramified. Clearly $(\pi_*\mathbb{C})_{\underline{o}} = \mathbb{C}$, and so $H^r(P^n,(\pi_*\mathbb{C})_{\underline{o}}) \approx H^r(\mathbb{P}^n,\mathbb{C})$ for all r. It follows that $H^r(P^n,(\pi_*\mathbb{C})_{\underline{a}}) = o$, $r \neq n$, $\underline{a} \neq \underline{o} = (o,\ldots,o)$, and so $(-1)^n \dim H^n(P^n,(\pi_*\mathbb{C})_{\underline{a}})$, $\underline{a} \neq \underline{o}$, is equal to the Euler-Poincaré characteristic of $(\pi_*\mathbb{C})_{\underline{a}}$. We have

$$EP(P^n,(\pi_*\mathbb{C})_{\underline{a}}) = EP(P^n - \bigcup_{a_i \neq o} H_i,\mathbb{C}) .$$

Suppose first that no a_i is zero. Then

$$(x_o: \ldots: x_n: -\Sigma x_i) \longleftrightarrow (x_o: \ldots: x_n): P^n \xleftrightarrow{\approx} \mathbb{P}^n$$

induces

$$P^n - \bigcup_{i=o}^{n+1} H_i \xleftrightarrow{\approx} \mathbb{P}^n - \bigcup_{i=o}^{n} H_i \cup P^{n-1} ,$$

where H_i denotes the coordinate hyperplane in $\mathbb{P}^{n+1}$ or $\mathbb{P}^n$. As

$$(\mathbb{P}^n - \cup H_i \cup P^{n-1}) \amalg (P^{n-1} - \cup H_i) = \mathbb{P}^n - \cup H_i ,$$

and $\mathbb{P}^n - \cup H_i$, being topologically isomorphic to $(\mathbb{C}^\times)^n$, has Euler-Poincaré characteristic zero, we have
$EP(P^n - \bigcup^{n+1} H_i) = -EP(P^{n-1} - \bigcup^{n} H_i) = \ldots = (-1)^n EP(P^\circ) = (-1)^n$.
If some, but not all, a_i are zero, then $P^n - \cup H_i \approx (\mathbb{C}^\times)^r \times \mathbb{C}^{n-r}$

with $r \geq 1$, and so $EP(P^n - \cup H_i) = 0^r \times 1^{n-r} = 0$.

Remark 7.5. The above proof shows also that the primitive cohomology of V ,

$$H^n(V,\mathbb{C})_{prim} = \bigoplus_{\underline{a}\neq \underline{o}} H^n(V,\mathbb{C})_{\underline{a}} .$$

The action of S on $H^n(V,\mathbb{C})$ respects the Hodge decomposition, and so $H^n(V,\mathbb{C})_{\underline{a}}$ is pure of bidegree (p,q) for some p , q with $p + q = n$.

Proposition 7.6. If no $a_i = o$, then $H^n(V,\mathbb{C})_{\underline{a}}$ is of bidegree (p,q) with $p = \langle a \rangle - 1$.

Proof: We apply the method of Griffiths [1,§8]. When V is a smooth hypersurface in $\mathbb{P}^{n+1}$, Griffiths shows that the maps in

$$H^{n+1}(\mathbb{P}^{n+1},\mathbb{C}) \xrightarrow{o} H^{n+1}(\mathbb{P}^{n+1} - V,\mathbb{C}) \to H^{n+2}_V(\mathbb{P}^{n+1},\mathbb{C}) \to H^{n+2}(\mathbb{P}^{n+1},\mathbb{C})$$
$$H^{n+2}_V(\mathbb{P}^{n+1},\mathbb{C}) \xrightarrow{\simeq} H^n(V)(-1)$$

induce an isomorphism

$$H^{n+1}(\mathbb{P}^{n+1} - V,\mathbb{C}) \xrightarrow{\approx} H^n(V)(-1)_{prim}$$

and that the Hodge filtration on $H^n(V)(-1)$ has the following explicit interpretation: identify $H^{n+1}(\mathbb{P}^{n+1} - V,\mathbb{C})$ with $\Gamma(\mathbb{P}^{n+1} - V, \Omega^{n+1})/d\Gamma(\mathbb{P}^{n+1} - V,\Omega^n)$ and let

$$\Omega^{n+1}_p(V) = \{\omega \in \Gamma(\mathbb{P}^{n+1} - V,\Omega^{n+1}) \mid \omega \text{ has a pole of order } \leq p \text{ on } V\} ;$$

then the map $R: \Omega_p^{n+1}(V) \to H^n(V,\mathbb{C})$ determined by

$$\langle \sigma, R(\omega) \rangle = \frac{1}{2\pi i} \int_\sigma \omega , \text{ all } \sigma \in H_n(V,\mathbb{C})$$

induces an isomorphism

$$\Omega_p^{n+1}(V)/d\Omega_{p-1}^n \xrightarrow{\approx} F^{n-p} H^n(V)(-1)_{prim} = F^{n-p+1} H^n(V)_{prim} .$$

(For example, if $p = 1$, R is the residue map $\Omega_1^{n+1}(V) \to F^n H^n(V) = H^0(V,\Omega^n)$).

Let f be the irreducible polynomial defining V. As $\Omega^{n+1}_{\mathbb{P}^{n+1}}(n+2) \approx O_{\mathbb{P}^{n+1}}$ has basis

$$\omega_0 = \Sigma(-1)^i X_i \, dX_0 \wedge \ldots \wedge \widehat{dX_i} \wedge \ldots \wedge dX_n ,$$

any differential form $\omega = P\omega_0/f^p$ with P a homogeneous polynomial of degree $p \deg(f) - (n+2)$ lies in $\Omega_p^{n+1}(V)$. In particular, when V is our Fermat hypersurface,

$$\omega = \frac{X_0^{\langle a_0\rangle -1} \cdots X_{n+1}^{\langle a_{n+1}\rangle -1}}{(X_0^d + \ldots + X_{n+1}^d)^{\langle \underline{a}\rangle}} \omega_0 = \frac{X_0^{\langle a_0\rangle} \cdots X_{n+1}^{\langle a_{n+1}\rangle}}{(X_0^d+\ldots+X_{n+1}^d)^{\langle \underline{a}\rangle}} \Sigma(-1)^i \frac{dX_0}{X_0} \wedge \ldots \wedge \widehat{\frac{dX_i}{X_i}} \wedge \ldots$$

lies in $\Omega^{n+1}_{\langle \underline{a}\rangle}(V)$. For $\underline{\zeta} \in S$, $\underline{\zeta} X_i = \zeta_i^{-1} X_i$, and so $\underline{\zeta}\omega = \underline{\zeta}^{-\underline{a}} \omega$. This shows that $H^n(V,\mathbb{C})_{-\underline{a}} \subset F^{n-\langle \underline{a}\rangle +1} H^n(V,\mathbb{C})$. Since $\langle -\underline{a}\rangle - 1 = n + 1 - \langle \underline{a}\rangle$, we can rewrite this inclusion as $H^n(V,\mathbb{C})_{\underline{a}} \subset F^{\langle \underline{a}\rangle -1} H^n(V,\mathbb{C})$. Thus $H^n(V,\mathbb{C})_{\underline{a}}$ is of bidegree (p,q) with $p \geq \langle \underline{a}\rangle - 1$. The complex conjugate of $H^n(V,\mathbb{C})_{\underline{a}}$ is $H^n(V,\mathbb{C})_{-\underline{a}}$, and is of bidegree (q,p). Hence

$$n - p = q \geq \langle -\underline{a}\rangle - 1 = n + 1 - \langle \underline{a}\rangle$$

and so $p \leq \langle \underline{a}\rangle - 1$.

Recall that $H^n_B(V)_{[\underline{a}]} = \bigoplus_{\underline{a}' \in [\underline{a}]} H^n_B(V)_{\underline{a}'}$; thus (7.4) shows that $H^n_B(V)_{[\underline{a}]}$ has dimension 1 over $\mathbb{Q}[\underline{a}]$ when no a_i is zero and otherwise $H^n_B(V)_{[\underline{a}]} \cap H^n_B(V)_{prim} = o$.

Corollary 7.7. Assume no $a_i = o$; $H^n_B(V)_{[\underline{a}]}$ is purely of bidegree $(\frac{n}{2},\frac{n}{2})$ if and only if $\langle u\underline{a}\rangle$ is independent of u .

Proof: As $\langle \underline{a}\rangle + \langle -\underline{a}\rangle = n + 2$, $\langle u\underline{a}\rangle$ is constant if and only if $\langle u\underline{a}\rangle = \frac{n}{2} + 1$ for all $u \in (\mathbb{Z}/d\,\mathbb{Z})^{\times}$, i.e. if and only if $\langle \underline{a}'\rangle = n/2 + 1$ for all $\underline{a}' \in [\underline{a}]$. Thus the corollary follows from the proposition.

Corollary 7.8. If no $a_i = o$ and $\langle u\underline{a}\rangle$ is constant, then $C^n_{AH}(\bar{V})_{[\underline{a}]}$ has dimension one over $\mathbb{Q}[\underline{a}]$.

Proof: This follows immediately from (7.7) since, as we have remarked, all Hodge cycles on V are absolutely Hodge.

The action of $Gal(\bar{\mathbb{Q}}/k)$ on the étale cohomology.

Let $\mathfrak{p}$ be a prime ideal of k not dividing d , and let $\mathbb{F}_q$ be the residue field of $\mathfrak{p}$. Then $d|q-1$ and reduction modulo $\mathfrak{p}$ defines an isomorphism $\mu_d \xrightarrow{\approx} \mathbb{F}_d^{\times}$ whose inverse we denote t . Fix an $\underline{a} = (a_o,\ldots,a_{n+1}) \in X(S)$ with all a_i nonzero, and define a character $\varepsilon_i \colon \mathbb{F}_q^{\times} \to \mu_d$ by

$$\varepsilon_i(x) = t(x^{(1-q)/d})^{a_i} \ , \ x \neq 0 \ .$$

As $\Pi\, \varepsilon_i = 1$, $\Pi\, \varepsilon_i(x_i)$ is well-defined for $\underline{x} = (x_o:\ldots:x_{n+1}) \in \mathbb{P}^{n+1}(\mathbb{F}_q)$, and we define a Jacobi sum

$$J(\varepsilon_o,\ldots,\varepsilon_{n+1}) = (-1)^n \sum_{\underline{x}\in P^n(\mathbb{F}_q)} \prod_{i=o}^{n+1} \varepsilon_i(x_i)$$

where P^n is the hyperplane $\Sigma X_i = o$ in $\mathbb{P}^{n+1}$. (As usual, we set $\varepsilon_i(o) = o$.) Let ψ be a nontrivial additive character $\psi: \mathbb{F}_q \to \mathbb{C}^\times$ and define Gauss sums

$$g(\mathscr{Y},a_i,\psi) = -\sum_{x\in\mathbb{F}_q} \varepsilon_i(x)\ \psi(x)$$

$$g(\mathscr{Y},\underline{a}) = q^{-\langle\underline{a}\rangle} \prod_{i=o}^{n+1} g(\mathscr{Y},a_i,\psi) .$$

<u>Lemma</u> 7.9. The Jacobi sum $J(\varepsilon_o,\ldots,\varepsilon_{n+1}) = q^{\langle\underline{a}\rangle -1}\, g(\mathscr{Y},\underline{a})$.

<u>Proof</u>: $q^{\langle\underline{a}\rangle} g(\mathscr{Y},\underline{a}) = \prod_{i=o}^{n+1} (-\sum_{x\in\mathbb{F}_q} \varepsilon_i(x)\ \psi(x))$

$$= (-1)^n \sum_{\underline{x}\in\mathbb{F}_q^{n+1}} ((\prod_{i=o}^{n+1} \varepsilon_i(x_i))\ \psi(\Sigma x_i)),\ \underline{x} = (x_o,\ldots),$$

$$= (-1)^n \sum_{\underline{x}\in P^n(\mathbb{F}_q)} \sum_{\lambda\in\mathbb{F}_q^\times} (\prod_{i=o}^{n+1} \varepsilon_i(\lambda x_i)\ \psi(\lambda\Sigma x_i)) .$$

We can omit the λ from $\Pi\,\varepsilon_i(\lambda x_i)$, and so obtain

$$q^{\langle\underline{a}\rangle} g(\mathscr{Y},\underline{a}) = (-1)^n \sum_{\underline{x}} (\prod_{i=o}^{n+1} \varepsilon_i(x_i)) \sum_{\lambda\in\mathbb{F}_q^\times} \psi(\lambda\Sigma x_i) .$$

Since $\sum_{\underline{x}} \prod_{i=o}^{n+1} \varepsilon_i(x_i) = \prod_{i=o}^{n+1} (\sum_{x\in\mathbb{F}_q} \varepsilon_i(x)) = o$, we can replace the sum over $\lambda\in\mathbb{F}_q^\times$ by a sum over $\lambda\in\mathbb{F}_q$. From

$$\sum_{\lambda \in \mathbb{F}_q} \psi(\lambda \Sigma x_i) = \begin{cases} q & \text{if } \Sigma x_i = 0 \\ 0 & \text{if } \Sigma x_i \neq 0 \end{cases}$$

we deduce finally that

$$q^{\langle \underline{a} \rangle} g(\wp, \underline{a}) = (-1)^n q \sum_{\underline{x} \in P^n(\mathbb{F}_q)} \left(\prod_{i=0}^{n+1} \varepsilon_i(x_i) \right)$$

$$= q \, J(\varepsilon_0, \ldots, \varepsilon_n) \, .$$

Note that this shows that $g(\wp, \underline{a})$ is independent of ψ and lies in k.

Let ℓ be a prime such that $\ell \nmid d$, $\wp \nmid \ell$, and $d | \ell - 1$. Then $\mathbb{Q}_\ell$ contains a primitive d^{th} root of 1 and so, after choosing an embedding $k \hookrightarrow \mathbb{Q}_\ell$, we can assume $g(\wp, \underline{a}) \in \mathbb{Q}_\ell$.

Proposition 7.10. Let $F_\wp \in \mathrm{Gal}(\mathbb{Q}/k)^{ab}$ be a geometric Frobenius element of $\wp \nmid d$; for any $v \in H^n(\bar{V}_{et}, \mathbb{Q}_\ell)_{\underline{a}}$,

$$F_\wp v = q^{\langle \underline{a} \rangle - 1} g(\wp, \underline{a}) v$$

Proof: As $\wp \nmid d$, V reduces to a smooth variety $V_\wp$ over $\mathbb{F}_q$ and the proper-smooth base change theorem shows that there is an isomorphism $H^n(\bar{V}, \mathbb{Q}_\ell) = H^n(\bar{V}_\wp, \mathbb{Q}_\ell)$ compatible with the action of S and carrying the action of $F_\wp$ on $H^n(\bar{V}, \mathbb{Q}_\ell)$ into the action of the Frobenius endomorphism Frob on $H^n(\bar{V}_\wp, \mathbb{Q}_\ell)$. The comparison theorem shows that $H^n(\bar{V}, \mathbb{Q}_\ell)_{\underline{a}}$ has dimension 1, and so it remains to compute

$$\mathrm{Tr}(F_\wp | H^n(\bar{V}, \mathbb{Q}_\ell)_{\underline{a}}) = \mathrm{Tr}(\mathrm{Frob} | H^n(\bar{V}_\wp, \mathbb{Q}_\ell)_{\underline{a}}) \, .$$

Let $\pi: V_{\mathfrak{X}} \to P^n$ be as before. Then $H^n(\bar{V}_{\mathfrak{X}}, \mathbb{Q}_\ell)_{\underline{a}} = H^n(P^n, (\pi_*\mathbb{Q}_\ell)_{\underline{a}})$, and the Lefschetz trace formula shows that

$$(-1)^n \operatorname{Tr}(\operatorname{Frob}|H^n(P^n, (\pi_*\mathbb{Q}_\ell)_{\underline{a}}) = \sum_{\underline{x} \in P^n(\mathbb{F}_q)} \operatorname{Tr}(\operatorname{Frob}|((\pi_*\mathbb{Q}_\ell)_{\underline{a}})_{\underline{x}}) \qquad (7.10.1)$$

where $((\pi_*\mathbb{Q}_\ell)_{\underline{a}})_{\underline{x}}$ is the stalk of $(\pi_*\mathbb{Q}_\ell)_{\underline{a}}$ at $\underline{x}$.

Fix an $\underline{x} \in P^n(\mathbb{F}_q)$ with no x_i zero, and let $\underline{y} \in V_{\mathfrak{X}}(\bar{\mathbb{F}}_q)$ map to $\underline{x}$; thus $y_i^d = x_i$ all i. Then $\pi^{-1}(\underline{x}) = \{\underline{\zeta}\underline{y} | \underline{\zeta} \in S\}$, and $(\pi_*\mathbb{Q}_\ell)_{\underline{x}}$ is the vector space $\mathbb{Q}_\ell^{\pi^{-1}(x)}$.

If ϕ denotes the arithmetic Frobenius automorphism (i.e., the generator $z \mapsto z^q$ of $\operatorname{Gal}(\bar{\mathbb{F}}_q/\mathbb{F}_q)$) then

$$\phi(y_i) = y_i^q = x_i^{\frac{q-1}{d}} y_i = t(x_i^{\frac{q-1}{d}}) y_i, \qquad 0 \le i \le n+1$$

and so

$$\operatorname{Frob}(\underline{y}) = \underline{\eta}\, \underline{y} \quad \text{where} \quad \underline{\eta} = (\ldots : t(x_i^{\frac{1-q}{d}}) : \ldots) \in S .$$

Thus Frob acts on $(\pi_*\mathbb{Q}_\ell)_{\underline{x}}$ as $\underline{\eta}$, and for $v \in ((\pi_*\mathbb{Q}_\ell)_{\underline{a}})_{\underline{x}}$ we have

$$\operatorname{Frob}(v) = \underline{\eta}\, v = \underline{\eta}^{\underline{a}} v, \quad \underline{\eta}^{\underline{a}} = \prod_{i=o}^{n+1} \varepsilon_i(x_i) \in k \subset \mathbb{Q}_\ell .$$

Consequently

$$\operatorname{Tr}(\operatorname{Frob}|((\pi_*\mathbb{Q}_\ell)_{\underline{a}})_{\underline{x}}) = \prod_{i=o}^{n+1} \varepsilon_i(x_i) .$$

If some $x_i = o$ then both sides are zero ($(\pi_*\mathbb{Q}_\ell)_{\underline{a}}$ is ramified over the coordinate hyperplanes), and so on summing over $\underline{x}$ and applying (7.10.1) and (7.9), we obtain the proposition.

Corollary 7.11. Assume that no a_i is zero and that $\langle u\underline{a}\rangle$ is constant. Then, for any $v \in H^n(\bar{V}_{et}, \mathbb{Q}_\ell)_{\underline{a}}(\frac{n}{2})$,

$$F_{\mathfrak{p}} v = g(\mathfrak{p}, \underline{a})\, v .$$

Proof: The hypotheses on $\underline{a}$ imply that $\langle \underline{a}\rangle = n/2 + 1$; therefore, if we write $v = v_o \otimes 1$ with $v_o \in H^n(\bar{V}_{et}, \mathbb{Q}_\ell)_{\underline{a}}$, then

$$F_{\mathfrak{p}} v = F_{\mathfrak{p}} v_o \otimes F_{\mathfrak{p}} 1 = q^{n/2} g(\mathfrak{p}, \underline{a})\, v_o \otimes q^{-n/2} = g(\mathfrak{p}, \underline{a})\, v .$$

Calculation of the periods

Recall that the Γ-function is defined by

$$\Gamma(s) = \int_o^\infty e^{-t}\, t^s \frac{dt}{t}, \quad s > o ,$$

and satisfies the following equations

$$\Gamma(s)\,\Gamma(1-s) = \pi(\sin \pi s)^{-1}$$
$$\Gamma(1+s) = s\,\Gamma(s) .$$

The last equation shows that, for $s \in \mathbb{Q}^\times$, the class of $\Gamma(s)$ in $\mathbb{C}/\mathbb{Q}^\times$ depends only on the class of s in $\mathbb{Q}/\mathbb{Z}$. Thus, for $\underline{a} \in X(S)$, we can define

$$\tilde{\Gamma}(\underline{a}) = (2\pi i)^{-\langle\underline{a}\rangle} \prod_{i=o}^{n+1} \Gamma(a_i/d) \in \mathbb{C}/\mathbb{Q}^\times .$$

Let V^o denote the open affine subvariety of V with equation

$$Y_1^d + \ldots + Y_{n+1}^d = -1 \qquad (\text{so } Y_i = X_i/X_o) .$$

Denote by Δ the n-simplex $\{(t_1,\ldots,t_{n+1}) \in \mathbb{R}^{n+1} \mid t_i \geq o,\ \Sigma\, t_i = 1\}$ and define $\sigma_o : \Delta \to V^o(\mathbb{C})$ to be

$$(t_1,\ldots,t_{n+1}) \mapsto (\varepsilon t_1^{1/d},\ldots,\varepsilon t_{n+1}^{1/d}) , \quad \varepsilon = e^{2\pi i/2d} = \sqrt[d]{-1} , \quad t_i^{1/d} > o .$$

<u>Lemma</u> 7.12, Let a_o , ..., a_{n+1} be positive integers such that $\Sigma a_i = o$. Then

$$\int_{\sigma_o(\Delta)} Y_1^{a_1} \cdots Y_{n+1}^{a_{n+1}} \frac{dY_1}{Y_1}\wedge\ldots\wedge\frac{dY_n}{Y_n} = \frac{1}{2\pi i}(1-\xi^{-a_o}) \prod_{i=o}^{n+1} \Gamma(\frac{a_i}{d})$$

where $\xi = e^{2\pi i/d}$.

<u>Proof</u>: Write ω_o for the integrand. Then

$$\int_{\sigma_o(\Delta)} \omega_o = \int_\Delta \sigma_o^*(\omega_o) = \int_\Delta (\varepsilon t_1^{1/d})^{a_1} \cdots (\varepsilon t_{n+1}^{1/d})^{a_{n+1}} d^{-n} \frac{dt_1}{t_1}\wedge\ldots\wedge\frac{dt_n}{t_n}$$

$$= c \int_\Delta t_1^{b_1} \cdots t_{n+1}^{b_{n+1}} \frac{dt_1}{t_1}\wedge\ldots\wedge\frac{dt_n}{t_n}$$

where $b_i = a_i/d$ and $c = \varepsilon^{a_1+\ldots+a_{n+1}}(\frac{1}{d})^n$. On multiplying by $\Gamma(1-b_o) = \Gamma(1+b_1+\ldots+b_{n+1}) = \int_o^\infty e^{-t} t^{b_1+\ldots+b_{n+1}} dt$ we obtain

$$\Gamma(1-b_o) \int_{\sigma_o(\Delta)} \omega_o = c \int_o^\infty \int_\Delta e^{-t} t^{b_1+\ldots+b_{n+1}} t_1^{b_1}\ldots t_{n+1}^{b_{n+1}} \frac{dt_1}{t_1}\wedge\ldots\wedge\frac{dt_n}{t_n}\wedge dt .$$

If, on the inner integral, we make the change of variables $s_i = tt_i$, the integral becomes

$$c \int_o^\infty \int_{\Delta(t)} e^{-t} s_1^{b_1}\ldots s_{n+1}^{b_{n+1}} \frac{ds_1}{s_1}\wedge\ldots\wedge\frac{ds_n}{s_n}\wedge dt$$

where $\Delta(t) = \{(s_1,\ldots,s_{n+1}) \mid s_i \geq o , \Sigma s_i = t\}$. We now let $t = \Sigma s_i$, and obtain

$$\Gamma(1-b_o) \int_{\sigma_o(\Delta)} \omega = c \int_o^\infty \cdots \int_o^\infty e^{-s_1-s_2\ldots-s_{n+1}} s_1^{b_1} \cdots s_{n+1}^{1+b_{n+1}} \frac{ds_\cdot}{s_1}\wedge\ldots\wedge\frac{ds_{n+1}}{s_{n+1}}$$

$$= c\,\Gamma(b_1)\,\Gamma(b_2)\ldots\Gamma(b_n)\,\Gamma(1+b_{n+1})$$

$$= c\,b_{n+1}\,\Gamma(b_1)\ldots\Gamma(b_{n+1}) .$$

The formula recalled above shows that $\Gamma(1-b_o) = \pi/(\sin\pi b_o)\,\Gamma(b_o)$,

and so $c\,\Gamma(1-b_o)^{-1} = \varepsilon^{-a_o}\dfrac{\sin\pi b_o}{\pi}\,\Gamma(b_o) \pmod{\mathbb{Q}^\times}$

$$= \frac{1}{\pi}\,e^{-2\pi i b_o/2}\left(\frac{e^{\pi i b_o} - e^{-\pi i b_o}}{2i}\right)\Gamma(b_o)$$

$$= \frac{1}{2\pi i}\,(1-\varepsilon^{-2a_o})\,\Gamma(b_o)\ .$$

The lemma is now obvious.

The group algebra $\mathbb{Q}[S]$ acts on the $\mathbb{Q}$-space of differentiable n-simplices in $V(\mathbb{C})$. For $\underline{a} \in X(S)$ and $\underline{\xi}_i = (1,\ldots,\xi,\ldots)$ ($\xi = e^{2\pi i/d}$ in i^{th} position), define

$$\sigma = \prod_{i=1}^{n+1}(1-\underline{\xi}_i)^{-1}\sigma_o(\Lambda) \subset V^o(\mathbb{C})$$

where σ_o and Λ are as above.

<u>Proposition</u> 7.13. Let $\underline{a} \in X(S)$ be such that no a_i is zero, and let ω^o be the differential

$$Y_1^{a'_1}\cdots Y_{n+1}^{a'_{n+1}}\,\frac{dY_1}{Y_1}\wedge\cdots\wedge\frac{dY_n}{Y_n}$$

on V^o, where a'_i represents $-a_i$, and $a'_i \gg o$. Then

(a) $\underline{\xi}\omega^o = \underline{\xi}^{\underline{a}}\,\omega^o$;

(b) $\displaystyle\int_\sigma \omega^o = \frac{1}{2\pi i}\prod_{i=o}^{n+1}(1-\xi^{a_i})\,\Gamma\!\left(\frac{-a_i}{d}\right)$.

<u>Proof</u>: (a) This is obvious since $\underline{\zeta}\,Y_i = (\zeta_i/\zeta_o)^{-1}\,Y_i$.

(b) $\displaystyle\int_\sigma \omega^o = \int_{\sigma_o(\Delta)}\prod_{i=1}^{n+1}(1-\xi_i)\omega^o = \prod_{i=1}^{n+1}(1-\xi^{a_i})\int_{\sigma_o(\Lambda)}\omega^o$

$$= \frac{1}{2\pi i}\prod_{i=o}^{n+1}(1-\xi^{a_i})\,\Gamma\!\left(-\frac{a_i}{d}\right)$$

Remark 7.14. From the Gysin sequence

$$(\mathbb{C}\approx)\ H^{n-2}(V-V^o,\mathbb{C}) \to H^n(V,\mathbb{C}) \to H^n(V^o,\mathbb{C}) \to o$$

we obtain an isomorphism $H^n(V,\mathbb{C})_{prim} \xrightarrow{\approx} H^n(V^o,\mathbb{C})$, which shows that there is an isomorphism

$$H^n_{DR}(V/k)_{prim} \xrightarrow{\approx} H^n_{DR}(V^o/k) = \Gamma(V^o,\Omega^n)/d\Gamma(V^o,\Omega^{n+1}) \ .$$

The class $[\omega^o]$ of the differential ω^o lies in $H^n_{DR}(V^o/k)_{\underline{a}}$. Correspondingly we get a C^∞ differential n-form ω on $V(\mathbb{C})$ such that

(a) the class $[\omega]$ of ω in $H^n_{DR}(V/\mathbb{C})$ lies in $H^n_{DR}(V/k)_{\underline{a}}$,

and

(b) $\int_\sigma \omega = \frac{1}{2\pi i} \prod_{i=o}^{n+1} (1-\xi^{a_i})\ \Gamma\left(-\frac{a_i}{d}\right)$, where $\sigma = \prod_{i=1}^{n+1} (1-\underline{\xi}_i)^{-1}\sigma_o(\Delta)$.

Note that, if we regard V as a variety over $\mathbb{Q}$, then $[\omega]$ even lies in $H^n_{DR}(V/\mathbb{Q})$.

The theorem

Recall that for $\underline{a} \in X(S)$, we set

$$\tilde{\Gamma}(\underline{a}) = (2\pi i)^{-\langle\underline{a}\rangle} \prod_{i=o}^{n+1} \Gamma(a_i/d) \qquad (\in \mathbb{C}/\mathbb{Q}^\times)$$

and for $\mathfrak{p}$ a prime of k not dividing d , we set

$$g(\mathfrak{p},\underline{a}) = q^{-\langle\underline{a}\rangle} \prod_i g(\mathfrak{p},a_i,\psi)\ ,\ g(\mathfrak{p},a_i,\psi) = -\sum_{x\in\mathbb{F}_q} t\left(x^{\frac{1-q}{d}}\right)^{a_i} \psi(x)$$

where q is the order of the residue field of $\mathfrak{p}$.

Theorem 7.15. Let $\underline{a} \in X(S)$ have no $a_i = o$ and be such that $\langle u\underline{a}\rangle = \langle\underline{a}\rangle$ $(= n/2 + 1)$ for all $u \in (\mathbb{Z}/d\,\mathbb{Z})^\times$.

(a) Then $\tilde{\Gamma}(\underline{a}) \in \bar{\mathbb{Q}}$ and generates an abelian extension of $k = \mathbb{Q}(e^{2\pi i/d})$.

(b) If $F_{\mathfrak{P}} \in \mathrm{Gal}(\bar{\mathbb{Q}}/k)^{ab}$ is the geometric Frobenius element at $\mathfrak{P}$, then

$$F_{\mathfrak{P}}(\tilde{\Gamma}(\underline{a})) = g(\mathfrak{P},\underline{a})\,\tilde{\Gamma}(\underline{a}) .$$

(c) For any $\tau \in \mathrm{Gal}(\bar{\mathbb{Q}}/\mathbb{Q})$, $\lambda_{\underline{a}}(\tau) \overset{df}{=} \tilde{\Gamma}(\underline{a})/\tau\tilde{\Gamma}(\underline{a})$ lies in k ; moreover, for any $u \in (\mathbb{Z}/d\,\mathbb{Z})^{\times}$,

$$\tau_u(\lambda_{\underline{a}}(\tau)) = \lambda_{u\underline{a}}(\tau)$$

where τ_u is the element of $\mathrm{Gal}(k/\mathbb{Q})$ defined by u .

<u>Proof</u>: Choose $\sigma \in H_n^B(V)$ and ω as in (7.14). Then all the conditions of (7.2) are fulfilled with α the character $\underline{a}$. Moreover, (7.14) and (7.11) show respectively that

$$P(\sigma,\omega) = \xi(\underline{a})\,\tilde{\Gamma}(-\underline{a}) \text{ , where } \xi(\underline{a}) = \prod_{i=o}^{n+1}(1-\xi^{a_i})$$

and

$$\chi(F_{\mathfrak{P}}) = g(\mathfrak{P},\underline{a})^{-1} .$$

As $\xi(\underline{a}) \in k$, (7.2) shows that $\tilde{\Gamma}(-\underline{a})$ generates an abelian algebraic extension of k and that $F_{\mathfrak{P}}\tilde{\Gamma}(-\underline{a}) = g(\mathfrak{P},\underline{a})^{-1}\tilde{\Gamma}(-\underline{a})$. It is clear from this equation that $g(\mathfrak{P},\underline{a})$ has absolute value 1 (in fact, it is a root of 1); thus $g(\mathfrak{P},\underline{a})^{-1} = \overline{g(\mathfrak{P},\underline{a})} = g(\mathfrak{P},-\underline{a})$. This proves (a) and (b) for $-\underline{a}$ and hence for $\underline{a}$.

To prove (c) we have to regard V as a variety over $\mathbb{Q}$. If S is interpreted as an algebraic group, then its action on V is rational over $\mathbb{Q}$. This means that

$$\tau(\underline{\zeta}\,\underline{x}) = \tau(\underline{\zeta})\ \tau(\underline{x})\ ,\ \tau \in Gal(\bar{\mathbb{Q}}/\mathbb{Q})\ ,\underline{\zeta} \in S(\bar{\mathbb{Q}})\ ,\ \underline{x} \in V(\bar{\mathbb{Q}})$$

and implies that

$$\tau(\underline{\zeta}\,\gamma) = \tau(\underline{\zeta})\ \tau(\gamma)\ ,\ \tau \in Gal(\bar{\mathbb{Q}}/\mathbb{Q})\ ,\ \underline{\zeta} \in S(\bar{\mathbb{Q}}),\ \gamma \in C^n_{AH}(\bar{V})\ .$$

Therefore $Gal(\bar{\mathbb{Q}}/\mathbb{Q})$ stabilizes $C^n_{AH}(\bar{V})_{[\underline{a}]}$ and, as this is a one-dimensional vector space over $\mathbb{Q}[\underline{a}]$, there exists for any $\gamma \in C^n_{AH}(\bar{V})_{[\underline{a}]}$ a crossed homomorphism $\lambda\colon Gal(\bar{\mathbb{Q}}/\mathbb{Q}) \to \mathbb{Q}[\underline{a}]^\times$ such that $\tau(\gamma) = \lambda(\tau)\gamma$ for all τ. On applying the canonical map $C^n_{AH}(\bar{V})_{[\underline{a}]} \to (C^n_{AH}(\bar{V}) \otimes k)_{\underline{a}}$ to this equality, we obtain $\tau(\gamma \otimes 1) = \lambda(\tau)^{\underline{a}}\ (\gamma \otimes 1)$.

We take γ to be the image of $\sigma \otimes (2\pi i)^{-n/2} \in H^B_n(V)(-\frac{n}{2})$ in $C^n_{AH}(\bar{V})_{[\underline{a}]}$. Then (cf. 7.3), $(\gamma \otimes 1)_{DR} = P(\sigma,\omega)^{-1}[\omega]$, if $[\omega]$ is as in (7.14). Hence

$$\lambda(\tau)^{\underline{a}} = P(\sigma,\omega)/\tau\ P(\sigma,\omega) = \lambda_{-\underline{a}}(\tau)\,(\xi(\underline{a})/\tau\ \xi(\underline{a}))\ .$$

On comparing

$$\lambda_{\underline{a}}(\tau) = \lambda(\tau)^{-\underline{a}}\ (\tau\ \xi(-\underline{a})/\ \xi(-\underline{a}))\quad \text{and}$$

$$\lambda_{u\underline{a}}(\tau) = \lambda(\tau)^{-u\underline{a}}(\tau\ \xi(-u\underline{a})/\xi(-u\underline{a}))\ ,$$

and using that $\tau(\xi(-u\underline{a})) = \tau(\tau_u(\xi(-\underline{a})) = \tau_u(\tau\ \xi(-\underline{a}))$, one obtains (c) of the theorem.

Remark 7.16 (a) The first statement of the theorem, that $\tilde{\Gamma}(\underline{a})$ is algebraic, has an elementary proof; see the appendix by Koblitz and Ogus to Deligne [7].

(b) Part (b) of the theorem has been proved up to sign by Gross and Koblitz [1, 4.5] using p-adic methods.

Remark 7.17 Let I_d be the group of ideals of k prime to d, and consider the character

$$\mathfrak{a} = \Pi \mathfrak{p}_i^{r_i} \longmapsto g(\mathfrak{a},\underline{a}) \overset{\mathrm{df}}{=} \Pi\, g(\mathfrak{p}_i,\underline{a})^{r_i} : I_d \to k^\times .$$

When $\underline{a}$ satisfies the conditions in the theorem, then this is an algebraic Hecke character (Weil [1], [3]; see also Deligne [5, §6]). This means that there exists an ideal m of k (dividing a power of d) and a homomorphism $\chi_{alg}: k^\times \longrightarrow k^\times$ that is algebraic (i.e., defined by a map of tori) and such that, for all $x \in k^\times$ totally positive and $\equiv 1 (\mathrm{mod}\ m)$, $g((x), \underline{a}) = \chi_{alg}(x)$. There is then a unique character $\chi_{\underline{a}}: \mathrm{Gal}(\bar{\mathbb{Q}}/k)^{ab} \to k^\times$ such that $\chi_{\underline{a}}(F_{\mathfrak{p}}) = g(\mathfrak{p},\underline{a})$ for all $\mathfrak{p}$ prime to d. Part (b) of the theorem can be stated as

$$\sigma(\tilde{\Gamma}(\underline{a})) = \chi_{\underline{a}}(\sigma)\, \tilde{\Gamma}(\underline{a}), \text{ all } \sigma \in \mathrm{Gal}(\bar{k}/k) .$$

(There is an elegant treatment of algebraic Hecke characters in Serre [2, II]. Such a character with conductor dividing a modulus m corresponds to a character χ of the torus S_m (loc. cit. p II-17). The map χ_{alg} is $k^\times \xrightarrow{\pi} T_m \hookrightarrow S_m \xrightarrow{\chi} k^\times$. One defines from χ a character χ_∞ of the idèle class group as in (loc. cit., II 2.7). Weil's determination of χ_{alg} shows that χ_∞ is of finite order; in particular it is trivial on the connected component of the idèle class group, and so gives rise to a character $\chi_{\underline{a}}: \mathrm{Gal}(\bar{\mathbb{Q}}/k)^{ab} \to k^\times$.)

Restatement of the theorem

For $b \in d^{-1}\mathbb{Z}/\mathbb{Z}$, we write $\langle b\rangle$ for the representative of b in $d^{-1}\mathbb{Z}$ with $1/d \leq \langle b\rangle \leq 1$. Let $\underline{b} = \Sigma\, n(b)\,\delta_b$ be an element of the free abelian group generated by the set $d^{-1}\mathbb{Z}/\mathbb{Z} - \{o\}$, and assume that $\Sigma\, n(b) \langle ub\rangle = c$ is an integer independent of $u \in \mathbb{Z}/d\,\mathbb{Z}$. Define

$$\tilde{\Gamma}(\underline{b}) = \frac{1}{(2\pi i)^c} \prod_b \Gamma(\langle b\rangle)^{n(b)} .$$

Let $\mathfrak{p}$ be a prime of k, not dividing d, and let $\mathbb{F}_q$ be the residue field at $\mathfrak{p}$. For ψ a non-trivial additive character of $\mathbb{F}_q$, define

$$g(\mathfrak{p},\underline{b}) = \frac{1}{q^c} \prod_b g(\mathfrak{p},b,\psi)^{n(b)} , \quad \text{where} \quad g(\mathfrak{p},b,\psi) = -\sum_{x\in\mathbb{F}_q} t(x^{b(1-q)})\psi(x) .$$

As in (7.17), $\mathfrak{p} \mapsto g(\mathfrak{p},\underline{b})$ defines an algebraic Hecke character of k and a character $\chi_{\underline{b}}\colon \mathrm{Gal}(\bar{\mathbb{Q}}/k) \to \mathbb{C}^\times$ such that $\chi_{\underline{b}}(F_{\mathfrak{p}}) = g(\mathfrak{p},\underline{b})$ for all $\mathfrak{p} \nmid b$.

Theorem 7.18. Assume that $\underline{b} = \Sigma\, n(b)\,\delta_b$ satisfies the condition above.

(a) Then $\tilde{\Gamma}(\underline{b}) \in k^{ab}$, and for all $\sigma \in \mathrm{Gal}(\bar{\mathbb{Q}}/k)^{ab}$,

$$\sigma\,\tilde{\Gamma}(\underline{b}) = \chi_{\underline{b}}(\sigma)\;\tilde{\Gamma}(\underline{b}) .$$

(b) For $\tau \in \mathrm{Gal}(\bar{\mathbb{Q}}/\mathbb{Q})$, let $\lambda_{\underline{b}}(\tau) = \tilde{\Gamma}(\underline{b})/\tau\,\tilde{\Gamma}(\underline{b})$; then $\lambda_{\underline{b}}(\tau) \in k$, and, for any $u \in (\mathbb{Z}/d\,\mathbb{Z})^\times$,

$$\tau_u(\lambda_{\underline{b}}(\tau)) = \lambda_{u\underline{b}}(\tau) .$$

Proof: Suppose first that $n(b) \geq o$ for all b. Let $n + 2 = \Sigma\, n(b)$, and let $\underline{a}$ be an $(n+2)$-triple in which each $a \in \mathbb{Z}/d\,\mathbb{Z}$ occurs exactly $n(a/d)$ times. Write

$\underline{a} = (a_o, \ldots, a_{n+1})$. Then $\Sigma a_i = d(\Sigma n(b)b) = dc \pmod d = o$, and so $\underline{a} \in X(S)$. Moreover,

$$\langle u\underline{a}\rangle \overset{df}{=} \frac{1}{d} \Sigma \langle ua_i\rangle = \Sigma n(b) \langle ub\rangle = c$$

for all $u \in \mathbb{Z}/d\mathbb{Z}$. Thus $\langle u\underline{a}\rangle$ is constant, and $c = \langle \underline{a}\rangle$. We deduce that $\tilde{\Gamma}(\underline{a}) = \tilde{\Gamma}(\underline{b})$, $g(\boldsymbol{\gamma},\underline{a}) = g(\boldsymbol{\gamma},\underline{b})$, and $\chi_{\underline{a}} = \chi_{\underline{b}}$. Thus in this case, (7.18) follows immediately from (7.15) and (7.17).

Let $\underline{b}$ be arbitrary. For some N , $\underline{b} + N\underline{b}_o$ has positive coefficients, where $\underline{b}_o = \Sigma \delta_b$. Thus (7.18) is true for $\underline{b} + N\underline{b}_o$. Since $\tilde{\Gamma}(\underline{b}_1 + \underline{b}_2) = \tilde{\Gamma}(\underline{b}_1)\,\tilde{\Gamma}(\underline{b}_2) \pmod{\mathbb{Q}^\times}$ and $g(\boldsymbol{\gamma},\underline{b}_1 + \underline{b}_2) = g(\boldsymbol{\gamma},\underline{b}_1)\, g(\boldsymbol{\gamma},\underline{b}_2)$ this completes the proof.

<u>Remark</u> 7.19. (a) Part (b) of the theorem determines $\Gamma(u\underline{b})$ (up to multiplication by a rational number) starting from $\Gamma(\underline{b})$.

(b) Conjecture 8.11 of Deligne [7] is a special case of part (a) of the above theorem. The more precise form of the conjecture, Deligne [7, 8.13], can be proved by a modification of the above methods.

<u>Final Note</u>. The original seminar of Deligne comprised fifteen lectures, given between 29/10/78 and 15/5/79. The first six sections of these notes are based on the first eight lectures of the seminar, and the final section on the last two lectures. The remaining five lectures (which the writer of these notes was unable to attend) were on the following topics:

(6/3/79) review of the proof that Hodge cycles on abelian varieties are absolutely Hodge; discussion of the expected action of the Frobenius endomorphism on the image of an absolute Hodge cycle in crystalline cohomology;

(13/3/79) definition of the category of motives using absolute Hodge cycles; semisimplicity of the category; existence of the motivic Galois group G;

(20/3/79) fibre functors in terms of torsors; the motives of Fermat hypersurfaces and K 3-surfaces are contained in the category generated by abelian varieties;

(27/3/79) Artin motives; the exact sequence

$$1 \to G^{\circ} \to G \xrightarrow{\pi} \mathrm{Gal}(\bar{\mathbb{Q}}/\mathbb{Q}) \to 1 \ ;$$

identification of G° with the Serre group, and description of the G°-torsor $\pi^{-1}(\tau)$;

(3/4/79) action of $\mathrm{Gal}(\bar{\mathbb{Q}}/\mathbb{Q})$ on G°; study of $G \otimes_{\mathbb{Q}} \mathbb{Q}_{\ell}$; Hasse principle for $H^1(\mathbb{Q},G^{\circ})$.

Most of the material in these five lectures is contained in the remaining articles of this volume (especially IV).

The writer of these notes is indebted to P. Deligne and A. Ogus for their criticisms of the first draft of the notes and to Ogus for his notes on which the final section is largely based.

REFERENCES

Artin, M., Grothendieck, A., and Verdier, J.

1. SGA4, Théorie des topos et cohomologie étale des schémas. Lecture Notes in Math. 269, 270, 305. Springer, Heidelberg, 1972-73.

Baily, W. and Borel, A.

1. Compactification of arithmetic quotients of bounded symmetric domains, Ann. of Math. 84(1966) 442-528.

Borel, A.

1. Linear Algebraic Groups (Notes by H. Bass). Benjamin, New York, 1969.

2. Some metric properties of arithmetic quotients of symmetric spaces and an extension theorem. J. Diffl. Geometry 6(1972) 543-560.

Borel, A. and Spinger, T.

1. Rationality properties of algebraic semisimple groups, Proc. Symp. Pure Math, A.M.S. 9(1966) 26-32.

Borovoǐ, M.

1. The Shimura-Deligne schemes $M_{\mathbb{C}}(G,h)$ and the rational cohomology classes of type (p,p). Voprosy Teorii Grupp i Gomologičeskoǐ Algebry, vypusk I, Jaroslavl, 1977, pp 3-53.

Deligne, P.

1. Théorème de Lefschetz et critères de dégénérescence de suite spectrales, Publ. math. I.H.E.S. 35(1968) 107-126.

2. Travaux de Griffiths, Sém. Bourbaki 1969/70, Exposé 376 Lecture Notes in Math 180, Springer, Heidelberg, 1971.

3. Travaux de Shimura, Sém. Bourbaki 1970/71, Exposé 389 Lecture Notes in Math 244, Springer, Heidelberg, 1971.

4. Théorie de Hodge II, Publ. Math. I.H.E.S. 40(1972) 5-57.

5. Applications de la formule des traces aux sommes trigonométriques, SGA4 $\frac{1}{2}$, Lecture Notes in Math 569, Springer, Heidelberg, 1977.

6. Variétés de Shimura: interpretation modulaire, et techniques de construction de modèles canoniques, Proc. Symp. Pure Math., A.M.S. 33(1979) Part 2, 247-290.

7. Valeurs de fonctions L et périodes d'intégrales. Proc. Symp. Pure Math., A.M.S. 33(1979) Part 2, 313-346.

Demazure, M.

1. Démonstration de la conjecture de Mumford (d'après W. Haboush), Sém Bourbaki 1974/75, Exposé 462, Lecture Notes in Math. 514, Springer, Heidelberg, 1976.

Griffiths, P.

1. On the periods of certain rational integrals: I, Ann. of Math. 90(1969) 460-495.

Gross, B.

1. On the periods of abelian integrals and a formula of Chowla and Selberg, Invent. Math. 45(1978) 193-211.

Gross, B. and Koblitz, N.

1. Gauss sums and the p-adic Γ-function, Ann. of Math. 109(1979) 569-581.

Grothendieck, A.

1. La théorie des classes de Chern, Bull. Soc. Math. France 86(1958) 137-154.

2. On the de Rham cohomology of algebraic varieties, Publ. Math. I.H.E.S. 29(1966) 95-103.

Katz, N.

1. Nilpotent connections and the monodromy theorem: applications of a result of Turritten, Publ. Math. I.H.E.S. 39(1970) 175-232.

Katz, N. and Oda, T.

1. On the differentiation of de Rham cohomology classes with respect to parameters. J. Math. Kyoto Univ. 8(1968) 199-213.

Knapp, A.

1. Bounded symmetric domains and holomorphic discrete series 211-246; in Symmetric Spaces (Boothby, W. and Weiss, G., Ed) Dekker, New York 1972.

Landherr, W.

1. Äquivalenz Hermitescher Formen über einem beliebigen algebraischen Zahlkörper. Abh. Math. Semin. Hamburg Univ. 11(1936) 245-248.

Milne, J.

1. Étale Cohomology, Princeton U.P., Princeton, 1980.

Piatetski-Shapiro, I.

1. Interrelation between the conjectures of Hodge and Tate for abelian varieties, Mat. Sb. 85(127) (1971) 610-620.

Pohlmann, H.

1. Algebraic cycles on abelian varieties of complex multiplication type, 88(1968) 161-180.

Saavedra Rivano, N.

1. Catégories Tannakiennes, Lecture Notes in Math 265, Springer, Heidelberg, 1972.

Serre, J-P.

1. Géométrie algébriques et géométrie analytique, Ann. Inst. Fourier, Grenoble 6(1955-56) 1-42.

2. Abelian ℓ-Adic Representations and Elliptic Curves, Benjamin, New York, 1968.

Singer, I. and Thorpe, J.

1. Lecture Notes on Elementary Topology and Geometry, Scott-Foresman, Glenview, 1967.

Warner, F.

1. Introduction to Manifolds, Scott-Foresman, Glenview, 1971.

Waterhouse,

1. Introduction to Affine Group Schemes, Springer, Heidelberg, 1979.

Weil, A.

1. Jacobi sums as grössencharaktere, Trans. A.M.S. 73(1952) 487-495.

2. Introduction à l'Étude des Variétés Kählériennes, Hermann, Paris, 1958.

3. Sommes de Jacobi et caractères de Hecke, Nachr. Akad. Wiss. Göttingen Math. Phys. Kl. II 1(1974) 1-14.

II TANNAKIAN CATEGORIES

by

P. Deligne and J. S. Milne

Introduction:

In the first section it is shown how to introduce on an abstract category operations of tensor products and duals having properties similar to the familiar operations on the category $\underline{\text{Vec}}_k$ of finite-dimensional vector spaces over a field k. What complicates this is the necessity of including enough constraints so that, whenever an obvious isomorphism (e.g., $U \otimes (V \otimes W) \xrightarrow{\approx} (V \otimes U) \otimes W)$ exists in $\underline{\text{Vec}}_k$, a unique isomorphism is constrained to exist also in the abstract setting.

The next section studies the category $\underline{\text{Rep}}_k(G)$ of finite-dimensional representations of an affine group scheme G over k and demonstrates necessary and sufficient conditions for a category $\underline{C}$ with a tensor product to be isomorphic to $\underline{\text{Rep}}_k(G)$ for G; such a category $\underline{C}$ is then called a neutral Tannakian category.

A fibre functor on a Tannakian category $\underline{C}$ with values in a field $k' \supset k$ is an exact k-linear functor $\underline{C} \to \underline{Vec}_{k'}$ that commutes with tensor products. For example, the forgetful functor is a fibre functor on $\underline{Rep}_k(G)$. In the third section it is shown that fibre functors on $\underline{Rep}_k(G)$ are in one-to-one correspondence with the torsors of G. Also, the notion of a (non-neutral) Tannakian category as introduced.

The fourth section studies the notion of a polarization (compatible families of sesquilinear forms having certain positivity properties) on a Tannakian category, and the fifth studies the notion of a graded Tannakian category.

In the sixth section, motives are defined using absolute Hodge cycles, and the related motivic Galois groups discussed. In an appendix, some terminology from non-abelian cohomology is reviewed.

We note that the introduction of Saavedra [1] is an excellent summary of Tannakian categories, except that two changes are necessary: Théorème 3 is, unfortunately, only a conjecture; in Théorème 4 the requirement that G be abelian or connected can be dropped.

Notations: Functors between additive categories are assumed to be additive. In general, rings are commutative with 1 except in §2. A morphism of functors is also called a functorial or natural morphism. A strictly full subcategory is a full subcategory containing with any X, all objects isomorphic to X. The empty set is denoted by Φ.

Our notations agree with those of Saavedra [1] except that we have made some simplifications: What would be called a $\otimes$-widget ACU by Saavedra, here becomes a tensor widget, and $\underline{\mathrm{Hom}}^{\otimes,1}$ becomes $\underline{\mathrm{Hom}}^{\otimes}$.

$\underline{\mathrm{Vec}}_k$: Category of finite-dimensional vector spaces over k ;

$\underline{\mathrm{Rep}}_k(G)$: Category of finite-dimensional representation of G over k ;

$\underline{\mathrm{Mod}}_R$: Category of finitely generated R-modules;

$\underline{\mathrm{Proj}}_R$: Category of finitely generated projective R-modules;

$\underline{\mathrm{Set}}$: Category of sets

§1. Tensor categories

Let $\underline{C}$ be a category and

$$\otimes: \underline{C} \times \underline{C} \to \underline{C}\ ,\ (X,Y) \mapsto X \otimes Y$$

a functor. An associativity constraint for $(\underline{C},\otimes)$ is a functorial isomorphism

$$\phi_{X,Y,Z}: X \otimes (Y \otimes Z) \xrightarrow{\approx} (X \otimes Y) \otimes Z$$

such that, for all objects X,Y,Z,T, the diagram

$$\begin{array}{ccccc} X \otimes (Y\otimes(Z\otimes T)) & \xrightarrow{\phi} & (X\otimes Y) \otimes (Z\otimes T) & \xrightarrow{\phi} & ((X\otimes Y)\otimes Z) \otimes T \\ \downarrow {\scriptstyle 1 \otimes \phi} & & & & \uparrow {\scriptstyle \phi \otimes 1} \\ X \otimes ((Y\otimes Z)\otimes T) & & \xrightarrow{\phi} & & (X\otimes(Y\otimes Z)) \otimes T \end{array} \qquad (1.0.1)$$

is commutative (this is the pentagon axiom). Here, as in subsequent diagrams, we have omitted the obvious subscripts on the maps; for example, the ϕ at top-left is $\phi_{X,Y,Z\otimes T}$. A commutativity constraint for $(\underline{C},\otimes)$ is a functorial isomorphism

$$\psi_{X,Y}: X \otimes Y \xrightarrow{\approx} Y \otimes X$$

such that, for all objects $X,Y, \psi_{Y,X} \circ \psi_{X,Y} = id_{X\otimes Y}: X \otimes Y \to X \otimes Y$. An associativity constraint ϕ and a commutativity constraint ψ are compatible if, for all objects X,Y,Z, the diagram

$$\begin{array}{ccccc} X \otimes (Y\otimes Z) & \xrightarrow{\phi} & (X\otimes Y) \otimes Z & \xrightarrow{\psi} & Z \otimes (X\otimes Y) \\ \downarrow {\scriptstyle 1 \otimes \psi} & & & & \downarrow {\scriptstyle \phi} \\ X \otimes (Z\otimes Y) & \xrightarrow{\phi} & (X\otimes Z) \otimes Y & \xrightarrow{\psi\otimes 1} & (Z\otimes X) \otimes Y \end{array} \qquad (1.0.2)$$

is commutative (hexagon axiom). A pair (U,u) comprising an object U of $\underline{C}$ and an isomorphism $u\colon U \to U \otimes U$ is an <u>identity object</u> of $(\underline{C},\otimes)$ if $X \longmapsto U \otimes X\colon \underline{C} \to \underline{C}$ is an equivalence of categories.

<u>Definition</u> 1.1. A system $(\underline{C},\otimes,\phi,\psi)$, in which ϕ and ψ are compatible associativity and commutativity constraints, is a <u>tensor category</u> if it has an identity object.

<u>Example</u> 1.2. The category $\underline{\text{Mod}}_R$ of finitely generated modules over a commutative ring R becomes a tensor category with the usual tensor product and the obvious constraints. (If one perversely takes ϕ to be the negative of the obvious isomorphism, then the pentagon (1.0.1) fails to commute by a sign.) A pair (U,u_0) comprising a free R-module U of rank 1 and a basis element u_0 determines an identity object (U,u) of $\underline{\text{Mod}}_R$ — take u to be the unique isomorphism $U \to U \otimes U$ mapping u_0 to $u_0 \otimes u_0$. Every identity element is of this form.

(For other examples, see the end of this section.)

<u>Proposition</u> 1.3. Let (U,u) be an identity object of the tensor category $(\underline{C},\otimes)$.

(a) There exists a unique functorial isomorphism

$$\ell_X\colon X \xrightarrow{\approx} U \otimes X$$

such that ℓ_U is u and the diagrams

$$\begin{array}{ccc} X \otimes Y & \xrightarrow{\ell} & U \otimes (X \otimes Y) \\ \| & & \downarrow{\scriptstyle\phi} \\ X \otimes Y & \xrightarrow{\ell \otimes 1} & (U \otimes X) \otimes Y \end{array} \qquad \begin{array}{ccc} X \otimes Y & \xrightarrow{\ell \otimes 1} & (U \otimes X) \otimes Y \\ \downarrow{\scriptstyle 1 \otimes \ell} & & \downarrow{\scriptstyle \psi \otimes 1} \\ X \otimes (U \otimes Y) & \xrightarrow{\phi} & (X \otimes U) \otimes Y \end{array}$$

are commutative.

(b) If (U',u') is a second identity object of $(\underline{C},\otimes)$ then there is a unique isomorphism $a\colon U \xrightarrow{\approx} U'$ making

$$\begin{array}{ccc} U & \xrightarrow{u} & U \otimes U \\ \downarrow a & & \downarrow a \otimes a \\ U' & \xrightarrow{u'} & U' \otimes U' \end{array}$$

commute

Proof (a) We confine ourselves to defining ℓ_X . (See Saavedra [1,I2.2.5.1,2.4.1] for details.) As $X \longmapsto U \otimes X$ is an equivalence of categories, it suffices to define $1 \otimes \ell_X\colon U \otimes X \to U \otimes (U \otimes X)$; this we take to be

$$U \otimes X \xrightarrow{u \otimes 1} (U \otimes U) \otimes X \xrightarrow{\phi^{-1}} U \otimes (U \otimes X) .$$

(b) The map $U \xrightarrow{\ell_U} U' \otimes U \xrightarrow{\psi} U \otimes U' \xrightarrow{\ell_{U'}^{-1}} U'$ has the required properties.

The functorial isomorphism $r_X \overset{\text{df}}{=} \psi_{U,X} \circ \ell_X : X \to X \otimes U$ has analogous properties to ℓ_X . We shall often use $(\underline{1},e)$ to denote a (the) identity object of $(\underline{C},\otimes)$.

Remark 1.4. Our notion of a tensor category is the same as that of a "$\otimes$-catégorie AC unifère" in (Saavedra [1]) and, because of (1.3), is essentially the same as the notion of a "$\otimes$-catégorie ACU" defined in (Saavedra [1, I.2.4.1]) (cf. Saavedra [1,I.2.4.3]).

Extending $\otimes$

Let ϕ be an associativity constraint for $(\underline{C},\otimes)$. Any functor $\underline{C}^n \to \underline{C}$ defined by repeated application of $\otimes$ is called an iterate of $\otimes$. If $F,F' : \underline{C}^n \to \underline{C}$ are iterates of $\otimes$, then it is possible to construct an isomorphism of functors $\tau: F \xrightarrow{\approx} F'$ out of ϕ and ϕ^{-1} . The significance of the pentagon axiom is that it implies that τ is unique: any two iterates of $\otimes$ to C^n are isomorphic by a unique isomorphism of functors constructed out of ϕ and ϕ^{-1} (MacLane [1], [2,VII.2]). In other words, there is an essentially unique way of extending $\otimes$ to a functor $\bigotimes_{i=1}^{n} : \underline{C}^n \to \underline{C}$ when $n \geq 1$. Similarly, if $(\underline{C},\otimes)$ is a tensor category, then it is possible to extend $\otimes$ in essentially one way to a functor $\bigotimes_{i \in I} : \underline{C}^I \to \underline{C}$ where I is any finite set: the tensor product of any finite family of objects of $\underline{C}$ is well-defined up to a unique isomorphism (MacLane [1]). We can make this more precise.

Proposition 1.5. The tensor structure on a tensor category $(\underline{C},\otimes)$ can be extended as follows. For each finite set I there is to be a functor

$$\bigotimes_{i \in I} : \underline{C}^I \to \underline{C} ,$$

and for each map $\alpha: I \to J$ of finite sets there is to be a functorial isomorphism

$$\chi(\alpha): \bigotimes_{i \in I} X_i \xrightarrow{\approx} \bigotimes_{j \in J} \left(\bigotimes_{i \mapsto j} X_i \right)$$

satisfying the following conditions:

(a) if I consists of a single element, then $\bigotimes_{i \in I}$ is the identity functor $X \longmapsto X$; if α is a map between single-element sets, then $\chi(\alpha)$ is the identity automorphism of the identity functor;

(b) the isomorphisms defined by maps $I \xrightarrow{\alpha} J \xrightarrow{\beta} K$ give rise to a commutative diagram

$$\begin{array}{ccc}
\bigotimes_{i\in I} X_i & \xrightarrow{\chi(\alpha)} & \bigotimes_{j\in J} \left(\bigotimes_{i\mapsto j} X_i \right) \\
\Big\downarrow \chi(\beta\alpha) & & \Big\downarrow \chi(\beta) \\
\bigotimes_{k\in K} \left(\bigotimes_{i\mapsto k} X_i \right) & \xrightarrow{\otimes(\chi(\alpha|I_k))} & \bigotimes_{k\in K} \left(\bigotimes_{j\mapsto k} \left(\bigotimes_{i\mapsto j} X_i \right) \right) .
\end{array}$$

where $I_k = (\beta\alpha)^{-1}(k)$.

<u>Proof</u>: Omitted.

By $(\bigotimes_{i\in I}, \chi)$ being an extension of the tensor structure on $\underline{C}$ we mean that $\bigotimes_{i\in I} X_i = X_1 \otimes X_2$ when $I = \{1,2\}$ and that the isomorphisms $X \otimes (Y \otimes Z) \longrightarrow (X \otimes Y) \otimes Z$ and $X \otimes Y \longrightarrow Y \otimes X$ induced by χ are equal to ϕ and ψ respectively. It is automatic that $(\bigotimes_{\phi} X_i, \chi(\phi \to \{1,2\}))$ is an identity object and that $\chi(\{2\} \hookrightarrow \{1,2\})$ is $\ell_X: X \to \underline{1} \otimes X$. If $(\bigotimes'_{i\in I}, \chi')$ is a second such extension, then

there is a unique system of isomorphisms $\bigotimes_{i\in I} X_i \to \bigotimes'_{i\in I} X_i$ compatible with χ and χ' and such that, when $I = \{i\}$, the isomorphism is id_{X_i} .

When a tensor category $(\underline{C}, \otimes)$ is given, we shall always assume that an extension as in (1.5) has been made. (We could, in fact, have defined a tensor category to be a system as in (1.5).)

Invertible objects

Let $(\underline{C},\otimes)$ be a tensor category. An object L of $\underline{C}$ is invertible if $X \longmapsto L \otimes X\colon \underline{C} \to \underline{C}$ is an equivalence of categories. Thus, if L is invertible, there exists an L' such that $L \otimes L' \approx \underline{1}$; the converse assertion is also true. An inverse of L is a pair (L^{-1},δ) where $\delta\colon \bigotimes_{i\in\{\pm\}} X_i \xrightarrow{\approx} \underline{1}$, $X_+ = L$, $X_- = L^{-1}$. Note that this definition is symmetric: (L,δ) is an inverse of L^{-1}. If (L_1,δ_1) and (L_2,δ_2) are both inverses of L, then there is a unique isomorphism $\alpha\colon L_1 \xrightarrow{\approx} L_2$ such that $\delta_1 = \delta_2 \circ (1\otimes\alpha)\colon L \otimes L_1 \to L \otimes L_2 \to \underline{1}$.

An object L of $\underline{\mathrm{Mod}}_R$ is invertible if and only if it is projective of rank 1. (Saavedra [1,0.2.2.2]).

Internal Hom

Let $(\underline{C},\otimes)$ be a tensor category.

Definition 1.6. If the functor $T \longmapsto \mathrm{Hom}(T\otimes X,Y)\colon \underline{C}^{\circ} \to \underline{\mathrm{Set}}$ is representable, then we denote by $\underline{\mathrm{Hom}}(X,Y)$ the representing object and by $ev_{X,Y}\colon \underline{\mathrm{Hom}}(X,Y) \otimes X \to Y$ the morphism corresponding to $id_{\underline{\mathrm{Hom}}(X,Y)}$.

Thus, to a g there corresponds a unique f such that $ev \circ (f\otimes id) = g$:

$$\begin{array}{ccc} T & \qquad\qquad & T \otimes X \\ \Big\downarrow{\scriptstyle f} & & \Big\downarrow{\scriptstyle f\otimes id} \searrow{\scriptstyle g} \\ \underline{\mathrm{Hom}}(X,Y) & & \underline{\mathrm{Hom}}(X,Y) \otimes X \xrightarrow{ev} Y \end{array} \qquad (1.6.1)$$

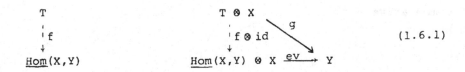

For example, in $\underline{\mathrm{Mod}}_R$, $\underline{\mathrm{Hom}}(X,Y) = \mathrm{Hom}_R(X,Y)$ regarded as an R-module, and ev is $f \otimes x \longmapsto f(x)$, whence its name.

Assume that in $(\underline{C},\otimes)$, $\underline{\mathrm{Hom}}(X,Y)$ exists for every pair (X,Y) . Then there is a composition map

$$\underline{\mathrm{Hom}}(X,Y) \otimes \underline{\mathrm{Hom}}(Y,Z) \to \underline{\mathrm{Hom}}(X,Z) \qquad (1.6.2)$$

(corresponding to $\underline{\mathrm{Hom}}(X,Y) \otimes \underline{\mathrm{Hom}}(Y,Z) \otimes X \xrightarrow{ev} \underline{\mathrm{Hom}}(Y,Z) \otimes Y \xrightarrow{ev} Z$) and an isomorphism

$$\underline{\mathrm{Hom}}(Z,\underline{\mathrm{Hom}}(X,Y)) \xrightarrow{\approx} \underline{\mathrm{Hom}}(Z\otimes X,Y) \qquad (1.6.3)$$

(inducing, for any object T,

$$\mathrm{Hom}(T,\underline{\mathrm{Hom}}(Z,\underline{\mathrm{Hom}}(X,Y))) \xrightarrow{\approx} \mathrm{Hom}(T\otimes Z,\underline{\mathrm{Hom}}(X,Y)) \xrightarrow{\approx} \mathrm{Hom}(T\otimes Z\otimes X,Y) \xrightarrow{\approx} \mathrm{Hom}(T,\underline{\mathrm{Hom}}(Z\otimes X,Y)))$$

Note that

$$\mathrm{Hom}(\underline{1},\underline{\mathrm{Hom}}(X,Y)) = \mathrm{Hom}(\underline{1}\otimes X,Y) = \mathrm{Hom}(X,Y) \qquad (1.6.4)$$

The <u>dual</u> $X^\vee$ of an object X is defined to be $\underline{\mathrm{Hom}}(X,\underline{1})$. There is therefore a map $ev_X\colon X^\vee \otimes X \to \underline{1}$ inducing a functorial isomorphism

$$\mathrm{Hom}(T,X^\vee) \xrightarrow{\approx} \mathrm{Hom}(T\otimes X,\underline{1}) \qquad (1.6.5)$$

The map $X \longmapsto X^\vee$ can be made into a contravariant functor: to $f\colon X \to Y$ we associate the unique map ${}^t f\colon Y^\vee \to X^\vee$ rendering commutative

$$\begin{array}{ccc} Y^\vee \otimes X & \xrightarrow{{}^t f \otimes id} & X^\vee \otimes X \\ \Big\downarrow id \otimes f & & \Big\downarrow ev_X \\ Y^\vee \otimes Y & \xrightarrow{ev_Y} & \underline{1} \end{array} \tag{1.6.6}$$

For example, in $\underline{\mathrm{Mod}}_R$, $X^\vee = \mathrm{Hom}_R(X,R)$ and ${}^t f$ is determined by the equation $\langle {}^t f(y), x\rangle_X = \langle y, f(x)\rangle_Y$, $y \in Y^\vee$, $x \in X$, where we have written $\langle\ ,\ \rangle_X$ and $\langle\ ,\ \rangle_Y$ for ev_X and ev_Y.

If f is an isomorphism, we let $f^\vee = ({}^t f)^{-1}: X^\vee \to Y^\vee$, so that

$$ev_Y \circ (f^\vee \otimes f) = ev_X: X^\vee \otimes X \to \underline{1}. \tag{1.6.7}$$

(E.g. in $\underline{\mathrm{Mod}}_R$, $\langle f^\vee(x'), f(x)\rangle_Y = \langle x', x\rangle_X$, $x' \in X^\vee, x \in X$.)

Let $i_X: X \to X^{\vee\vee}$ be the map corresponding in (1.6.5) to $ev_X \circ \psi : X \otimes X^\vee \to \underline{1}$. If i_X is an isomorphism then X is said to be <u>reflexive</u>. If X has an inverse $(X^{-1}, \delta: X^{-1} \otimes X \xrightarrow{\approx} \underline{1})$ then X is reflexive and δ determines an isomorphism $X^{-1} \xrightarrow{\approx} X^\vee$ as in (1.6.1).

For any finite families of objects $(X_i)_{i \in I}$ and $(Y_i)_{i \in I}$ there is a morphism

$$\bigotimes_{i \in I} \underline{\mathrm{Hom}}(X_i, Y_i) \to \underline{\mathrm{Hom}}(\bigotimes_{i \in I} X_i, \bigotimes_{i \in I} Y_i) \tag{1.6.8}$$

corresponding in (1.6.1) to

$$(\bigotimes_{i \in I} \underline{\mathrm{Hom}}(X_i,Y_i)) \otimes (\bigotimes_{i \in I} X_i) \xrightarrow{\approx} \bigotimes_{i \in I} (\underline{\mathrm{Hom}}(X_i,Y_i) \otimes X_i) \xrightarrow{\otimes ev} \bigotimes_{i \in I} Y_i .$$

In particular, there are morphisms

$$\underset{i\in I}{\otimes} X_i^{\vee} \longrightarrow (\underset{i\in I}{\otimes} X_i)^{\vee} \tag{1.6.9}$$

and

$$X^{\vee} \otimes Y \longrightarrow \underline{\mathrm{Hom}}(X,Y) \tag{1.6.10}$$

obtained respectively by taking $Y_i = \underline{1}$ all i, and $X_1 = X,\ X_2 = \underline{1} = Y_1,\ Y_2 = Y$.

Rigid tensor categories

Definition 1.7. A tensor category $(\underline{C},\otimes)$ is rigid if $\underline{\mathrm{Hom}}(X,Y)$ exists for all objects X and Y, the maps (1.6.8) are isomorphisms for all finite families of objects, and all objects of $\underline{C}$ are reflexive.

In fact, it suffices to require that the maps (1.6.8) be isomorphisms in the special case that $I = \{1,2\}$.

Let $(\underline{C},\otimes)$ be a rigid tensor category. The functor

$$(X,f) \longmapsto (X^{\vee}, {}^tf):\ \underline{C}^0 \longrightarrow \underline{C}$$

is an equivalence of categories because its composite with itself is isomorphic to the identity functor. (It is even an equivalence of tensor categories in the sense defined below—note that $\underline{C}^\circ$ has an obvious tensor structure for which $\otimes X_i^\circ = (\otimes X_i)^\circ$.) In particular

$$f \longmapsto {}^tf:\quad \mathrm{Hom}(X,Y) \longrightarrow \mathrm{Hom}(Y^{\vee},X^{\vee}) \tag{1.7.1}$$

is an isomorphism. There is also a canonical isomorphism

$$\underline{\mathrm{Hom}}(X,Y) \xrightarrow{\approx} \underline{\mathrm{Hom}}(Y^\vee,X^\vee)\ , \qquad (1.7.2)$$

namely $\underline{\mathrm{Hom}}(X,Y) \xleftarrow[\approx]{1.6.10} X^\vee \otimes Y \xrightarrow{\approx} X^\vee \otimes Y^{\vee\vee} \xrightarrow{\approx}$ $Y^{\vee\vee} \otimes X^\vee \xrightarrow[\approx]{1.6.10} \underline{\mathrm{Hom}}(Y^\vee,X^\vee)$.

For any object X of $\underline{C}$, there is a morphism

$$\underline{\mathrm{Hom}}(X,X) \xrightarrow{1.6.10} X^\vee \otimes X \xrightarrow{ev} \underline{1}\ .$$

On applying the functor $\mathrm{Hom}(\underline{1},-)$ to this we obtain (see 1.6.4)) a morphism

$$\mathrm{Tr}_X\colon\ \mathrm{End}(X) \to \mathrm{End}(\underline{1}) \qquad (1.7.3)$$

called the <u>trace</u> morphism. The <u>rank</u>, $\mathrm{rk}(X)$, of X is defined to be $\mathrm{Tr}_X(\mathrm{id}_X)$. There are the formulas (Saavedra [1,I 5.1.4]):

$$\mathrm{Tr}_{X\otimes X'}(f\otimes f') = \mathrm{Tr}_X(f)\mathrm{Tr}_{X'}(f')$$
$$\mathrm{Tr}_{\underline{1}}(f) = f \qquad (1.7.4)$$

In particular,

$$\mathrm{rk}(X\otimes X') = \mathrm{rk}(X)\mathrm{rk}(X')$$
$$\mathrm{rk}(\underline{1}) = \mathrm{id}_{\underline{1}}\ . \qquad (1.7.5)$$

<u>Tensor functors</u>

Let $(\underline{C},\otimes)$ and $(\underline{C}',\otimes')$ be tensor categories.

<u>Definition</u> 1.8. A <u>tensor functor</u> $(\underline{C},\otimes) \to (\underline{C}',\otimes')$ is a pair (F,c) comprising a functor $F\colon\ \underline{C} \to \underline{C}'$ and a functorial isomorphism $c_{X,Y}\colon\ F(X)\otimes F(Y) \xrightarrow{\approx} F(X\otimes Y)$ with the properties:

(a) for all $X,Y,Z \in ob(\underline{C})$, the diagram

$$\begin{array}{ccccc} FX \otimes (FY \otimes FZ) & \xrightarrow{id \otimes c} & FX \otimes F(Y \otimes Z) & \xrightarrow{c} & F(X \otimes (Y \otimes Z)) \\ \downarrow \phi' & & & & \downarrow F(\phi) \\ (FX \otimes FY) \otimes FZ & \xrightarrow{c \otimes id} & F(X \otimes Y) \otimes FZ & \xrightarrow{c} & F((X \otimes Y) \otimes Z) \end{array}$$

is commutative;

(b) for all $X,Y \in ob(\underline{C})$, the diagram

$$\begin{array}{ccc} FX \otimes FY & \xrightarrow{c} & F(X \otimes Y) \\ \downarrow \psi' & & \downarrow F(\psi) \\ FY \otimes FX & \xrightarrow{c} & F(Y \otimes X) \end{array}$$

is commutative;

(c) if (U,u) is an identity object of $\underline{C}$ then $(F(U),F(u))$ is an identity object of $\underline{C}'$.

In (Saavedra [1,I4.2.4]) a tensor functor is called a "$\otimes$-foncteur ACU".

Let (F,c) be a tensor functor $\underline{C} \to \underline{C}'$. The conditions (a), (b), (c) imply that, for any finite family $(X_i)_{i \in I}$ of objects of $\underline{C}$, c gives rise to a well-defined isomorphism

$$c: \bigotimes_{i \in I} F(X_i) \xrightarrow{\approx} F(\bigotimes_{i \in I} X_i) ;$$

moreover, for any map $\alpha: I \to J$, the diagram

$$\begin{array}{ccc}
\bigotimes_{i\in I} F(X_i) & \xrightarrow[\approx]{\ c\ } & F(\bigotimes_{i\in I} X_i) \\
\approx \Big\downarrow \chi'(\alpha) & & \approx \Big\downarrow F(\chi(\alpha)) \\
\bigotimes_{j\in J}(\bigotimes_{i\mapsto j} F(X_i)) \xrightarrow[\approx]{c} & \bigotimes_{j\in J}(F(\bigotimes_{i\mapsto j} X_i)) & \xrightarrow[\approx]{c} F(\bigotimes_{j\in J}(\bigotimes_{i\mapsto j} X_i))
\end{array}$$

is commutative. In particular, (F,c) maps inverse objects to inverse objects. Also, the morphism
$F(ev): F(\underline{Hom}(X,Y)) \otimes F(X) \to F(Y)$ gives rise to morphisms
$F_{X,Y}: F(\underline{Hom}(X,Y)) \to \underline{Hom}(FX,FY)$ and $F_X: F(X^\vee) \to F(X)^\vee$.

<u>Proposition 1.9.</u> Let $(F,c): \underline{C} \to \underline{C}'$ be a tensor functor. If $\underline{C}$ and $\underline{C}'$ are rigid, then $F_{X,Y}: F(\underline{Hom}(X,Y)) \to \underline{Hom}(FX,FY)$ is an isomorphism for all $X,Y \in ob(\underline{C})$.

<u>Proof</u>: It suffices to show that F preserves duality, but this is obvious from the following characterization of the dual of X: it is a pair $(Y, Y\otimes X \xrightarrow{ev} \underline{1})$, for which there exists $\varepsilon: \underline{1} \to X \otimes Y$ such that $X = \underline{1} \otimes X \xrightarrow{\varepsilon\otimes id} (X\otimes Y)\otimes X = X\otimes(Y\otimes X) \xrightarrow{id\otimes ev} X$, and the same map with X and Y interchanged, are identity maps.

<u>Definition 1.10.</u> A tensor functor $(F,c): \underline{C} \to \underline{C}'$ is a <u>tensor equivalence</u> (or an <u>equivalence of tensor categories</u>) if $F: \underline{C} \to \underline{C}'$ is an equivalence of categories.

The definition is justified by the following proposition.

Proposition 1.11. Let (F,c): $\underline{C} \to \underline{C}'$ be a tensor equivalence; then there is a tensor functor (F',c'): $\underline{C}' \to \underline{C}$ and isomorphisms of functors $F' \circ F \xrightarrow{\approx} id_{\underline{C}}$ and $F \circ F' \xrightarrow{\approx} id_{\underline{C}'}$ commuting with tensor products (i.e. isomorphisms of tensor functors; see below).

Proof: Saavedra [1, I4.4].

A tensor functor F: $\underline{C} \to \underline{C}'$ of rigid tensor categories induces a morphism F: $End(\underline{1}) \to End(\underline{1}')$. The following formulas hold:

$$Tr_{F(X)} F(f) = F(Tr_X(f))$$

$$rk(F(X)) = F(rk(X)) .$$

Morphisms of tensor functors

Definition 1.12. Let (F,c) and (G,d) be tensor functors $\underline{C} \to \underline{C}'$; a morphism of tensor functors $(F,c) \longrightarrow (G,d)$ is a morphism of functors λ: $F \to G$ such that, for all finite families $(X_i)_{i \in I}$ of objects in $\underline{C}$, the diagram

$$\begin{array}{ccc} \bigotimes_{i\in I} F(X_i) & \xrightarrow{c} & F(\bigotimes_{i\in I} X_i) \\ \downarrow \otimes\lambda_{X_i} & & \downarrow \lambda_{\otimes X_i} \\ \bigotimes_{i\in I} G(X_i) & \xrightarrow{d} & G(\bigotimes_{i\in I} X_i) \end{array} \qquad (1.12.1)$$

is commutative.

In fact, it suffices to require that the diagram (1.12.1) be commutative when I is $\{1,2\}$ or the empty set. For the empty set, (1.12.1) becomes

$$\begin{array}{ccc} \underline{1}' & \xrightarrow{\approx} & F(\underline{1}) \\ \| & & \downarrow \lambda_{\underline{1}} \\ \underline{1}' & \xrightarrow{\approx} & G(\underline{1}) \end{array} \tag{1.12.2}$$

in which the horizontal maps are the unique isomorphisms compatible with the structures of $\underline{1}'$, $F(\underline{1})$, and $G(\underline{1})$ as identity objects of $\underline{C}'$. In particular, when (1.12.2) commutes, $\lambda_{\underline{1}}$ is an isomorphism.

We write $\mathrm{Hom}^{\otimes}(F,F')$ for the set of morphisms of tensor functors $(F,c) \to (G,d)$.

<u>Proposition</u> 1.13. Let (F,c) and (G,d) be tensor functors $\underline{C} \to \underline{C}'$. If $\underline{C}$ and $\underline{C}'$ are rigid, then any morphism of tensor functors λ: $F \to G$ is an isomorphism.

<u>Proof</u>: The morphism μ: $G \to F$, making the diagrams

$$\begin{array}{ccc} F(X^{\vee}) & \xrightarrow{\lambda_{X^{\vee}}} & G(X^{\vee}) \\ \downarrow \approx & & \downarrow \approx \\ F(X)^{\vee} & \xrightarrow{t(\mu_X)} & G(X)^{\vee} \end{array}$$

commutative for all $X \in \mathrm{ob}(\underline{C})$, is an inverse for λ.

For any field k and k-algebra R, there is a canonical tensor functor ϕ_R: $\underline{\mathrm{Vec}}_k \to \underline{\mathrm{Mod}}_R$, $V \longmapsto V \otimes_k R$. If (F,c) and (G,d) are tensor functors $\underline{C} \to \underline{\mathrm{Vec}}_k$, then we define $\underline{\mathrm{Hom}}^{\otimes}(F,G)$ to be the functor of k-algebras such that

$$\underline{\mathrm{Hom}}^{\otimes}(F,G)(R) = \mathrm{Hom}^{\otimes}(\phi_R \circ F, \phi_R \circ G) \tag{1.13.1}$$

Tensor subcategories

Definition 1.14. Let $\underline{C}'$ be a strictly full subcategory of a tensor category $\underline{C}$. We say $\underline{C}'$ is a tensor subcategory of $\underline{C}$ if it is closed under the formation of finite tensor products (equivalently, if it contains an identity object of $\underline{C}$ and if $X_1 \otimes X_2 \in ob(\underline{C}')$ whenever $X_1, X_2 \in ob(\underline{C}')$) . A tensor subcategory of rigid tensor category is said to be a rigid tensor subcategory if it contains $X^{\vee}$ whenever it contains X .

A tensor subcategory becomes a tensor category under the induced tensor structure, and similarly for rigid tensor subcategories.

When $(\underline{C},\otimes)$ is abelian (see below), then we say that a family $(X_i)_{i\in I}$ of objects of $\underline{C}$ is a tensor generating family for $\underline{C}$ if every object of $\underline{C}$ is isomorphic to a subquotient of $P(X_i)$, $P(t_i) \in \mathbb{N}[t_i]_{i\in I}$, where in $P(X_i)$ multiplication is interpreted as $\otimes$ and addition as $\oplus$.

Abelian tensor categories; End $(\underline{1})$

Our convention, that functors between additive categories are to be additive, forces the following definition.

Defintion 1.15. An additive (resp. abelian) tensor category is a tensor category $(\underline{C},\otimes)$ such that $\underline{C}$ is an additive (resp. abelian) category and $\otimes$ is a bi-additive functor.

If $(C,\otimes)$ is such a category, then $R = \mathrm{End}(\underline{1})$ is a ring which acts, via $\ell_X : X \to \underline{1} \otimes X$, on each object X. The action of R on X commutes with endomorphisms of X and so, in particular, R is commutative. The category $\underline{C}$ is R-linear and $\otimes$ is R-bilinear. When $\underline{C}$ is rigid, the trace morphism is an R-linear map $\mathrm{Tr}: \mathrm{End}(X) \to R$.

<u>Proposition 1.16</u>. Let $(\underline{C},\otimes)$ be a rigid tensor category. If $\underline{C}$ is abelian then $\otimes$ is bi-additive and commutes with direct and inverse limits in each factor; in particular it is exact.

<u>Proof</u>: The functor $X \mapsto X \otimes Y$ has a right adjoint, namely $Z \mapsto \underline{\mathrm{Hom}}(Y,Z)$, and therefore commutes with direct limits and is additive. By considering the opposite category $\underline{C}^\circ$, one deduces that it also commutes with inverse limits. (In fact, $Z \mapsto \underline{\mathrm{Hom}}(Y,Z)$ is also a left adjoint for $X \mapsto X \otimes Y$).

<u>Proposition 1.17</u>. Let $(\underline{C},\otimes)$ be a rigid abelian tensor category. If U is a subobject of $\underline{1}$, then $\underline{1} = U \oplus U^\perp$ where $U^\perp = \ker(\underline{1} \to U^\vee)$. Consequently $\underline{1}$ is a simple object if $\mathrm{End}(\underline{1})$ is a field.

<u>Proof</u>: Let $V = \mathrm{coker}\,(U \to \underline{1})$. On tensoring $0 \to U \to \underline{1} \to V \to 0$ with itself, we obtain an exact commutative diagram

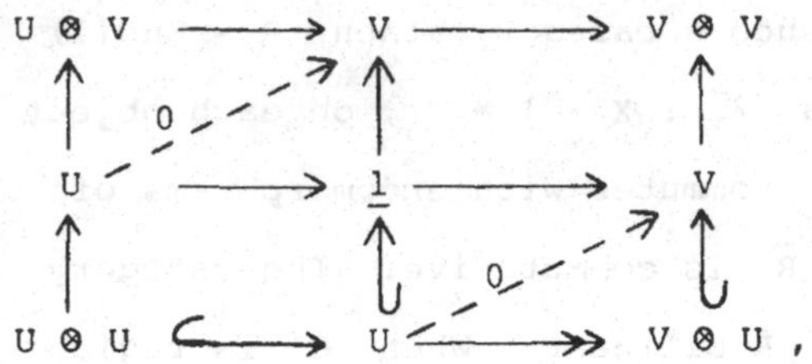

from which it follows that $U \otimes V = 0$ and that $U \otimes U = U$ as a subobject of $\underline{1} \otimes \underline{1} = \underline{1}$.

For any X, the largest subobject Y of X such that $U \otimes Y = 0$ is also the largest subobject for which the map $U \otimes Y \hookrightarrow Y$ $(=(U \hookrightarrow \underline{1}) \otimes Y)$ is zero or, equivalently, such that $Y \hookrightarrow Y \otimes U^{\vee}$ is zero; hence $Y = \ker(X \twoheadrightarrow X \otimes U^{\vee}) = X \otimes U^{\perp}$. On applying this remark with $X = V$, and using that $U \otimes V = 0$, we find that $V \otimes U^{\perp} = V$; on applying it with $X = U$, we find $U \otimes U^{\perp} = 0$. From

$$0 \to U \otimes U^{\perp} \to \underline{1} \otimes U^{\perp} \to V \otimes U^{\perp} \to 0$$

we deduce that $U^{\perp} \xrightarrow{\approx} V$, and that $\underline{1} = U \oplus U^{\perp}$.

Remark 1.18. The proposition shows that there is a one-to-one correspondence between subobjects of $\underline{1}$ and idempotents in $\mathrm{End}(\underline{1})$. Such an idempotent e determines a decomposition of tensor categories $\underline{C} = \underline{C}' \times \underline{C}''$ in which $ob(\underline{C}')$ is the set of objects of $\underline{C}$ on which e acts as the identity map.

Proposition 1.19. Let $\underline{C}$ and $\underline{C}'$ be rigid abelian tensor categories and assume that $End(\underline{1})$ is a field and that $\underline{1}' \neq 0$, where $\underline{1}$ and $\underline{1}'$ are identity objects in $\underline{C}$ and $\underline{C}'$. Then any exact tensor functor $F : \underline{C} \to \underline{C}'$ is faithful.

Proof: The criterion in $\underline{C}$,

$$X \neq 0 \Leftrightarrow X \otimes \overset{\vee}{X} \to \underline{1} \text{ surjective}$$

is respected by F.

A criterion to be a rigid tensor category

Proposition 1.20. Let $\underline{C}$ be a k-linear abelian category, where k is a field, and let $\otimes: \underline{C} \times \underline{C} \to \underline{C}$ be a k-bilinear functor. Suppose there are given a faithful exact k-linear functor $F: \underline{C} \to \underline{Vec}_k$, a functorial isomorphism $\phi_{X,Y,Z}: X \otimes (Y \otimes Z) \to (X \otimes Y) \otimes Z$, and a functorial isomorphism $\psi_{X,Y}: X \otimes Y \to Y \otimes X$ with the following properties:

(a) $F \circ \otimes = \otimes \circ F \times F$;

(b) $F(\phi_{X,Y,Z})$ is the usual associativity isomorphism in $\underline{Vec}_k$;

(c) $F(\psi_{X,Y})$ is the usual commutativity isomorphism in $\underline{Vec}_k$;

(d) there exists a $U \in ob(\underline{C})$ such that $k \to End(U)$ is an isomorphism and $F(U)$ has dimension 1 ;

(e) if $F(L)$ has dimension 1, then there exists an object L^{-1} in $\underline{C}$ such that $L \otimes L^{-1} \approx U$.

Then $(C, \otimes, \phi, \psi)$ is a rigid abelian tensor category.

Proof: It is not difficult to prove this directly — essentially one only has to show that the object U of (d) is an identity object and that (e) is sufficient to show that $\underline{C}$ is rigid — but we shall indicate a more elegant approach in (2.18) below.

Examples

(1.21) $\underline{\text{Vec}}_k$, for k a field, is rigid abelian tensor category and $\text{End}(\underline{1}) = k$. All of the above definitions take on a familiar meaning when applied to $\underline{\text{Vec}}_k$. For example, Tr: $\text{End}(X) \to k$ is the usual trace map.

(1.22) $\underline{\text{Mod}}_R$ is an abelian tensor category and $\text{End}(\underline{1}) = R$. In general it will not be rigid because not all R-modules will be reflexive.

(1.23) The category $\underline{\text{Proj}}_R$ of projective modules of finite type over a commutative ring R is a rigid additive tensor category and $\text{End}(\underline{1}) = R$. The rigidity follows easily from considering the objects of $\underline{\text{Proj}}_R$ as locally-free modules of finite rank on spec(R) .

(1.24) Let G be an affine group scheme over a field k and let $\underline{\text{Rep}}_k(G)$ be the category of finite-dimensional representations of G over k . Thus an object of $\underline{\text{Rep}}_k(G)$ consists of a finite-dimensional vector space V over k and a homomorphism $g \longmapsto g_V : G \to GL(V)$ of affine group schemes over k . Then $\underline{\text{Rep}}_k(G)$ is a rigid abelian tensor category and $\text{End}(\underline{1}) = k$. These categories, and more generally the categories of representations of affine gerbs (see §3), are the main topic of study of this article.

(1.25) (Vector spaces graded by $\mathbb{Z}/2\mathbb{Z}$). Let $\underline{C}$ be the category whose objects are pairs (V^0, V^1) of finite dimensional vector spaces over a field k. We give $\underline{C}$ the tensor structure whose commutativity constraint is determined by the Koszul rule of signs, i.e., that defined by the isomorphisms

$$v \otimes w \mapsto (-1)^{ij}\, w \otimes v : V^i \otimes W^j \to W^j \otimes V^i .$$

Then $\underline{C}$ is a rigid abelian tensor category and $\mathrm{End}(\underline{1}) = k$, but it is not of the form $\underline{\mathrm{Rep}}_k(G)$ for any G because $\mathrm{rk}(V^\circ, V^1) = \dim(V^\circ) - \dim(V^1)$ may not be positive.

(1.26) The rigid additive tensor category freely generated by an object T is a pair $(\underline{C}, T)$ comprising a rigid additive tensor category $\underline{C}$ and that $\mathrm{End}(\underline{1}) = \mathbb{Z}[t]$ and an object T having the property that

$$F \mapsto F(T) : \mathrm{Hom}^{\otimes}(\underline{C}, \underline{C}') \to \underline{C}'$$

is an equivalence of categories for all rigid additive tensor categories $\underline{C}'$ (t will turn out to be the rank of T). We show how to construct such a pair $(\underline{C}, T)$ — clearly it is unique up to a unique equivalence of tensor categories preserving T.

Let V be a free module of finite rank over a commutative ring k and let $T^{a,b}(V)$ be the space $V^{\otimes a} \otimes V^{\vee \otimes b}$ of tensors with covariant degree a and contravariant degree b. A map $f : T^{a,b}(V) \to T^{c,d}(V)$ can be identified with a tensor "f" in $T^{b+c,a+d}(V)$. When $a+d = b+c$, $T^{b+c,a+d}(V)$ contains a special element, namely the $(a+d)^{th}$ tensor power of "id" $\in T^{1,1}(V)$, and other elements can be obtained by allowing an element of the symmetric group S_{a+d} to permute the contravariant components of this special element. We have therefore a map

$$\varepsilon : S_{a+d} \to \mathrm{Hom}(T^{a,b},T^{c,d}) \quad \text{(when } a+d = b+c\text{)} .$$

The induced map $k[S_{a+d}] \to \mathrm{Hom}(T^{a,b},T^{c,d})$ is injective provided $rk(V) \geq a+d$. One checks that the composite of two such maps $\varepsilon(\sigma) : T^{a,b}(V) \to T^{c,d}(V)$ and $\varepsilon(\tau) : T^{c,d}(V) \to T^{e,f}(V)$ is given by a universal formula

$$\varepsilon(\tau)\cdot\varepsilon(\sigma) = (rk\ V)^N\ \varepsilon(\rho) \qquad (1.26.1)$$

with ρ and N depending only on a,b,c,d,e,f,σ, and τ.

We define $\underline{C}'$ to be the category having as objects symbols $T^{a,b}$ $(a,b, \in \mathbb{N})$, and for which $\mathrm{Hom}(T^{a,b},T^{c,d})$ is the free $\mathbb{Z}[t]$-module with basis S_{a+d} if $a+d = b+c$ and is zero otherwise. Composition of morphisms is defined to be $\mathbb{Z}[t]$-bilinear and to agree on basis elements with the universal formula (1.26.1) with $rk\ V$ replaced by the inderterminate t. The associativity law holds for this composition because it does whenever t is replaced by a large enough positive integer (it becomes the associativity law in a category of modules). Tensor products are defined by $T^{a,b} \otimes T^{c,d} = T^{a+d,b+d}$ and by an obvious rule for morphisms. We define T to be $T^{1,0}$.

The category $\underline{C}$ is deduced from $\underline{C}'$ by formally adjoining direct sums of objects. Its universality follows from the fact that the formula (1.26.1) holds in any rigid additive category.

(1.27) ($\underline{GL}_t$) . Let n be an integer, and use $t \longmapsto n : \mathbb{Z}[t] \to \mathbb{C}$ to extend the scalars in the above example from $\mathbb{Z}[t]$ to $\mathbb{C}$. If V is an n-dimensional complex vector space, and if $a+d \leq n$, then

$$\mathrm{Hom}(T^{a,b}, T^{c,d}) \otimes_{\mathbb{Z}[t]} \mathbb{C} \to \mathrm{Hom}_{GL(V)}(T^{a,b}(V), T^{c,d}(V))$$

is an isomorphism. For any sum T' of $T^{a,b}$s and large enough integer n , $\mathrm{End}(T') \otimes_{\mathbb{Z}[t]} \mathbb{C}$ is therefore a product of matrix algebras. This implies that $\mathrm{End}(T') \otimes_{\mathbb{Z}[t]} \mathbb{Q}(t)$ is a semisimple algebra.

After extending the scalars in $\underline{C}$ to $\mathbb{Q}(t)$ (i.e., replacing $\mathrm{Hom}(T',T'')$ with $\mathrm{Hom}(T',T'') \otimes_{\mathbb{Z}[t]} \mathbb{Q}(t)$) and passing to the pseudo-abelian (Karoubian) envelope (i.e., formally adjoining images of idempotents), we obtain a semisimple rigid abelian tensor category $\underline{GL}_t$. The rank of T in $\underline{GL}_t$ is $t \notin \mathbb{N}$ and so, although $\mathrm{End}(\underline{1}) = \mathbb{Q}(t)$ is a field, $\underline{GL}_t$ is not of the form $\underline{\mathrm{Rep}}_k(G)$ for any group scheme (or gerb) G .

§2. Neutral Tannakian categories

Throughout this section, k will be a field.

Affine group schemes

Let $G = \mathrm{spec}\ A$ be an affine group scheme over k . The maps mult: $G \times G \to G$, identity $\{1\} \to G$, inverse: $G \to G$ induce maps of k-algebras

$$\Delta: \quad A \to A \otimes_k A , \qquad \varepsilon: \ A \to k , \qquad S: \ A \to A$$

(the comultiplication, coidentity, and coinverse maps) such that

$(id \otimes \Delta)\Delta = (\Delta \otimes id)\Delta: \ A \to A \otimes A \rightrightarrows A \otimes A \otimes A$

(coassociativity axiom),

$id = (\varepsilon \otimes id)\Delta: A \to A \otimes A \to k \otimes A = A$ (coidentity axiom), and

$(A \xrightarrow{\Delta} A \otimes A \xrightarrow{(S,id)} A) = (A \xrightarrow{\varepsilon} k \hookrightarrow A)$ (coinverse axiom).

We define a <u>bialgebra</u> over k to be a k-algebra A together with maps Δ, ε, and S satisfying the three axioms. (This terminology is not standard).

<u>Proposition</u> 2.1. The functor $A \longmapsto \operatorname{spec} A$ defines an equivalence between the category of k-bialgebras and the category of affine group schemes over k .

<u>Proof</u>: Obvious.

If A is finitely generated (as a k-algebra) we say that G is <u>algebraic</u> or that it is an <u>algebraic group</u>.

We define a <u>coalgebra</u> over k to be a vector space C over k together with k-linear maps $\Delta: \ C \to C \otimes_k C$ and $\varepsilon: \ C \to k$ satisfying the coassociativity and coidentity axioms.

A <u>comodule</u> over C is a vector space V over k together with a k-linear map $\rho: \ V \to V \otimes C$ such that $(id \otimes \varepsilon)\rho: \ V \to V \otimes C \to V \otimes k = V$ is the identity map and $(id \otimes \Delta)\rho = (\rho \otimes id)\rho : \ V \to V \otimes C \otimes C$.

For example, Δ defines an C-comodule structure on C .

<u>Proposition</u> 2.2. Let $G = \operatorname{spec} A$ be an affine group scheme over k and let V be a vector space over k . There is a canonical one-to-one correspondence between the A-comodule structures on V and the linear representations of G on V .

Proof. Let $G \to GL(V)$ be a representation. The element $id \in Mor(G,G) = G(A)$ maps to an element of $GL(V \otimes A)$ whose restriction to $V = V \otimes k \subset V \otimes A$ is a comodule structure on V. Conversely, a comodule structure ρ on V determines a representation of G on V such that, for R a k-algebra and $g \in G(R)$, the restriction of g_V: $V \otimes R \to V \otimes R$ to $V = V \otimes k \subset V \otimes R$ is

$$(id \otimes g)\rho: \quad V \to V \otimes A \to V \otimes R .$$

Proposition 2.3. Let C be a k-coalgebra and (V,ρ) a comodule over C. Any finite subset of V is contained in a sub-comodule of V having finite dimension over k .

Proof: Let $\{a_i\}$ be a basis for C over k . If v is in the finite subset, write $\rho(v) = \sum v_i \otimes a_i$ (finite sum). The k-space generated by the v and the v_i is a sub-comodule.

Corollary 2.4. Any k-rational representation of an affine group scheme is a directed union of finite-dimensional subrepresentations V_i .

Proof: Combine (2.2) and (2.3) .

Corollary 2.5. An affine group scheme G is algebraic if and only if it has a faithful finite-dimensional representation over k .

Proof: The sufficiency is obvious. For the necessity, let V be the regular representation of G , and write $V = \cup V_i$ with the V_i as in (2.4). Then $\bigcap_i Ker(G \to GL(V_i)) = \{1\}$ because V is a faithful representation, and it follows that

$\mathrm{Ker}(G \to GL(V_{i_0})) = \{1\}$ for some i_0 because G is Noetherian as a topological space.

Proposition 2.6. Let A be a k-bialgebra. Any finite subset of A is contained in a sub-bialgebra that is finitely generated as an algebra over k.

Proof: According to (2.3), the finite subset is contained in a finite-dimensional subspace V of A such that $\Delta(V) \subset V \otimes A$. Let $\{v_i\}$ be a basis for V and let $\Delta(v_j) = \sum v_i \otimes a_{ij}$. The subalgebra $k[v_j, a_{ij}, Sv_j, Sa_{ij}]$ of A is a sub-bialgebra. (See Waterhouse [1,3.3]).

Corollary 2.7. Any affine group scheme G over k is a directed inverse limit $G = \varprojlim G_i$ of affine algebraic groups over k in which the transition maps $G_i \leftarrow G_j$, $i \leq j$, are surjective.

Proof: The functor spec transforms a direct limit $A = \cup A_i = \varinjlim A_i$ into an inverse limit $G = \varprojlim G_i$. The transition map $G_i \leftarrow G_j$ is surjective because A_j is faithfully flat over its subalgebra A_i (Waterhouse [1, 14.1]).

The converse to (2.7) is also true; in fact the inverse limit of any family of affine group schemes is again an affine group scheme.

Determining a group scheme from its representations.

Let G be an affine group scheme over k and let ω, or ω^G, be the forgetful functor $\underline{\mathrm{Rep}}_k(G) \to \underline{\mathrm{Vec}}_k$. For R a k-algebra, $\underline{\mathrm{Aut}}^{\otimes}(\omega)(R)$ consists of families (λ_X), $X \in \mathrm{ob}(\underline{\mathrm{Rep}}_k(G))$, where λ_X is an R-linear automorphism of $X \otimes R$ such that $\lambda_{X_1 \otimes X_2} = \lambda_{X_1} \otimes \lambda_{X_2}$, $\lambda_{\underline{1}}$ is the identity map (on R), and

$$\lambda_Y \circ (\alpha \otimes 1) = (\alpha \otimes 1) \circ \lambda_X : X \otimes R \to Y \otimes R$$

for all G-equivariant k-linear maps $\alpha: X \to Y$ (see 1.12). Clearly any $g \in G(R)$ defines an element of $\underline{\mathrm{Aut}}^{\otimes}(\omega)(R)$.

<u>Proposition</u> 2.8. The natural map $G \to \underline{\mathrm{Aut}}^{\otimes}(\omega)$ is an isomorphism of functors of k-algebras.

<u>Proof</u>: Let $X \in \underline{\mathrm{Rep}}_k(G)$ and let $\underline{C}_X$ be the strictly full subcategory of $\underline{\mathrm{Rep}}_k(G)$ of objects isomorphic to a subquotient of $P(X, X^{\vee})$, $P \in \mathbb{N}[t,s]$ (cf. the discussion following (1.14)). The map $\lambda \longmapsto \lambda_X$ identifies $\underline{\mathrm{Aut}}^{\otimes}(\omega|\underline{C}_X)(R)$ with a subgroup of $GL(X \otimes R)$. Let G_X be the image of G in $GL(X)$; it is a closed algebraic subgroup of G, and clearly

$$G_X(R) \subset \underline{\mathrm{Aut}}^{\otimes}(\omega|\underline{C}_X)(R) \subset GL(X \otimes R).$$

If $V \in \mathrm{ob}(\underline{C}_X)$ and $t \in V$ is fixed by G, then $a \longmapsto at: k \xrightarrow{\alpha} V$ is G-equivariant, and so $\lambda_V(t \otimes 1) = (\alpha \otimes 1)\lambda_{\underline{1}}(1) = t \otimes 1$. Now (I.3.2) shows that $G_X = \underline{\mathrm{Aut}}^{\otimes}(\omega|\underline{C}_X)$.

If $X' = X \oplus Y$ for some representation Y of G, then $\underline{C}_X \subset \underline{C}_{X'}$, and there is a commutative diagram

$$\begin{array}{ccc} G_{X'} & \xrightarrow{\approx} & \underline{\mathrm{Aut}}^{\otimes}(\omega|\underline{C}_{X'}) \\ \downarrow & & \downarrow \\ G_X & \xrightarrow{\approx} & \underline{\mathrm{Aut}}^{\otimes}(\omega|\underline{C}_X) \end{array}.$$

It is clear from (2.5) and (2.7) that $G = \varprojlim G_{X'}$, and so, on passing to the inverse limit over these diagrams, we obtain an isomorphism $G \xrightarrow{\approx} \underline{Aut}^{\otimes}(\omega)$.

A homomorphism $f: G \to G'$ defines a functor ω^f: $\underline{Rep}_k(G') \to \underline{Rep}_k(G)$, namely $(G' \to GL(V)) \mapsto (G \xrightarrow{f} G' \to GL(V))$, such that $\omega^G \circ \omega^f = \omega^{G'}$.

Corollary 2.9. Let G and G' be affine group schemes over k and let $F: \underline{Rep}_k(G') \to \underline{Rep}_k(G)$ be a tensor functor such that $\omega^G \circ F = \omega^{G'}$. Then there is a unique homomorphism $f: G \to G'$ such that $F = \omega^f$.

Proof: For $\lambda \in \underline{Aut}^{\otimes}(\omega^G)(R)$, R a k-algebra, define $F^*(\lambda) \in \underline{Aut}^{\otimes}(\omega^{G'})(R)$ by the rule $F^*(\lambda)_{X'} = \lambda_{F(X')}$. The proposition allows us to regard F^* as a homomorphism $G \to G'$, and clearly $F \mapsto F^*$ and $f \mapsto \omega^f$ are inverse maps.

Remark 2.10. Proposition 2.8 shows that G is determined by the triple $(\underline{Rep}_k(G), \otimes, \omega^G)$; it can be shown that the coalgebra of G is already determined by $(\underline{Rep}_k(G), \omega^G)$ (cf. the proof of Theorem 2.11).

The main theorem

Theorem 2.11. Let $\underline{C}$ be a rigid abelian tensor category such that $k = \underline{End}(\underline{1})$, and let $\omega: \underline{C} \to \underline{Vec}_k$ be an exact faithful k-linear tensor functor. Then,

(a) the functor $\underline{Aut}^{\otimes}(\omega)$ of k-algebras is representable by an affine group scheme G;

(b) ω defines an equivalence of tensor categories $\underline{C} \to \underline{\mathrm{Rep}}_k(G)$.

<u>Proof</u>: We first construct the coalgebra A of G without using the tensor structure on $\underline{C}$. The tensor structure then enables us to define an algebra structure on A , and the rigidity of $\underline{C}$ implies that spec A is a group scheme (rather than a monoid scheme). The following easy observation will allow us to work initially with algebras rather than coalgebras: for a finite-dimensional (not necessarily commutative) k-algebra A and its dual coalgebra $A^{\vee} \overset{\mathrm{df}}{=} \mathrm{Hom}(A,k)$, the bijection

$$\mathrm{Hom}(A \otimes_k V,\ V) \xleftarrow{\ \approx\ }\!\!\!\!\rightarrow \mathrm{Hom}(V, A^{\vee} \otimes_k V)$$

determines a one-to-one correspondence between the A-module structures on a vector space V and the $A^{\vee}$-comodule structure on V .

We begin with some constructions that are valid in any k-linear abelian category $\underline{C}$. For V a finite-dimensional vector space over k and X and object of $\underline{C}$, we define $V \otimes X$ to be the system $((X^n)_\alpha, \phi_{\beta,\alpha})$ where α runs through the isomorphisms $k^n \xrightarrow{\approx} V$, $(X^n)_\alpha = X^n \overset{\mathrm{df}}{=} X \oplus \ldots \oplus X$ (n copies), and $\phi_{\beta,\alpha}\colon (X^n)_\alpha \to (X^n)_\beta$ is defined by $\beta^{-1} \circ \alpha \in GL_n(k)$. Note that $\phi_{\gamma,\beta} \circ \phi_{\beta,\alpha} = \phi_{\gamma,\alpha}$. A morphism $V \otimes X \to T$ or $T \to V \otimes X$, where $T \in \mathrm{ob}(\underline{C})$, is a family of morphisms compatible with the $\phi_{\beta,\alpha}$. There is a canonical k-linear map $V \to \mathrm{Hom}(X, V \otimes X)$ under which $v \in V$ maps to $(X \xrightarrow{\psi_\alpha} (X^n)_\alpha)$ where ψ_α is defined by $\alpha^{-1}(v) \in k^n$. This map induces a functorial isomorphism

$\underline{\mathrm{Hom}}(V \otimes X, T) \xrightarrow{\approx} \mathrm{Hom}(V, \mathrm{Hom}(X,T))$, $T \in \mathrm{ob}(\underline{C})$. Any k-linear functor $F\colon \underline{C} \to \underline{C}'$ has the property that $F(V \otimes X) = V \otimes F(X)$. When $\underline{C}$ is $\underline{\mathrm{Vec}}_k$, $V \otimes X$ can be identified with the usual object.

For $V \in \mathrm{ob}(\underline{\mathrm{Vec}}_k)$ and $X \in \mathrm{ob}(\underline{C})$, we define $\underline{\mathrm{Hom}}(V,X)$ to be $V^{\vee} \otimes X$. If $W \subset V$ and $Y \subset X$, then the subobject of $\underline{\mathrm{Hom}}(V,X)$ mapping W into Y is defined to be

$$(Y:W) = \mathrm{Ker}(\underline{\mathrm{Hom}}(V,X) \to \underline{\mathrm{Hom}}(W,X/Y)) .$$

Lemma 2.12. Let $\underline{C}$ be a k-linear abelian category and $\omega\colon \underline{C} \to \underline{\mathrm{Vec}}_k$ a k-linear exact faithful functor. Then, for any $X \in \mathrm{ob}(\underline{C})$, the following two objects are equal:

(a) the largest subobject P of $\underline{\mathrm{Hom}}(\omega(X),X)$ whose image in $\underline{\mathrm{Hom}}(\omega(X)^n,X^n)$ (embedded diagonally) is contained in $(Y\colon \omega(Y))$ for all $Y \subset X^n$;

(b) the smallest subobject P' of $\underline{\mathrm{Hom}}(\omega(X),X)$ such that the subspace $\omega(P')$ of $\mathrm{Hom}(\omega(X),\omega(X))$ contains $\mathrm{id}\colon \omega(X) \to \omega(X)$.

Proof: Clearly $\omega(X) = 0$ implies $\mathrm{End}(X) = 0$, which implies $X = 0$. Thus if $X \subset Y$ and $\omega(X) = \omega(Y)$ then $X = Y$, and it follows that all objects of $\underline{C}$ are both Artinian and Noetherian. The objects P and P' therefore obviously exist.

The functor ω maps $\underline{\mathrm{Hom}}(V,X)$ to $\mathrm{Hom}(V,\omega(X))$ and $(Y:W)$ to $(\omega(Y):W)$ for all $W \subset V \in \mathrm{ob}(\underline{\mathrm{Vec}}_k)$ and $Y \subset X \in \mathrm{ob}(\underline{C})$. It therefore maps $P \overset{\mathrm{df}}{=} \cap\, (\underline{\mathrm{Hom}}(\omega(X),X) \cap (Y:\omega(Y)))$ to $\cap\, (\mathrm{End}(\omega(X)) \cap (\omega Y:\omega Y))$. This means ωP is the largest subring of $\mathrm{End}(\omega(X))$ stabilizing $\omega(Y)$ for all $Y \subset X^n$. Hence $\mathrm{id} \in \omega(P)$ and $P \supset P'$.

Let V be a finite-dimensional vector space over k ; there is an obvious map $\underline{\mathrm{Hom}}(\omega(X),X) \to \underline{\mathrm{Hom}}(\omega(V\otimes X),V\otimes X)$ (inducing $f \mapsto 1 \otimes f\colon \mathrm{End}(\omega(X)) \to \mathrm{End}(V\otimes\omega(X))$ and $\omega(P) \subset \mathrm{End}(\omega(X))$ stabilizes $\omega(Y)$ for all $Y \subset V \otimes X$. On applying this remark to a $Q \subset \underline{\mathrm{Hom}}(\omega(X),X) = \omega(X)^{\vee} \otimes X$, we find that $\omega(P)$, when acting by left multiplication on $\mathrm{End}(\omega(X))$, stablizes $\omega(Q)$. Therefore, if $\omega(Q)$ contains 1, then $\omega(P) \subset \omega(Q)$, and $P \subset Q$; this shows that $P \subset P'$.

Let $P_X \subset \underline{\mathrm{Hom}}(\omega(X),X)$ be the subobject defined in (a) (or (b)) of the lemma, and let $A_X = \omega(P_X)$; it is the largest subalgebra of $\mathrm{End}(\omega(X))$ stabilizing $\omega(Y)$ for all $Y \subset X^n$. Let $\langle X \rangle$ be the strictly full subcategory of $\underline{C}$ such that $\mathrm{ob}(\langle X \rangle)$ consists of the objects of $\underline{C}$ that are isomorphic to subquotients of X^n , $n \in \mathbb{N}$. Then $\omega|\langle X \rangle\colon \langle X \rangle \to \underline{\mathrm{Vec}}_k$ factors through $\underline{\mathrm{Mod}}_{A_X}$.

Lemma 2.13. Let $\omega\colon \underline{C} \to \underline{\mathrm{Vec}}_k$ be as in (2.12). Then ω defines an equivalence of categories $\langle X \rangle \to \underline{\mathrm{Mod}}_{A_X}$ carrying $\omega|\langle X \rangle$ into the forgetful functor. Moreover $A_X = \mathrm{End}(\omega|\langle X \rangle)$.

Proof: The right action $f \mapsto f \circ a$ of A_X on $\underline{\mathrm{Hom}}(\omega(X),X)$ stabilizes P_X because obviously $(Y{:}\omega(Y))(\omega(Y){:}\omega(Y)) \subset (Y{:}\omega(Y))$. If M is an A_X-module we define

$$P_X \otimes_{A_X} M = \mathrm{Coker}(P_X \otimes A_X \otimes M \rightrightarrows P_X \otimes M) .$$

Then $\omega(P_X \otimes_{A_X} M) = \omega(P_X) \otimes_{A_X} M = A_X \otimes_{A_X} M = M$. Recall that

$P_X \otimes_{A_X} M$ is a family (Y_α) of objects of $\underline{C}$ with given compatible isomorphisms $Y_\alpha \to Y_\beta$. If we choose one α, then $\omega(Y_\alpha) \approx M$, which shows that ω is essentially surjective. A similar argument shows that $\langle X \rangle \to \underline{\mathrm{Mod}}_{A_X}$ is full.

Clearly any element of A_X defines an endomorphism of $\omega | \langle X \rangle$. On the other hand an element λ of $\mathrm{End}(\omega | \langle X \rangle)$ is determined by $\lambda_X \in \mathrm{End}(\omega(X))$; thus $\mathrm{End}(\omega(X)) \supset \mathrm{End}(\omega | \langle X \rangle) \supset A_X$. But λ_X stabilizes $\omega(Y)$ for all $Y \subset X^n$, and so $\mathrm{End}(\omega | \langle X \rangle) \subset A_X$. This completes the proof of the lemma.

Let $B_X = A_X^\vee$. The remark at the start of the proof allows us to restate (2.13) as follows: ω defines an equivalence

$$(\langle X \rangle, \omega | \langle X \rangle) \to (\underline{\mathrm{Comod}}_{B_X}, \mathrm{forget})$$

where $\underline{\mathrm{Comod}}_{B_X}$ is the category of B_X-comodules of finite dimension over k.

On passing to the inverse limit over X (cf. the proof of (2.8)), we obtain the following result.

<u>Proposition</u> 2.14. Let $(\underline{C}, \omega)$ be as in (2.12) and let $B = \varinjlim \mathrm{End}(\omega | \langle X \rangle)^\vee$. Then ω defines an equivalence of categories $\underline{C} \to \underline{\mathrm{Comod}}_B$ carrying ω into the forgetful functor.

<u>Example</u> 2.15. Let A be a finite-dimensional k-algebra and let ω be the forgetful functor $\underline{\mathrm{Mod}}_A \to \underline{\mathrm{Vec}}_k$. For R

a commutative k-algebra, let ϕ_R be the functor $R \otimes -: \underline{Vec}_k \to \underline{Mod}_R$. There is a canonical map $\alpha: R \otimes_k A \to \mathrm{End}(\phi_R \circ \omega)$, which we shall show to be an isomorphism by defining an inverse β. For $\lambda \in \mathrm{End}(\phi_R \circ \omega)$, set $\beta(\lambda) = \lambda_A(1)$. Clearly $\beta\alpha = id$, and so we have to show $\alpha\beta = id$. For $M \in ob(\underline{Mod}_A)$, let $M_0 = \omega(M)$. The A-module $A \otimes_k M_0$ is a direct sum of copies of A, and the additivity of λ shows that $\lambda_{A \otimes M_0} = \lambda_A \otimes id_{M_0}$. The map $a \otimes m \longmapsto am: A \otimes_k M_0 \to M$ is A-linear, and hence

$$\begin{array}{ccc} R \otimes A \otimes M_0 & \longrightarrow & R \otimes M \\ \downarrow \lambda & & \downarrow \lambda \\ R \otimes A \otimes M_0 & \longrightarrow & R \otimes M \end{array}$$

is commutative. Therefore $\lambda_M(m) = \lambda_A(1)m = (\alpha\beta(\lambda))_M(m)$ for $m \in R \otimes M$.

In particular, $A \xrightarrow{\approx} \mathrm{End}(\omega)$, and it follows that, if in (2.13) we take $\underline{C} = \underline{Mod}_A$ so that $\underline{C} = \langle A \rangle$, then the equivalence of categories obtained is the identity functor.

Let B be a coalgebra over k and let ω be the forgetful functor $\underline{Comod}_B \to \underline{Vec}_k$. The above discussion shows that $B = \varinjlim \mathrm{End}(\omega | \langle X \rangle)^{\vee}$. We deduce, as in (2.9), that every functor $\underline{Comod}_B \to \underline{Comod}_{B'}$, carrying the forgetful functor into the forgetful functor arises from a unique homomorphism $B \to B'$.

Again let B be a coalgebra over k. A homomorphism $u: B \otimes_k B \to B$ defines a functor

$$\phi^u: \underline{\text{Comod}}_B \times \underline{\text{Comod}}_B \to \underline{\text{Comod}}_B$$

sending (X,Y) to $X \otimes_k Y$ with the B-comodule structure

$$X \otimes Y \xrightarrow{\rho_X \otimes \rho_Y} X \otimes B \otimes Y \otimes B \xrightarrow{1 \otimes u} X \otimes Y \otimes B .$$

Proposition 2.16. The map $u \longmapsto \phi^u$ defines a one-to-one correspondence between the set of homomorphisms $B \otimes_k B \to B$ and the set of functors $\phi: \underline{\text{Comod}}_B \times \underline{\text{Comod}}_B \to \underline{\text{Comod}}_B$ such that $\phi(X,Y) = X \otimes_k Y$ as k vector spaces. The natural associativity and commutativity constraints on $\underline{\text{Vec}}_k$ induce similar contraints on $(\underline{\text{Comod}}_B, \phi^u)$ if and only if the multiplication defined by u on B is associative and commutative; there is an identity object in $(\underline{\text{Comod}}_B, \phi^u)$ with underlying vector space k if and only if B has an identity element.

Proof: The pair $(\underline{\text{Comod}}_B \times \underline{\text{Comod}}_B, \omega \otimes \omega)$, with $(\omega \otimes \omega)(X \otimes Y)$ $\omega(X) \otimes \omega(Y)$ (as a k vector space), satisfies the conditions of (2.14), and $\varinjlim \text{End}(\omega \otimes \omega | < (X,Y) >)^\vee = B \otimes B$. Thus the first statement of the proposition follows from (2.15). The remaining statements are easy.

Let $(\underline{C}, \omega)$ and B be as in (2.14) except now assume that $\underline{C}$ is a tensor category and ω is a tensor functor. The tensor structure on $\underline{C}$ induces a similar structure on

$\underline{\text{Comod}}_B$ and hence, because of (2.16), the structure of an associative commutative k-algebra with identity element on B. Thus B lacks only a coinverse map S to be a bialgebra, and G = spec B is an affine monoid scheme. Using (2.15) we find that, for any k-algebra R, $\underline{\text{End}}(\omega)(R) \overset{\text{df}}{=} \text{End}(\phi_R \circ \omega) = \varprojlim \text{Hom}_{k\text{-lin}}(B_X, R) = \text{Hom}_{k\text{-lin}}(B,R)$. An element $\lambda \in \text{Hom}_{k\text{-lin}}(B,R)$ corresponds to an element of $\underline{\text{End}}(\omega)(R)$ commuting with the tensor structure if and only if λ is a k-algebra homomorphism; thus $\underline{\text{End}}^{\otimes}(\omega)(R) = \text{Hom}_{k\text{-alg}}(B,R) = G(R)$. We have shown that if in the statement of (2.11) the rigidity condition is omitted, then one can conclude that $\underline{\text{End}}^{\otimes}(\omega)$ is representable by an affine monoid scheme G = spec B and ω defines an equivalence of tensor categories $\underline{C} \to \underline{\text{Comod}}_B = \underline{\text{Rep}}_k(G)$. If we now assume that $(\underline{C},\otimes)$ is rigid, then (1.13) shows that $\underline{\text{End}}^{\otimes}(\omega) = \underline{\text{Aut}}^{\otimes}(\omega)$, and the theorem follows.

Remark 2.17. Let $(\underline{C},\omega)$ be $(\text{Rep}_k(G),\omega^G)$. On following through the proof of (2.11) in this case one recovers (2.8): $\underline{\text{Aut}}^{\otimes}(\omega^G)$ is represented by G.

Remark 2.18. Let $(\underline{C},\otimes,\phi,\psi,F)$ satisfy the conditions of (1.20). Then $(\underline{C},\otimes,\phi,\psi)$ is obviously a tensor category, and the proof of (2.11) shows that F defines an equivalence of tensor categories $\underline{C} \to \underline{\text{Rep}}_k(G)$ where G is an affine group monoid representing $\underline{\text{End}}^{\otimes}(\omega)$. We can assume that $\underline{C} = \underline{\text{Rep}}_k(G)$.

Let $\lambda \in G(R)$. If $L \subset \underline{\mathrm{Rep}}_k(G)$ has dimension 1 , then $\lambda_L: R \otimes L \to R \otimes L$ is invertible, as follows easily from the existence of a G-isomorphism $L \otimes L^{-1} \xrightarrow{\approx} k$. It follows that λ_X is invertible for any $X \in \mathrm{ob}(\underline{\mathrm{Rep}}_k(G))$ because $\det(\lambda_X) \underline{\underline{\mathrm{df}}} \Lambda^d\lambda_X = \lambda_{\Lambda^d X}$, where $d = \dim X$, is invertible.

Definition 2.19. A neutral Tannakian category over k is a rigid abelian k-linear tensor category $\underline{C}$ for which there exists an exact faithful k-linear tensor functor ω: $\underline{C} \to \underline{\mathrm{Vec}}_k$. Any such functor ω is said to be a fibre functor for $\underline{C}$.

Thus (2.11) shows that any neutral Tannakian category is equivalent (in possibly many different ways) to the category of finite-dimensional representations of an affine group scheme.

Properties of G and of $\underline{\mathrm{Rep}}_k(G)$.

In view of the last remark, it is natural to ask how properties of G are reflected in $\underline{\mathrm{Rep}}_k(G)$.

Proposition 2.20. Let G be an affine group scheme over k .

(a) G is finite if and only if there exists an object X of $\underline{\mathrm{Rep}}_k(G)$ such that every object of $\underline{\mathrm{Rep}}_k(G)$ is isomorphic to a subquotient of X^n , some $n \geq 0$.

(b) G is algebraic if and only if there exists an object X of $\underline{\mathrm{Rep}}_k(G)$ that is a tensor generator for $\underline{\mathrm{Rep}}_k(G)$.

Proof (a). If G is finite then the regular representation of G has the required properties. Conversely if, with the notations of the proof of (2.11), $\underline{\mathrm{Rep}}_k(G) = \langle X \rangle$, then $G = \mathrm{spec}\, B$ where B is the linear dual of the finite k-algebra A_X .

(b) If G is algebraic, then it has a finite-dimensional faithful representation X (2.5), and one shows as in (I.3.1a) that $X \oplus X^{\vee}$ is a tensor generator for $\underline{\mathrm{Rep}}_k(G)$. Conversely, if X is a tensor generator for $\underline{\mathrm{Rep}}_k(G)$ then it is a faithful representation of G.

Proposition 2.21. Let $f: G \to G'$ be a homomorphism of affine group schemes over k, and let ω^f be the corresponding functor $\underline{\mathrm{Rep}}_k(G') \to \underline{\mathrm{Rep}}_k(G)$.

(a) f is faithfully flat if and only if ω^f is fully faithful and every subobject of $\omega^f(X')$, for $X' \in ob(\underline{\mathrm{Rep}}_k(G'))$, is isomorphic to the image of a subobject of X'.

(b) f is a closed immersion if and only if every object of $\underline{\mathrm{Rep}}_k(G)$ is isomorphic to a subquotient of an object of the form $\omega^f(X')$, $X' \in ob(\underline{\mathrm{Rep}}_k(G'))$.

Proof (a). If $G \xrightarrow{f} G'$ is faithfully flat, and therefore is an epimorphism, then $\underline{\mathrm{Rep}}_k(G')$ can be identified with the subcategory of $\underline{\mathrm{Rep}}_k(G)$ of representations $G \to GL(V)$ factoring through G'. It is therefore obvious that ω^f is fully faithful etc. Conversely, if ω^f is fully faithful, it defines an equivalence of $\mathrm{Rep}_k(G')$ with a full subcategory of $\underline{\mathrm{Rep}}_k(G)$, and the second condition shows that, for $X' \in ob(\mathrm{Rep}_k(G'))$, $\langle X' \rangle$ is equivalent to $\langle \omega^f(X') \rangle$. Let $G = \mathrm{spec}\, B$ and $G' = \mathrm{spec}\, B'$; then (2.15) shows that

$$B' = \varinjlim \mathrm{End}(\omega' | \langle X' \rangle)^{\vee} = \varinjlim \mathrm{End}(\omega | \langle \omega^f(X') \rangle)^{\vee} \subset \varinjlim \mathrm{End}(\omega | \langle X \rangle)^{\vee} = B,$$

and $B' \to B$ being injective implies that $G \to G'$ is faithfully flat (Waterhouse [1,14]).

(b) Let $\underline{C}$ be the strictly full subcategory of $\underline{Rep}_k(G)$ whose objects are isomorphic to subquotients of objects of the form $\omega^f(X')$. The functors

$$\underline{Rep}_k(G') \to \underline{C} \to \underline{Rep}_k(G)$$

correspond (see (2.14,2.15)) to homomorphisms of k-coalgebras

$$B' \to B'' \to B$$

where $G = \text{spec } B$ and $G' = \text{spec } B'$. An argument as in the above proof shows that $B'' \to B$ is injective. Moreover, for $X' \in \text{ob}(\underline{Rep}_k(G'))$, $\text{End}(\omega| < \omega^f(X) >) \to \text{End}(\omega' | < X' >)$ is injective, and so $B' \to B''$ is surjective. If f is a closed immersion, then $B' \to B$ is surjective and it follows that $B'' \xrightarrow{\approx} B$, and $\underline{C} = \underline{Rep}_k(G)$. Conversely, if $\underline{C} = \underline{Rep}_k(G)$, then $B'' = B$ and $B' \to B$ is surjective.

Corollary 2.22 Assume k has characteristic zero; then G is connected if and only if, for any representation X of G on which G acts non-trivially, the strictly full subcategory of $\underline{Rep}_k(G)$ whose objects are isomorphic to subquotients of X^n, $n \geq 0$, is not stable under $\otimes$.

Proof: G is connected if and only if there is no non-trivial epimorphism $G \to G'$ with G' finite. According to (2.21a), this is equivalent to $\underline{Rep}_k(G)$ having no non-trivial subcategory of the type described in (2.20a).

<u>Proposition</u> 2.23. Assume k has characteristic zero and that G is connected; then G is pro-reductive if and only if $\underline{\mathrm{Rep}}_k(G)$ is semisimple.

<u>Proof</u>: As every finite-dimensional representation $G \to GL(V)$ of G factors through an algebraic quotient of G, we can assume that G itself is an algebraic group.

<u>Lemma</u> 2.24. Let X be a representation of G; a subspace $Y \subset X$ is stable under G if and only if it is stable under $\mathrm{Lie}(G)$.

<u>Proof</u>: Standard.

<u>Lemma</u> 2.25. Let $\bar{k}$ be the algebraic closure of k; then $\underline{\mathrm{Rep}}_k(G)$ is semisimple if and only if $\underline{\mathrm{Rep}}_{\bar{k}}(G_{\bar{k}})$ is semisimple.

<u>Proof</u>: Let $U(G)$ be the universal enveloping algebra of $\mathrm{Lie}(G)$, and let X be a finite-dimensional representation of G. The last lemma shows that X is semisimple as a representation of G if and only if it is semisimple as a representation of $\mathrm{Lie}(G)$, or of $U(G)$. But X is a semisimple $U(G)$-module if and only if $\bar{k} \otimes_k X$ is a semisimple $\bar{k} \otimes U(G)$-module (Bourbaki [1,13.4]). Since $\bar{k} \otimes U(G) = U(G_{\bar{k}})$, this shows that if $\underline{\mathrm{Rep}}_{\bar{k}}(G_{\bar{k}})$ is semisimple then so is $\underline{\mathrm{Rep}}_k(G)$. For the converse, let $\bar{X}$ be an object of $\underline{\mathrm{Rep}}_{\bar{k}}(G_{\bar{k}})$. There is a finite extension k' of k and a representation X' of $G_{k'}$ over k' giving $\bar{X}$ by extension of scalars. If we regard X' as a vector space over k then we obtain a k-representation X of G. By assumption, X is semisimple and, as was observed above, this implies that $\bar{k} \otimes_k X$ is semisimple. Since $\bar{X}$ is a quotient of $\bar{k} \otimes_k X$, $\bar{X}$ is semisimple.

Lemma 2.26 (Weyl). Let L be a semisimple Lie algebra over an algebraically closed field k (of characteristic zero). Any finite-dimensional representation of L is semisimple.

Proof: For an algebraic proof, see for example (Humphries [1, 6.3]). Weyl's original proof was as follows: we can assume $k = \mathbb{C}$; let L_0 be a compact real form of L and G_0 a connected simply-connected real Lie group with Lie algebra L_0 ; as G_0 is compact, any finite-dimensional representation of it carries a positive-definite form (see (I3.6)) and therefore is semisimple; thus any finite-dimensional (real or complex) representation of L_0 is semisimple, and it is then obvious that any (complex) representation of L is semisimple.

For the remainder of the proof, we assume that k is algebraically closed.

Lemma 2.27. If N is a normal closed subgroup of G and $\rho : G \to GL(X)$ is semisimple, then $\rho|N$ is semisimple.

Proof: We can assume X is a simple G-module. Let Y be a nonzero simple N-submodule of X. For any $g \in G(k)$, gY is an N-module and is simple because $S \mapsto g^{-1}S$ maps N-submodules of gY to N-submodules of Y. The sum $\Sigma\, gY$, $g \in G(k)$, is G-stable and nonzero, and therefore equals X. Thus X, being a sum of simple N-submodules, is semisimple.

We now prove the proposition. If G is reductive, then $G = Z.G'$ where Z is the centre of G and G' is the derived subgroup of G. Let $\rho: G \to GL(X)$ be a finite-dimensional representation of G. As Z is a torus, $\rho|Z$ is diagonalizable: $X = \oplus X_i$ as a Z-module, where any $z \in Z$ acts on X_i as a scalar $\chi_i(z)$. Each X_i is G'-stable and, as G' is semisimple, is a direct sum of simple G'-modules. It is now clear that X is semisimple as a G-module.

Conversely, assume that $\underline{\mathrm{Rep}}_k(G)$ is semisimple and choose a faithful representation X of G. Let N be the unipotent radical of G. Lemma 2.27 shows that X is semisimple as an N-module: $X = \oplus X_i$ where each X_i is a simple N-module. As N is solvable, the Lie-Kolchin theorem shows that each X_i has dimension one, and as N is unipotent, it has a fixed vector in each X_i. Therefore N acts trivially on each X_i, and on X, and, as X is faithful, this shows that $N = \{1\}$.

Remark 2.28. The proposition can be strengthened as follows: assume that k has characteristic zero; then the identity component G^0 of G is pro-reductive if and only if $\underline{\mathrm{Rep}}_k(G)$ is semisimple.

To prove this one has to show that $\underline{\mathrm{Rep}}_k(G)$ is semisimple if and only if $\underline{\mathrm{Rep}}_k(G^\circ)$ is semisimple. The necessity follows from (2.27). For the sufficiency, let X be a representation of G (where G is assumed to be algebraic) and let Y be a G-stable subspace of X. By assumption, there is a G°-equivariant map $p : X \to Y$ such that $p|Y = id$. Define

$$q : \bar{k} \otimes X \to \bar{k} \otimes Y, \quad q = \frac{1}{n} \sum_g g_Y p g_X^{-1}$$

where $n = (G(\bar{k}) : G^\circ(\bar{k}))$ and g runs over a set of coset representatives for $G^\circ(\bar{k})$ in $G(\bar{k})$. One checks easily that q has the following properties:

(i) it is independent of the choice of the coset representatives;

(ii) for all $\sigma \in Gal(\bar{k}/k)$, $\sigma(q) = q$;

(iii) for all $y \in \bar{k} \otimes Y$, $q(y) = q$;

(iv) for all $g \in G(\bar{k})$, $g_Y \cdot q = q \cdot g_X$.

Thus q is defined over k, restricts to the identity map on Y, and is G-equivariant.

Remark 2.29. When, as in the above remark, $\underline{Rep}_k(G)$ is semisimple, the second condition in (2.21a) is superfluous; thus $f : G \to G'$ is faithfully flat if and only if ω^f is fully faithful.

Examples.

(2.30) (Graded vector spaces) Let $\underline{C}$ be the category whose objects are families $(V^n)_{n \in \mathbb{Z}}$ of vector spaces over k with finite-dimensional sum $V = \oplus V^n$. There is an obvious rigid tensor structure on $\underline{C}$ for which $End(\underline{1}) = k$ and $\omega : (V^n) \longmapsto \oplus V^n$ is a fibre functor. Thus, according to (2.11), there is an equivalence of tensor categories $\underline{C} \to \underline{Rep}_k(G)$ for some G. This equivalence is easy to describe: Take $G = \mathbb{G}_m$ and make (V^n) correspond to the representation of $\mathbb{G}_m$ on $\oplus V^n$ for which $\mathbb{G}_m$ acts on V^n through the character $\lambda \longmapsto \lambda^n$.

(2.31) A real Hodge structure is a finite dimensional vector space V over $\mathbb{R}$ together with a decomposition $V \otimes \mathbb{C} = \bigoplus_{p,q} V^{p,q}$ such that $V^{p,q}$ and $V^{q,p}$ are conjugate complex subspaces of $V \otimes \mathbb{C}$. There is an obvious rigid tensor structure on the category $\underline{\mathrm{Hod}}_{\mathbb{R}}$ of real Hodge structures and $\omega : (V,(V^{p,q})) \longmapsto V$ is a fibre functor. The group corresponding to $\underline{\mathrm{Hod}}_{\mathbb{R}}$ and ω is the real algebraic group $\mathbb{S}$ obtained from $\mathbb{G}_m$ by restriction of scalars from $\mathbb{C}$ to $\mathbb{R}$: $\mathbb{S} = \mathrm{Res}_{\mathbb{C}/\mathbb{R}} \mathbb{G}_m$. The real Hodge structure $(V,(V^{p,q}))$ corresponds to the representation of $\mathbb{S}$ on V such that an element $\lambda \in \mathbb{S}(\mathbb{R}) = \mathbb{C}^{\times}$ acts on $V^{p,q}$ as $\lambda^{-p}\bar{\lambda}^{-q}$. We can write $V = \oplus V^n$ where $V^n \otimes \mathbb{C} = \bigoplus_{p+q=n} V^{p,q}$. The functor $(V,(V^{p,q})) \longmapsto (V^n)$ from $\underline{\mathrm{Hod}}_{\mathbb{R}}$ to the category of real graded vector spaces corresponds to the homomorphism $\mathbb{G}_m \to \mathbb{S}$ which, on real points, is $t \longmapsto t^{-1}$: $\mathbb{R}^{\times} \to \mathbb{C}^{\times}$.

(2.32) The preceding examples have a common generalization. Recall that an algebraic group G is of <u>multiplicative type</u> if $G_{\bar{k}}$, where $\bar{k}$ is the separable algebraic closure of k, is diagonalizable in some faithful representation, and that the character group $X(G) \overset{df}{=} \mathrm{Hom}(G_{\bar{k}},\mathbb{G}_m)$ of such a G is a finitely generated abelian group on which $\Gamma = \mathrm{Gal}(\bar{k}/k)$ acts continuously. Write $M = X(G)$, and let $k' \subset \bar{k}$ be a Galois extension of k over which all elements of M are defined. For any finite-dimensional representation V of G, $V \otimes_k k' = \bigoplus_{m \in M} V^m$ where $V^m = \{v \in V \otimes_k k' \mid gv = m(g)v \text{ , all } g \in G(k')\}$. A finite-dimensional vector space V over k together with a decomposition $k' \otimes V = \oplus V^m$ arises from a representation of G if and only if $V^{\sigma(m)} = \sigma V^m (\overset{\mathrm{df}}{=} V^m_{\otimes_{k',\sigma}} k')$ for all $m \in M$ and $\sigma \in \Gamma$. Thus an object of $\underline{\mathrm{Rep}}_k(G)$ can be identified with a finite-dimensional vector space V over k together with an M-grading on $V \otimes_k k'$ that is compatible with the action of the Galois group.

(2.33) (Tannakian duality) Let K be a topological group. The category $\underline{\mathrm{Rep}}_{\mathbb{R}}(K)$ of continuous representations of K on finite-dimensional real vector spaces is, in a natural way, a neutral Tannakian category with the forgetful functor as fibre functor. There is therefore a real affine algebraic group $\tilde{K}$, called the <u>real algebraic envelope</u> of K, for which there exists an equivalence $\underline{\mathrm{Rep}}_{\mathbb{R}}(K) \tilde{\to} \underline{\mathrm{Rep}}_{\mathbb{R}}(\tilde{K})$. There is also a map $K \to \tilde{K}(\mathbb{R})$, which is an isomorphism when K is compact.

In general, a real algebraic group G is said to be <u>compact</u> if $G(\mathbb{R})$ is compact and the natural functor $\underline{\mathrm{Rep}}_{\mathbb{R}}(G(\mathbb{R})) \to \underline{\mathrm{Rep}}_{\mathbb{R}}(G)$ is an equivalence. The second condition is equivalent to each connected component of $G(\mathbb{C})$ containing a real point (or to $G(\mathbb{R})$ being Zariski dense in G). We note for reference that Deligne [1, 2.5] shows that a subgroup of a compact real algebraic group is compact.

(2.34) (The true fundamental group). Recall that a vector bundle E on a curve C is <u>semi-stable</u> if for every sub-bundle $E' \subset E$, $(\deg E')/(\operatorname{rank} E') \leq (\deg E)/(\operatorname{rank} E)$. Let X be a complete connected reduced k-scheme, where k is assumed to be perfect. A vector bundle E on X will be said to be semi-stable if for every nonconstant morphism $f : C \to X$ with C a projective smooth connected curve, f^*E is semi-stable of degree zero. Such a bundle E is <u>finite</u> if there exist polynomials $g,h \in \mathbb{N}[t]$, $g \neq h$, such that $g(E) \approx h(E)$. Let $\underline{C}$ be the category of semi-stable vector bundles on X that are isomorphic to a subquotient of a finite vector bundle. If X has a k-rational point x then $\underline{C}$ is a neutral Tannakian category over k with fibre functor $\omega : E \mapsto E_x$. The group associated with $(\underline{C},\omega)$ is a pro-finite group scheme over k, called the <u>true fundamental group</u> $\pi_1(X,x)$ of X, which classifies all G-coverings of X with G a finite group scheme over k. The maximal pro-étale quotient of $\pi_1(X,x)$ is the usual étale fundamental group of X. See Nori [1].

(2.35) Let K be a field of characteristic zero, complete with respect to a discrete valuation, whose residue field is algebraically closed of characteristic $p \neq 0$. The Hodge-Tate modules for K from a neutral Tannakian category over $\mathbb{Q}_p$ (see Serre [2]).

§3. <u>Fibre Functors; the general notion of a Tannakian category</u>

Throughout this section, k denotes a field

<u>Fibre functors</u>

Let G be an affine group scheme over k and let $U = \operatorname{spec} R$ be an affine k-scheme. A G-<u>torsor</u> over U (for the f.p.q.c. topology) is an affine scheme T, faithfully flat over S, together with a morphism $T \times_U G \to T$ such that $(t,g) \longmapsto (t,tg) : T \times_U G \to T \times T$ is an isomorphism. Such a scheme T is determined by its points functor, $h_T = (R' \longmapsto T(R'))$. A non-vacuous set-valued functor h of R-algebras with functorial pairing $h(R') \times G(R') \to h(R')$ arises from a G-torsor if

(3.1a) For each R-algebra R' such that $h(R')$ is non-empty, $G(R')$ acts simply transitively on $h(R')$, and

(3.1b) h is respectable by an affine scheme faithfully flat over U. Descent theory shows that (3.1b) can be replaced by the condition that h be a sheaf for the f.p.q.c. topology on U (see Waterhouse [1,V]). There is an obvious notion of a morphism of G-torsors.

Let $\underline{C}$ be a k-linear abelian tensor category; a <u>fibre functor</u> on $\underline{C}$ with values in a k-algebra R is a k-linear exact faithful tensor functor $\eta : \underline{C} \to \underline{\text{Mod}}_R$ that takes values in the subcategory $\underline{\text{Proj}}_R$ of $\underline{\text{Mod}}_R$. Assume now that $\underline{C}$ is a neutral Tannakian category over k. There then exists a fibre functor ω with values in k and we proved in the last section that if we let $G = \underline{\text{Aut}}^{\otimes}(\omega)$, ω defines an equivalence $\underline{C} \xrightarrow{\sim} \underline{\text{Rep}}_k(G)$. For any fibre functor η with values in R, composition defines a pairing

$$\underline{\text{Hom}}^{\otimes}(\omega,\eta) \times \underline{\text{Aut}}^{\otimes}(\omega) \to \underline{\text{Hom}}^{\otimes}(\omega,\eta)$$

of functors of R-algebras. Proposition 1.13 shows that

$\underline{\mathrm{Hom}}^{\otimes}(\omega,\eta) = \underline{\mathrm{Isom}}^{\otimes}(\omega,\eta)$, and therefore that $\underline{\mathrm{Hom}}^{\otimes}(\omega,\eta)$ satisfies (3.1a).

Theorem 3.2. Let $\underline{C}$ be a neutral Tannakian category over k .
(a) For any fibre functor η on $\underline{C}$ with values in R , $\underline{\mathrm{Hom}}^{\otimes}(\omega,\eta)$ is representable by an affine scheme faithfully flat over spec R ; it is therefore a G-torsor.
(b) The functor $\eta \mapsto \underline{\mathrm{Hom}}^{\otimes}(\omega,\eta)$ determines an equivalence between the category of fibre functors on $\underline{C}$ with values in R and the category of G-torsors over R .

Proof: Let $X \in \mathrm{ob}(\underline{C})$, and, with the notations of the proof of (2.11), define

$$\begin{cases} A_X \subset \mathrm{End}(\omega(X)) , & A_X = \bigcap_Y (\omega(Y) : \omega(Y)) , \quad Y \subset X^n , \\ P_X \subset \underline{\mathrm{Hom}}(\omega(X),X) , & P_X = \bigcap_Y (Y : \omega(Y)) , \quad Y \subset X^n . \end{cases}$$

Then $\omega(P_X) = A_X$ and $P_X \in \mathrm{ob}(\langle X\rangle)$. For any R-algebra R' , $\underline{\mathrm{Hom}}(\omega|\langle X\rangle,\eta|\langle X\rangle)(R')$ is the subspace of $\mathrm{Hom}(\omega(P_X)\otimes_k R',\eta(P_X)\otimes_R R')$ of maps respecting all $Y \subset X^n$; it therefore equals $\eta(P_X) \otimes R'$. Thus

$$\underline{\mathrm{Hom}}(\omega|\langle X\rangle,\eta|\langle X\rangle)(R') \xrightarrow{\approx} \mathrm{Hom}_{R\text{-lin}}(\eta(P_X^{\vee}),R') .$$

Let Q be the ind-object $(P_X^{\vee})_X$, and let $B = \varinjlim A_X^{\vee}$. As we saw in the last section, the tensor structure on $\underline{C}$ defines an algebra structure on B ; it also defines a ring structure on Q (i.e., a map $Q \otimes Q \to Q$ in $\mathrm{Ind}(\underline{C})$) making $\omega(Q) \overset{\approx}{\to} B$ into an isomorphism of k-algebras. We have

$$\begin{aligned}\underline{\mathrm{Hom}}(\omega,\eta)(R') &= \varprojlim \underline{\mathrm{Hom}}(\omega|\langle X\rangle, \eta|\langle X\rangle)(R') \\ &= \varprojlim \mathrm{Hom}_{R\text{-lin}}(\eta(P_X^\vee), R') \\ &= \mathrm{Hom}_{R\text{-lin}}(\eta(Q), R)\end{aligned}$$

where $\eta(Q) \overset{\mathrm{df}}{=} \varinjlim \eta(P_X^\vee)$. Under this correspondence,

$$\underline{\mathrm{Hom}}^{\otimes}(\omega,\eta)(R') = \mathrm{Hom}_{R\text{-alg}}(\eta(Q), R') ,$$

and so $\underline{\mathrm{Hom}}^{\otimes}(\omega,\eta)$ is represented by $\eta(Q)$. By definition $\eta(P_X^\vee)$ is a projective R-module, and so $\eta(Q) = \varinjlim \eta(P_X^\vee)$ is flat over R . For each X there is a surjection $P_X \twoheadrightarrow \underline{1}$, and the exact sequence

$$0 \to \underline{1} \to P_X^\vee \to P_X^\vee/\underline{1} \to 0$$

gives rise to an exact sequence

$$0 \to \eta(\underline{1}) \to \eta(\underline{P}_X^\vee) \to \eta(\underline{P}_X^\vee/\underline{1}) \to 0$$

As $\eta(\underline{1}) = R$ and $\eta(\underline{P}_X^\vee/\underline{1})$ is flat, this shows that $\eta(\underline{P}_X^\vee)$ is a faithfully flat R-module. Hence $\eta(Q)$ is faithfully flat over R , which completes the proof that $\underline{\mathrm{Hom}}^{\otimes}(\omega,\eta)$ is a G-torsor.

To show that $\eta \mapsto \underline{Hom}^{\otimes}(\omega,\eta)$ is an equivalence, we construct a quasi-inverse. Let T be a G-torsor over R. For a fixed X, define $R' \mapsto \eta_T(X)(R')$ to be the sheaf associated with

$$R' \mapsto (\omega(X) \otimes R') \times T(R')/G(R') .$$

Then $X \mapsto \eta_T(X)$ is a fibre functor on $\underline{C}$ with values in R.

<u>Remark</u> 3.3

(a) Define

$$A_X \subset \underline{Hom}(X,X), \ \underline{A}_X = \bigcap (Y:Y) \ , \ Y \subset X^n .$$

Then $\underline{A}_X$ is a ring in $\underline{C}$ such that $\omega(\underline{A}_X) = A_X$ (as k-algebras). Let $\underline{B}$ be the ind-object $(\underline{A}_X^{\vee})$. Then

$$\underline{End}^{\otimes}(\omega) = \text{spec}\ \omega(\underline{B}) = G$$
$$\underline{End}^{\otimes}(\eta) = \text{spec}\ \eta(\underline{B}) .$$

(b) The proof of (3.2) can be made more concrete by using (2.11) to replace $(\underline{C},\omega)$ with $(\underline{Rep}_k(G),\omega^G)$.

<u>Remark</u> 3.4. The situation described in the theorem is analogous to the following. Let X be a connected topological space and let $\underline{C}$ be the category of locally constant sheaves of $\mathbb{Q}$ vector spaces on X. For any $x \in X$, there is a fibre functor $\omega_x : \underline{C} \to \underline{Vec}_{\mathbb{Q}}$, and ω_x defines an equivalence of categories

$\underline{C} \tilde{\to} \underline{\mathrm{Rep}}_{\mathbb{Q}}(\pi_1(X,x))$. Let $\Pi_{x,y}$ be the set of homotopy classes of paths from x to y ; then $\Pi_{x,y} \approx \underline{\mathrm{Isom}}(\omega_x,\omega_y)$, and $\Pi_{x,y}$ is a $\pi_1(X,x)$-torsor.

Question 3.5. Let $\underline{C}$ be a rigid abelian tensor category whose objects are of finite length and which is such that $\mathrm{End}(\underline{1})=k$ and $\otimes$ is exact. (Thus $\underline{C}$ lacks only a fibre functor with values in k to be a neutral Tannakian category). As in (3.3) one can define

$$\underline{A}_X \subset \underline{\mathrm{Hom}}\ (X,X),\ \underline{A}_X = \bigcap (Y:Y),\ Y \subset X^n$$

and hence obtain a bialgebra $B = \text{"}\varinjlim\text{"}\ A_X^\vee$ in $\mathrm{Ind}(\underline{C})$ which can be thought of as defining an affine group scheme G in $\mathrm{Ind}(\underline{C})$.

Is it true that for $X \subset X'$, $A_{X'} \to A_X$ is an epimorphism?

For any X in $\underline{C}$, there is a morphism $X \overset{\rho}{\to} X \otimes \underline{B}$, which can be regarded as a representation of G . Define X^G , the subobject fixed by G , to be the largest subobject of X such that $X^G \to X \otimes B_X$ factors through $X^G \otimes 1 \hookrightarrow X \otimes B_X$. Is it true that $\underline{\mathrm{Hom}}(\underline{1},X) \otimes_k \underline{1} \to X^G$ is an isomorphism?

If for all X there exists an N such that $\Lambda^N X = 0$, is $\underline{C}$ Tannakian in the sense of Definition 3.7 below? (See note at the end of the article.)

The general notion of a Tannakian category

In this subsection, we need to use some terminology from non-abelian 2-cohomology, for which we refer the reader to the

Appendix. In particular $\underline{Aff}_S$ or $\underline{Aff}_k$ denotes the category of affine schemes over $S = \operatorname{spec} k$ and PROJ is the stack over $\underline{Aff}_S$ such that $PROJ_U = \underline{Proj}_R$ for $R = \Gamma(U, O_U)$. For any gerb $\underline{G}$ over $\underline{Aff}_k$ (for the f.p.q.c. topology) we let $\underline{Rep}_k(\underline{G})$ denote the category of cartesian functors $\underline{G} \to PROJ$. Thus an object Φ of $\underline{Rep}_k(G)$ determines (and is determined by) functors $\Phi_R : \underline{G}_R \to \underline{Proj}_R$, one for each k-algebra R , and functorial isomorphisms $\Phi_{R'}(g^*Q) \xleftrightarrow{\approx} \Phi_R(Q) \otimes_R R'$ defined whenever $g : R \to R'$ is a homomorphism of k-algebras and $Q \in ob(\underline{G}_R)$. There is an obvious rigid tensor structure on $\underline{Rep}_k(G)$, and $End(\underline{1}) = k$.

Example 3.6. Let G be an affine group scheme over k , and let TORS(G) be the gerb over $\underline{Aff}_S$ such that $TORS(G)_U$ is the category of G-torsors over U . Let G_r be G regarded as a right G-torsor, and let Φ be an object of $\underline{Rep}_k(TORS(G))$. The isomorphism $G \xrightarrow{\approx} \underline{Aut}(G_r)$ defines a representation of G on the vector space $\Phi_k(G_r)$, and it is not difficult to show that $\Phi \longmapsto \Phi_k(G_r)$ extends to an equivalence of categories $\underline{Rep}_k(TORS(G)) \xrightarrow{\approx} \underline{Rep}_k(G)$.

Let $\underline{C}$ be a rigid abelian tensor category with $End(\underline{1}) = k$. For any k-algebra R , the fibre functors on $\underline{C}$ with values in R form a category $FIB(\underline{C})_R$, and the collection of these categories forms in a natural way a fibred category $FIB(\underline{C})$ over $\underline{Aff}_k$. Descent theory for projective modules shows that $FIB(\underline{C})$ is a stack, and (1.13) shows that its fibres are groupoids. There is a canonical k-linear tensor functor $\underline{C} \to \underline{\text{Rep}}_k(\text{FIB}(C))$ associating to $X \in \text{ob}(\underline{C})$ the family of functors $\omega \longmapsto \omega(X) : \text{FIB}(\underline{C})_R \to \underline{\text{Proj}}_R$.

Definition 3.7. A Tannakian category over k is a rigid abelian tensor category $\underline{C}$ with $\text{End}(\underline{1}) = k$ such that $\text{FIB}(\underline{C})$ is an affine gerb and $\underline{C} \to \underline{\text{Rep}}_k(\text{FIB}(\underline{C}))$ is an equivalence of categories.

Example 3.8. Let $\underline{C}$ be a neutral Tannakian category over k. Theorem 3.2 shows that the choice of a fibre functor ω with values in k determines an equivalence of fibred categories $\text{FIB}(\underline{C}) \overset{\approx}{\to} \text{TORS}(G)$ where G represents $\underline{\text{Aut}}^{\otimes}(\omega)$. Thus $\text{FIB}(\underline{C})$ is an affine gerb and the commutative diagram of functors

$$\begin{array}{ccc} \underline{C} & \to & \underline{\text{Rep}}_k(\text{FIB}(\underline{C})) \\ \wr\downarrow\omega & & \downarrow\wr \\ \underline{\text{Rep}}_k(G) & \overset{\approx}{\to} & \underline{\text{Rep}}_k(\text{TORS}(G)) \end{array}$$

shows that $\underline{C}$ is a Tannakian category. Thus a Tannakian category in the sense of (3.7) is a neutral Tannakian category in the sense of (2.19) if an only if it has a fibre functor with values in k.

Remark 3.9. The condition in (3.7) that $\text{FIB}(\underline{C})$ is a gerb means that $\underline{C}$ has a fibre functor ω with values in some field $k' \supset k$ and that any two fibre functors are locally isomorphic for the f.p.q.c. topology. The condition that the gerb $\text{FIB}(\underline{C})$ be affine means that $\underline{\text{Aut}}^{\otimes}(\omega)$ is representable by an affine group scheme over k'.

<u>Remark</u> 3.10. A Tannakian category $\underline{C}$ over k is said to be <u>algebraic</u> if FIB($\underline{C}$) is an algebraic gerb. There then exists a finite field extension k' of k and a fibre functor ω with values in k' (App., Proposition), and the algebraicity of $\underline{C}$ means that $G = \underline{\mathrm{Aut}}^{\otimes}(\omega)$ is an algebraic group over k'. As in the neutral case (2.20), a Tannakian category is algebraic if and only if it has a tensor generator. Consequently, any Tannakian category is a filtered union of algebraic Tannakian categories.

<u>Tannakian categories neutralized by a finite extension</u>

Let $\underline{C}$ be a k-linear category, and let A be a commutative k-algebra. An <u>A-module</u> in $\underline{C}$ is a pair (X,α_X) with X an object of $\underline{C}$ and α_X a homomorphism $A \to \mathrm{End}(X)$. For example, an A-module in $\underline{\mathrm{Vec}}_{k'}$, where $k' \supset k$, is simply an $A \otimes_k k'$-module that is of finite dimension over k'. With an obvious notion of morphism, the A-modules in $\underline{C}$ form an A-linear category $\underline{C}_{(A)}$. If $\underline{C}$ is abelian so also is $\underline{C}_{(A)}$, and if $\underline{C}$ has a tensor structure and its objects have finite length then we define $(X,\alpha_X) \otimes (Y,\alpha_Y)$ to be the A-module in $\underline{C}$ with object the largest quotient of $X \otimes Y$ to which $\alpha_X(a) \otimes \mathrm{id}$ and $\mathrm{id} \otimes \alpha_Y(a)$ agree for all $a \in A$.

Now let $\underline{C}$ be a Tannakian category over k, and let k' be a finite field extension of k. As the tensor operation on $\underline{C}$ commutes with direct limits (1.16), it extends to $\mathrm{Ind}(\underline{C})$, which is therefore an abelian tensor category. The functor $\underline{C} \to \mathrm{Ind}(\underline{C})$ defines an equivalence between $\underline{C}$ and the strictly full subcategory $\underline{C}^e$ of $\mathrm{Ind}(\underline{C})$ of essentially constant ind-objects. In $\underline{C}^e$ it is possible to define external tensor products with objects of $\underline{Vec}_k$ (cf. the proof of (2.11)) and hence a functor

$$X \mapsto i(X) = (k' \otimes_k X,\ a' \mapsto a' \otimes \mathrm{id}) : \underline{C}^e \to \underline{C}^e_{(k')} .$$

This functor is left adjoint to

$$(X,\alpha) \mapsto j(X,\alpha) = X : \underline{C}^e_{(k')} \to \underline{C}^e$$

and has the property that $k' \otimes_k \mathrm{Hom}(X,Y) \xrightarrow{\approx} \mathrm{Hom}(i(X),\ i(Y))$. Let ω be a fibre functor on $\underline{C}^e$ (or $\underline{C}$) with values in k' . For any $(X,\alpha) \in \mathrm{ob}(\underline{C}^e_{(k')})$, $(\omega(X),\ \omega(\alpha))$ is a k'-module in $\underline{Vec}_{k'}$, i.e., it is a $k' \otimes_k k'$-module. If we define

$$\omega'(X,\alpha) = k' \otimes_{k'\otimes k'} \omega(X) \qquad (3.10.1)$$

Then

$$\begin{array}{ccc} \underline{C}^e & \longrightarrow & \underline{C}^e_{(k')} \\ & \searrow^{\omega} & \downarrow{\scriptstyle \omega'} \\ & & \underline{Vec}_{k'} \end{array}$$

commutes up to a canonical isomorphism.

Proposition 3.11. Let $\underline{C}$ be a Tannakian category over k and let ω be a fibre functor on $\underline{C}$ with values in a finite field extension k' of k ; extend ω' to $\underline{C}_{(k')}$ using the formula (3.10.1) ; then ω' defines an equivalence of tensor categories $\underline{C}_{(k')} \tilde{\to} \underline{\mathrm{Rep}}_{k'}(G)$ where $G = \underline{\mathrm{Aud}}^{\otimes}(\omega)$. In particular, ω' is exact.

Proof: One has simply to compose the following functors:

$$\underline{C}_{(k')} \tilde{\to} \underline{\mathrm{Rep}}_k(G)_{(k')}$$

arising from the equivalence $\underline{C} \tilde{\to} \underline{\mathrm{Rep}}_k(\underline{G})$ ($\underline{G} = \mathrm{FIB}(\underline{C})$) in the definition (3.7);

$$\mathrm{Rep}_k(\underline{G})_{(k')} \tilde{\to} \underline{\mathrm{Rep}}_{k'}(\underline{G}/k')$$

where $\underline{G}/k'$ denotes the restriction of $\underline{G}$ to $\underline{\mathrm{Aff}}_{k'}$ (the functor sends $(\Phi,\alpha) \in \mathrm{ob}(\underline{\mathrm{Rep}}_k(\underline{G})_{(k')})$ to Φ' where, for any k'-algebra R and $Q \in \underline{G}_R$, $\Phi'_R(Q) = R \otimes_{k'\otimes R}\Phi_R(Q))$;

$$\underline{\mathrm{Rep}}_{k'}(\underline{G}/k') \tilde{\to} \underline{\mathrm{Rep}}_{k'}(\mathrm{TORS}(G))$$

arising from $\mathrm{TORS}(G) \tilde{\to} \underline{G}/k'$;

$$\underline{\mathrm{Rep}}_{k'}(\mathrm{TORS}(G)) \tilde{\to} \underline{\mathrm{Rep}}_{k'}(G) \qquad \text{(see 3.6).}$$

<u>Remark</u> 3.12. Let $\underline{C} = \underline{\mathrm{Rep}}_k(G)$ and let k' be a finite extension of k. Then $\underline{C}_{(k')} = \underline{\mathrm{Rep}}_{k'}(G)$ and $i : \underline{C} \to \underline{C}_{(k')}$ is $X \mapsto k' \otimes_k X$. Let ω be the fibre functor $X \mapsto k' \otimes_k X : \underline{\mathrm{Rep}}_k(G) \to \underline{\mathrm{Vec}}_{k'}$. Then $G_{k'} = \mathrm{Aut}^{\otimes}(\omega)$ and the equivalence $\underline{C}_{(k')} \xrightarrow{\sim} \underline{\mathrm{Rep}}_{k'}(G_{k'})$ defined in the proposition is

$$X \mapsto k' \otimes_{k' \otimes k'} X : \underline{\mathrm{Rep}}_{k'}(G) \to \underline{\mathrm{Rep}}_{k'}(G_{k'}) .$$

<u>Descent of Tannakian categories</u>

Let k'/k be a finite Galois extension with Galois group Γ, and let $\underline{C}'$ be a Tannakian category over k'. A <u>descent datum</u> on $\underline{C}'$ relative to k'/k is

(3.13a) a family $(\beta_\gamma)_{\gamma \in \Gamma}$ of equivalences of tensor categories $\beta_\gamma : \underline{C}' \to \underline{C}'$, β_γ being semi-linear relative to γ, together with

(3.13b) a family $(\mu_{\gamma',\gamma})$ of isomorphisms of tensor functors $\mu_{\gamma',\gamma} : \beta_{\gamma'\gamma} \xrightarrow{\approx} \beta_{\gamma'} \circ \beta_\gamma$ such that

$$\begin{array}{ccc}
\beta_{\gamma''\gamma'\gamma}(X) & \xrightarrow{\mu_{\gamma'',\gamma'\gamma}(X)} & \beta_{\gamma''}(\beta_{\gamma'\gamma}(X)) \\
\Big\downarrow{\scriptstyle \mu_{\gamma''\gamma',\gamma}(X)} & & \Big\downarrow{\scriptstyle \beta_{\gamma''}(\mu_{\gamma'\gamma}(X))} \\
\beta_{\gamma''\gamma'}(\beta_\gamma(X)) & \xrightarrow{\mu_{\gamma''\gamma'}(\beta_\gamma(X))} & \beta_{\gamma''}(\beta_{\gamma'}(\beta_\gamma(X)))
\end{array}$$

commutes for all $X \in \mathrm{ob}(\underline{C})$.

A Tannakian category $\underline{C}$ over k gives rise to a Tannakian category $\underline{C}' = \underline{C}_{(k')}$ over k' together with a descent datum for which $\beta_\gamma(X,\alpha_X) = (X,\alpha_X\circ\gamma^{-1})$. Conversely, a Tannakian category $\underline{C}'$ over k' together with a descent datum relative to k'/k gives rise to a Tannakian category $\underline{C}$ over k whose objects are pairs $(X,(a_\gamma))$, where $X \in ob(\underline{C}')$ and $(a_\gamma : X \to \beta_\gamma(X))_{\gamma\in\Gamma}$ is such that $(\mu_{\gamma',\gamma})_X\circ a_{\gamma'\gamma} = \beta_{\gamma'}(a_\gamma)\circ a_{\gamma'}$, and whose morphisms are morphisms in $\underline{C}'$ commuting with the a_γ. These two operations are quasi-inverse, so that to give a Tannakian category over k (up to a tensor equivalence, unique up to a unique isomorphism) is the same as to give a Tannakian category over k' together with a descent datum relative to k'/k (Saavedra [1, III 1.2]). On combining this statement with (3.11) we see that to give a Tannakian category over k together with a fibre functor with values in k' is the same as to give an affine group scheme G over k' together with a descent datum on the Tannakian category $\underline{Rep}_{k'}(G)$.

Questions

(3.14) Let $\underline{G}$ be an affine gerb over k . There is a morphism of gerbs

$$\underline{G} \to FIB(\underline{Rep}_k(\underline{G})) \qquad (3.14.1)$$

which, to an object Q of $\underline{G}$ over $S = spec\ R$, associates the fibre functor $F \mapsto F(Q)$ with values in R . Is (3.14.1) an equivalence of gerbs? If $\underline{G}$ is algebraic, or if the band of $\underline{G}$ is defined by an affine group scheme over k , then it is (Saavedra

[1, III 3.2.5]) but the general question is open. A positive answer would provide the following classification of Tannakian categories: The maps $\underline{C} \mapsto FIB(\underline{C})$ and $\underline{G} \mapsto \underline{Rep}_k(\underline{G})$ determine a one-to-one correspondence between the set of tensor equivalence classes of Tannakian categories over k and the set of equivalence classes of affine gerbs over k ; the affine gerbs bound by a given band B are classified by $H^2(S,B)$, and $H^2(S,B)$ is a pseudo-torsor over $H^2(S,Z)$ where Z is the centre of B .

(3.15) In [1, III 3.2.1] Saavedra defines a Tannakian category over k to be a k-linear rigid abelian tensor category $\underline{C}$ for which there exists a fibre functor with values in a field $k' \supset k$. He then claims to prove (ibid. 3.2.3.1) that $\underline{C}$ satisfies the conditions we have used to define a Tannakian category. This is false. For example, $\underline{Vec}_{k'}$ for k' a field containing k is a Tannakian category over k according to his definition but the fibre functors $V \mapsto \sigma V \overset{df}{=} V \otimes_{k',\sigma} k'$ for $\sigma \in Aut(k'/k)$ are not locally isomorphic for the f.p.q.c. topology on spec k' . There is an error in the proof (ibid. p. 197, ℓ.7) where it is asserted that "par définition" the objects of G_S are locally isomorphic.

The question remains of whether Saavedra's conditions plus the condition that $End(\underline{1}) = k$ imply our conditions. As we noted in (3.8), when there is a fibre functor with values in k they do, but the general question is open. The essential point is the following: Let $\underline{C}$ be a rigid abelian tensor category with $End(\underline{1}) = k$ and let ω be a fibre functor with values in a finite field extension k' of k ; is the functor ω' ,

$$X \longmapsto k' \otimes_{k' \otimes k'} \omega(X) : \underline{C}_{(k')} \to \underline{\mathrm{Vec}}_{k'} ,$$

exact? (See Saavedra [1, p. 195]; the proof there that ω' is faithful is valid.) The answer is yes if $\underline{C} = \underline{\mathrm{Rep}}_k(G)$, G an affine group scheme over k , but we know of no proof simpler than to say that ω' is defined by a G-torsor on k' , and $\underline{C}_{(k')} = \underline{\mathrm{Rep}}_{k'}(G)$. (See note at end.)

§4. Polarizations

Throughout this section $\underline{C}$ will be an algebraic Tannakian category over $\mathbb{R}$ and $\underline{C}'$ will be its extension to $\mathbb{C}$: $\underline{C}' = \underline{C}_{(\mathbb{C})}$.

Tannakian categories over $\mathbb{R}$

According to (3.13) and the paragraph following it, to give $\underline{C}$ is the same as to give the following data:

(4.1a) A Tannakian category $\underline{C}'$ over $\mathbb{C}$;

(4.1b) A semi-linear tensor functor $X \longmapsto \bar{X} : \underline{C}' \to \underline{C}'$;

(4.1c) A functorial tensor isomorphism $\mu_X : X \xrightarrow{\approx} \bar{\bar{X}}$ such that $\mu_{\bar{X}} = \bar{\mu}_X$.

An object of $\underline{C}$ can be identified with an object X of $\underline{C}'$ together with a descent datum (an isomorphism $a : X \xrightarrow{\approx} \bar{X}$ such that $\bar{a} \circ a = \mu_X$) . Note that $\underline{C}'$ is automatically neutral (3.10).

Example 4.2. Let G be an affine group scheme over $\mathbb{C}$ and let $\sigma : G \to G$ be a semi-linear isomorphism (meaning $f \longmapsto \sigma \circ f$: $\Gamma(G, O_G) \to \Gamma(G, O_G)$ is a semi-linear isomorphism). Assume there is given $c \in G(\mathbb{C})$ such that

$$\sigma^2 = \underline{ad}(c) \ , \ \sigma(c) = c \qquad (4.2.1).$$

From (G,σ,c) we can construct data as in (4.1):

(a) define $\underline{C}'$ to be $\underline{Rep}_{\mathbb{C}}(G)$;

(b) for any vector space V over $\mathbb{C}$ there is an (essentially) unique vector space $\bar{V}$ and semi-linear isomorphism $v \mapsto \bar{v} : V \xrightarrow{\approx} \bar{V}$; if V is a G-representation, we define a representation of G on $\bar{V}$ by the rule $\overline{gv} = \sigma(g)\bar{v}$;

(c) define μ_V to be the map $cv \mapsto \bar{\bar{v}} : V \xrightarrow{\approx} \bar{\bar{V}}$.

Let $m \in G(\mathbb{C})$. Then $\sigma' = \sigma \circ \underline{ad}(m)$ and $c' = \sigma(m)cm$ again satisfy (4.2.1). The element m defines an isomorphism of the functor $V \mapsto \bar{V}$ (rel. to (σ,c)) with the functor $V \mapsto \bar{V}$ (rel. to (σ',c')) by $\overline{mv} \mapsto \bar{v} : \bar{V}$ (rel. to (σ,c)) $\to \bar{V}$ (rel. to (σ',c')) . This isomorphism carries μ_V (rel. to (σ,c)) to μ_V (rel. to (σ',c')) , and hence defines an equivalence $\underline{C}$ (rel. to (σ,c)) with $\underline{C}$ (rel. to (σ',c')).

Proposition 4.3. Let $\underline{C}$ be an algebraic Tannakian category over $\mathbb{R}$, and let $\underline{C}' = \underline{C}_{(\mathbb{C})}$. Choose a fibre functor ω on $\underline{C}'$ with values in $\mathbb{C}$ and let $G = \underline{Aut}^{\otimes}(\omega)$.

(a) There exists a pair (σ,c) satisfying (4.2.1) and such that under the equivalence $\underline{C}' \xrightarrow{\approx} \underline{Rep}_{\mathbb{C}}(G)$ defined by ω , $X \mapsto \bar{X}$ corresponds to $V \mapsto \bar{V}$ and $\omega(\mu_X) = \mu_{\omega(X)}$.

(b) The pair (σ,c) in (a) is uniquely determined up to replacement by a pair (σ',c') with $\sigma' = \sigma \circ \underline{ad}(m)$ and $c' = \sigma(m)cm$, some $m \in G(\mathbb{C})$.

Proof: (a) Let $\bar{\omega}$ be the fibre functor $X \mapsto \overline{\omega(\bar{X})}$ and let $T = \underline{\mathrm{Hom}}^{\otimes}(\omega,\bar{\omega})$. According to (3.2), T is a G-torsor, and the Nullstellensatz shows that it is trivial. The choice of a trivialization provides us with a functorial isomorphism $\omega(X) \xrightarrow{\approx} \bar{\omega}(X)$ and therefore with a semi-linear functorial isomorphism $\lambda_X : \omega(X) \xrightarrow{\approx} \omega(\bar{X})$. Define σ by the condition that $\sigma(g)_{\bar{X}} = \lambda_X \circ g_X \circ \lambda_X^{-1}$ for all $g \in G(\mathbb{C})$, and let c be such that $c_X = \omega(\mu_X)^{-1} \circ \lambda_{\bar{X}} \circ \lambda_X$.

(b) The choice of a different trivialization of T replaces λ_X with $\lambda_X \circ m_X$ for some $m \in G(\mathbb{C})$, and σ with $\sigma \circ \underline{ad}(m)$ and c with $\sigma(m)\, c m$.

Sesquilinear forms

Let $\underline{1}$ (with $e : \underline{1} \otimes \underline{1} \xrightarrow{\approx} \underline{1}$) be an identity object for $\underline{C}'$. Then $\bar{\underline{1}}$ (with $\bar{e}$) is again an identity object, and the unique isomorphism of identity objects $a : \underline{1} \to \bar{\underline{1}}$ is a descent datum. It will be used to identify $\underline{1}$ and $\bar{\underline{1}}$.

A sesquilinear form on an object X of C' is a morphism

$$\phi : X \otimes \bar{X} \to \underline{1} .$$

On applying $-$, we obtain a morphism $\bar{X} \otimes \bar{\bar{X}} \to \bar{\underline{1}}$, which can be identified with a morphism

$$\bar{\phi} : \bar{X} \otimes X \to \underline{1} .$$

There are associated with ϕ two morphisms $\phi^{\sim}, {}^{\sim}\phi : X \to \bar{X}^{\vee}$ determined by

$$\phi^{\sim}(x)(y) = \phi(x \otimes y)$$
$$(4.3.1)$$
$${}^{\sim}\phi(x)(y) = \bar{\phi}(y \otimes x)$$

The form ϕ is said to be <u>non-degenerate</u> if $\phi^{\sim}$ (equivalently ${}^{\sim}\phi$) is an isomorphism. The <u>parity</u> of a non-degenerate sesquilinear form ϕ is the unique morphism $\varepsilon_{\phi} : X \to X$ such that

$$\phi^{\sim} = {}^{\sim}\phi \circ \varepsilon_{\phi} \quad , \quad \phi(x,y) = \bar{\phi}(y, \varepsilon_{\phi} x) \ . \qquad (4.3.2)$$

Note that

$$\phi \circ (\varepsilon_{\phi} \otimes \bar{\varepsilon}_{\phi}) = \phi, \ \phi(\varepsilon_{\phi} x, \bar{\varepsilon}_{\phi} y) = \phi(x,y). \qquad (4.3.3)$$

The <u>transpose</u> u^{ϕ} of $u \in \mathrm{End}(X)$ relative to ϕ is determined by

$$\phi \circ (u \otimes id_{\bar{X}}) = \phi \circ (id_X \otimes \overline{u^{\phi}}) \ , \ \phi(ux,y) = \phi(x, \overline{u^{\phi}} y) \ . \qquad (4.3.4)$$

There are the formulas

$$(uv)^{\phi} = v^{\phi} u^{\phi} \ , \ (id)^{\phi} = id, \ (u^{\phi})^{\phi} = \varepsilon_{\phi} u \varepsilon_{\phi}^{-1}, (\varepsilon_{\phi})^{\phi} = \varepsilon_{\phi}^{-1} \qquad (4.3.5)$$

and $u \mapsto u^{\phi}$ is a semi-linear bijection $End(X) \to End(X)$.

If ϕ is a non-degenerate sequilinear form on X , then any other non-degenerate sequilinear form can be written

$$\phi_{\alpha} = \phi\circ(\alpha\otimes id) \ , \ \phi_{\alpha}(x,y) = \phi(\alpha x,y) = \phi(x,\overline{\alpha^{\phi}y}) \qquad (4.3.6)$$

for a uniquely determined automorphism α of X . There are the formulas

$$u^{\phi\alpha} = (\alpha u\alpha^{-1})^{\phi} \ , \ \varepsilon_{\phi_{\alpha}} = (\alpha^{\phi})^{-1}\ \varepsilon_{\phi}\alpha \ . \qquad (4.3.7)$$

When ε_{ϕ} is in the centre of $End(X)$, ϕ_{α} has the same parity as ϕ if and only if $\alpha^{\phi} = \alpha$.

Remark 4.4. There is also the notion of a bilinear form on an object X of a tensor category: It is a morphism $X \otimes X \to \underline{1}$. Most of the notions associated with bilinear forms on vector spaces make sense in the context of Tannakian categories; see Saavedra [1, V 2.1].

Weil forms

A non-degenerate sesquilinear form ϕ on X is a Weil form if its parity ε_{ϕ} is in the centre of $End(X)$ and if for all nonzero $u \in End(X)$, $Tr_{X}(uu^{\phi})>0$.

Proposition 4.5. Let ϕ be a Weil form on X .

(a) The map $u \mapsto u^{\phi}$ is an involution on End(X) inducing complex conjugation on $\mathbb{C} = \mathbb{C}.id_X$, and $(u,v) \mapsto Tr(uv^{\phi})$ is a positive definite Hermitian form on End(X) .

(b) End(X) is a semisimple $\mathbb{C}$-algebra.

(c) Any commutative sub-$\mathbb{R}$-algebra A of End(X) composed of symmetric elements (i.e., such that $u^{\phi}=u$) is a product of copies of $\mathbb{R}$.

Proof. (a) is obvious.

(b) Let I be a nilpotent ideal in End(X) ; we have to show that $I = 0$. Suppose on the contrary that there is a $u \neq 0$ in I . Then $v \overset{df}{=} u u^{\phi} \in I$ and is nonzero because $Tr(v)>0$. As $v = v^{\phi}$, we have $Tr(v^2)>0$, $Tr(v^4)>0, \ldots$, contradicting the nilpotence of I .

(c) The argument used in (b) shows that A is semisimple and is therefore a product of fields. If $\mathbb{C}$ occurs as a factor of A , then $Tr_X|\mathbb{C}$ is a multiple of the identity map, and $Tr(u^2) = Tr(uu^{\phi})>0$ is impossible.

Two Weil forms, ϕ on X and ψ on Y , are said to be compatible if the sesquilinear form $\phi \oplus \psi$ on $X \oplus Y$ is a Weil form. Note that if $Hom(X,Y)=0=Hom(Y,X)$, then ϕ and ψ are automatically compatible.

Proposition 4.6. Let ϕ be a Weil form on X ; then $\alpha \mapsto \phi_{\alpha} \overset{df}{=} \phi \circ \alpha \otimes 1$ induces a bijection between

$\{\alpha \in \mathrm{Aut}(X) \mid \alpha^{\phi} = \alpha,\ \alpha \text{ is a square in } \mathbb{R}[\alpha] \subset \mathrm{End}(X)\}$

and the set of Weil forms on X that have the same parity as ϕ and are compatible with ϕ.

Proof: We saw in (4.3.6) that any non-degenerate sesquilinear form on X is of the form ϕ_α for a unique automorphism α of X. Moreover, ϕ_α has the same parity as ϕ if and only if $\alpha = \alpha^{\phi}$. Assume $\alpha = \alpha^{\phi}$ then $u^{\phi_\alpha} = \alpha u^{\phi} \alpha^{-1}$ and so ϕ_α is a Weil form if and only if $\mathrm{Tr}(u\alpha u^{\phi}\alpha^{-1}) > 0$ for all $u \neq 0$. Let $v = \begin{pmatrix} 0 & 0 \\ u & 0 \end{pmatrix} \in \mathrm{End}(X \oplus X)$; then $v^{\phi\oplus\phi_\alpha} = \begin{pmatrix} 0 & u^{\phi}\alpha \\ 0 & 0 \end{pmatrix}$ and $\mathrm{Tr}_{X\oplus X}(v^{\phi\oplus\phi_\alpha} v) = \mathrm{Tr}(u^{\phi}\alpha u)$. Therefore if ϕ_α is compatible with ϕ, then $\mathrm{Tr}_X(u^{\phi}\alpha u) > 0$ for all $u \neq 0$. One checks easily that the converse statement also holds.

Now assume α to be symmetric and equal to β^2 with $\beta \in \mathbb{R}[\alpha]$. Then $\mathrm{Tr}(u\alpha u^{\phi}\alpha^{-1}) = \mathrm{Tr}((u\beta)\beta u^{\phi}\alpha^{-1}) = \mathrm{Tr}(\beta u^{\phi}\alpha^{-1}(u\beta)) = \mathrm{Tr}((\beta^{-1}u\beta)^{\phi}\beta^{-1}u\beta) > 0$ for $u \neq 0$, and $\mathrm{Tr}(u^{\phi}\alpha u) = \mathrm{Tr}((\beta u)^{\phi}\beta u) > 0$ for $u \neq 0$. Hence ϕ_α is a Weil form and is compatible with ϕ. Conversely, if ϕ_α has the same parity as ϕ and is compatible with it, then α is symmetric and $\mathrm{Tr}_X(u^2\alpha) > 0$ for all $u \neq 0$ in $\mathbb{R}[\alpha]$; this last statement implies that α is a square in $\mathbb{R}[\alpha]$.

Corollary 4.7. Let ϕ and ϕ' be compatible Weil forms on X with the same parity, and let ψ be a Weil form on Y. If ϕ is compatible with ψ, then so also is ϕ'. In particular, compatibility is an equivalence relation for Weil forms on X having a given parity.

<u>Proof</u>: This follows easily from writing $\phi' = \phi_\alpha$.

<u>Example</u> 4.8. Let X be a simple object in $\underline{C}'$, so that $\mathrm{End}(X) = \mathbb{C}$, and let $\varepsilon \in \mathrm{End}(X)$. If $\bar{X}$ is isomorphic to $\check{X}$, then (4.3.6) shows that the sesquilinear forms on X form a complex line; (4.3.7) shows that if there is a nonzero such form with parity ε , then the set of sesquilinear forms on X with parity ε is a real line; (4.6) shows that if there is a Weil form with parity ε , then the set of such forms falls into two compatibility classes, each parametrized by $\mathbb{R}_{>0}$.

<u>Remark</u> 4.9. Let X_o be an object in $\underline{C}$ and let ϕ_o be a non-degenerate bilinear form $\phi_o : X_o \otimes X_o \to \underline{1}$ The <u>parity</u> of ϕ_o is defined by the equation $\phi_o(x,y) = \phi_o(y,\varepsilon x)$. The form ϕ_o is said to be a <u>Weil form</u> on X_o if ε is in the centre of $\mathrm{End}(X_o)$ and if for all nonzero $u \in \mathrm{End}(X_o)$, $\mathrm{Tr}(uu^{\phi_o})>0$. Two Weil forms ϕ_o and ψ_o are said to be compatible if $\phi_o \oplus \psi_o$ is also a Weil form.

Let X_o correspond to the pair (X,a) with $X \in \mathrm{ob}(\underline{C}')$. Then ϕ_o defines a bilinear form ϕ on X , and $\psi \overset{\mathrm{df}}{=} (X \otimes \bar{X} \xrightarrow{1 \otimes a^{-1}} X \otimes X \xrightarrow{\phi} \underline{1})$ is a non-degenerate sesquilinear form on X . If ϕ_o is a Weil form, then ψ is a Weil form on X which is compatible with its conjugate $\bar{\psi}$, and every such ψ arises from a ϕ_o; moreover $\varepsilon(\psi) = \varepsilon(\phi_o)$.

Polarizations.

Let Z be the centre of the band associated with $\underline{C}$ (see the appendix). Thus Z is a commutative algebraic group over $\mathbb{R}$ such that $Z(\mathbb{C})$ is the centre of $\mathrm{Aut}^{\otimes}(\omega)$ for any fibre functor on $\underline{C}'$ with values in $\mathbb{C}$. Moreover, Z represents $\underline{\mathrm{Aut}}^{\otimes}(\mathrm{id}_{\underline{C}})$.

Let $\varepsilon \in Z(\mathbb{R})$ and, for each $X \in \mathrm{ob}(\underline{C}')$, let $\pi(X)$ be an equivalence class (for the relation of compatibility) of Weil forms on X with parity ε; we say that π is a (homogeneous) polarization on $\underline{C}$ if

(4.10a) for all X, $\bar{\phi} \in \pi(X)$ whenever $\phi \in \pi(\bar{X})$, and

(4.10b) for all X and Y, $\phi \oplus \psi \in \pi(X \oplus Y)$ and

$\phi \otimes \psi \in \pi(X \otimes Y)$ whenever $\phi \in \pi(X)$ and $\psi \in \pi(Y)$.

We call ε the parity of π and say that ϕ is positive for π if $\phi \in \pi(X)$. Thus the conditions require that $\bar{\phi}$, $\phi \oplus \psi$, and $\phi \otimes \psi$ be positive for π whenever ϕ and ψ are.

Proposition 4.11. Let π be a polarization on $\underline{C}$.

(a) The categories $\underline{C}$ and $\underline{C}'$ are semisimple.

(b) If $\phi \in \pi(X)$ and $Y \subset X$ then $X = Y \oplus Y^{\perp}$ and the restriction ϕ_Y of ϕ to Y is in $\pi(Y)$.

Proof. (a) Let X be an object of $\underline{C}'$; let Y be a nonzero simple subobject of X and let $u : Y \hookrightarrow X$ denote the inclusion map. Choose $\phi \in \pi(Y)$ and $\psi \in \pi(X)$. Consider $v = \begin{pmatrix} 0 & u \\ 0 & 0 \end{pmatrix}$: $X \oplus Y \to X \oplus Y$ and let $u' : X \to Y$ be such that $v^{\psi \oplus \phi} = \begin{pmatrix} 0 & 0 \\ u' & 0 \end{pmatrix}$.

Then $Tr_Y(u'u) = Tr_{Y\oplus X}(v^{\psi\oplus\phi}v) > 0$, and so $u'u$ is an automorphism w of Y . The map $p = w^{-1}\circ u'$ projects X onto Y , which shows that Y is a direct summand of X , and X is semisimple.

The same argument, using the bilinear forms (4.9) shows that $\underline{C}$ is semisimple.

(b) Let $Y' = Y\cap Y^{\perp}$, where $Y^{\perp}$ is the largest subobject of X such that ϕ is zero on $Y \otimes \bar{Y}^{\perp}$, and let $p : X \to X$ project X onto Y' (by which we mean that $p(X)\subset Y'$ and $p|Y' = id$). As ϕ is zero on $Y' \otimes \bar{Y}'$, $0 = \phi\circ(p\otimes\bar{p}) = \phi\circ(id\otimes\overline{p^{\phi}p})$, and so $p^{\phi}p = 0$. Therefore $Tr(p^{\phi}p) = 0$ and so p , and Y', are zero. Thus $X = Y \oplus Y^{\perp}$ and $\phi = \phi_Y \oplus \phi_{Y^{\perp}}$. Let $\phi_1 \in \pi(Y)$ and $\phi_2 \in \pi(Y^{\perp})$. Then $\phi_1 \oplus \phi_2$ is compatible with ϕ , and this implies that ϕ_1 is compatible with ϕ_Y .

Remark 4.12. Suppose $\underline{C}$ is defined by a triple (G,σ,c) , as in (4.1), so that $\underline{C}' = \underline{Rep}_{\mathbb{C}}(G)$. A sesquilinear form $\phi: X \otimes \bar{X} \to \underline{1}$ defines a sesquilinear form ϕ' on X in the usual, vector space, sense by the formula

$$\phi'(x,y) = \phi(x \otimes \bar{y}) , \quad x,y \in X \qquad (4.12.1).$$

The conditions that ϕ be a G-morphism and have a parity $\varepsilon \in Z(\mathbb{R})$ become respectively

$$\phi'(x,y) = \phi'(gx,\sigma^{-1}(g)y) , \quad g \in G(\mathbb{C}) , \qquad (4.12.2)$$

$$\overline{\phi'(y,x)} = \phi'(x, \varepsilon c^{-1}y) \qquad (4.12.3).$$

When G acts trivially on X, then (4.12.3) becomes

$$\overline{\phi'(y,x)} = \phi'(x,y) ,$$

and so ϕ' is a Hermitian form in the usual sense on X. If X is one-dimensional and $\phi \in \pi(X)$, then ϕ' is positive-definite (for otherwise $\phi \otimes \phi \notin \pi(X)$). Now (4.11b) shows that the same is true for any X, and (4.6) shows that $\{\phi' | \phi \in \pi(X)\}$ is the complete set of positive-definite Hermitian forms on X (when G acts trivially on X). In particular, $\underline{\mathrm{Vec}}_{\mathbb{R}}$ has a unique polarization.

Remark 4.13. A polarization π on $\underline{C}$ with parity ε defines, for each simple object X of $\underline{C}'$, an orientation of the real line of sesquilinear forms on X with parity ε (see 4.8), and π is obviously determined by this family of orientations. Choose a fibre functor ω for $\underline{C}'$, and choose for each simple object X_i a $\phi_i \in \pi(X_i)$. Then

$$\pi(X_i) = \{r\,\phi_i | r \in \mathbb{R}_{>0}\} .$$

If X is isotypic of type X_i, so that $\omega(X) = W \otimes \omega(X_i)$ where $\underline{\mathrm{Aut}}^{\otimes}(\omega)$ acts trivially on W, then

$$\{\omega(\phi)' | \phi \in \pi(X)\} = \{\psi \otimes \omega(\phi_i)' | \psi \text{ Hermitian } \psi > 0\} .$$

If $X = \oplus X^{(i)}$ where the $X^{(i)}$ are the isotypic components of X, then

$$\pi(X) = \oplus\, \pi(X^{(i)}) .$$

Remark 4.14. Let $\varepsilon \in Z(\mathbb{R})$ and, for each $X_o \in ob(\underline{C})$, let $\pi(X_o)$ be an equivalence class of bilinear Weil forms on X_o with parity ε (see (4.9)). One says that π is a homogeneous polarization on $\underline{C}$ if $\phi_o \oplus \psi_o \in \pi(X \oplus Y)$ and $\phi_o \otimes \psi_o \in \pi(X \otimes Y)$ whenever $\phi_o \in \pi(X)$ and $\psi_o \in \pi(Y)$. As $\{X \mid (X,a) \in ob(\underline{C})\}$ generates $\underline{C}'$, the relation between bilinear and sesquilinear forms noted in (4.9) establishes a one-to-one correspondence between polarizations in this bilinear sense and in the sesquilinear sense of (4.10).

In the situation of (4.12), a bilinear form ϕ_o on X_o defines a sesquilinear form ψ' on $X = \mathbb{C} \otimes X_o$ (in the usual vector space sense) by the formula

$$\psi'(z_1 v_1, z_2 v_2) = z_1 \bar{z}_2\, \phi_o(v_1, v_2),\ v_1, v_2 \in X_o,\ z_1, z_2 \in \mathbb{C} .$$

Description of polarizations

Let $\underline{C}$ be defined by a triple (G,σ,c) satisfying (4.2.1), and let K be a maximal compact subgroup of $G(\mathbb{C})$. As all maximal compact subgroups of $G(\mathbb{C})$ are conjugate (Hochschild [1, XV. 3.1]), there exists $m \in G(\mathbb{C})$ such that $\sigma^{-1}(K) = m\, K\, m^{-1}$. Therefore, after replacing σ by $\sigma \circ \underline{ad}(m)$, we can assume that

$\sigma(K) = K$. Subject to this constraint, (σ, c) is determined up to modification by an element m in the normalizer of K .

Assume that $\underline{C}$ is polarizable. Then (4.11a) and (2.28) show that G^o is reductive, and it follows that K is a compact real form G , i.e., K has the structure of a compact real algebraic group in the sense of (2.33) and $K_{\mathbb{C}} = G$ (see Springer [1, 5.6]). Let σ_K be the semi-linear automorphism of G such that, for $g \in G(\mathbb{C})$, $\sigma_K(g)$ is the conjugate of g relative to the real structure on G defined by K ; note that σ_K determines K . The normalizer of K is $K.Z(\mathbb{C})$, and so $c \in K.Z(\mathbb{C})$.

Fix a polarization π on $\underline{C}$ with parity ε . If X is an irreducible representation of G and ψ is a positive-definite K-invariant Hermitian form on X , then for any $\phi \in \pi(X)$,

$$(\phi(x \otimes \bar{y} \stackrel{\mathrm{df}}{=\!=}) \phi'(x,y) = \psi(x, \beta y)$$

for some $\beta \in \mathrm{Aut}(X)$. Equations (4.12.2) and (4.12.3) can be re-written as

$$\beta g_X = \sigma(g)_X \beta \ , \ g \in K(\mathbb{R}) \qquad (4.14.1)$$

$$\beta^* = \beta \ \varepsilon_X \ c_X^{-1} \qquad (4.14.2)$$

where β^* is the adjoint of β relative to ψ : $\psi(\beta x, y) = \psi(x, \beta^* y)$. As $K(\mathbb{R})$ is Zariski dense in $K(\mathbb{C})$, X is also irreducible as a representation of $K(\mathbb{R})$, and so the set $c(X,\pi)$ of such β's is parametrized by $\mathbb{R}_{>0}$. An arbitrary finite-dimensional representation X of G can be written

$$X = \oplus_i W_i \otimes X_i$$

where the sum is over the non-isomorphic irreducible representations X_i of G, and G acts trivially on each W_i; let ψ_i and ψ_i' respectively be K-invariant positive-definite Hermitian forms on X_i and W_i, and let $\psi = \oplus \psi_i' \otimes \psi_i$; then for any $\phi \in \pi(X)$, $\phi'(x,y) = \psi(x,\beta y)$ where $\beta = \oplus \beta_i' \otimes \beta_i$ with $\beta_i \in c(X_i,\pi)$ and β_i' is positive-definite and Hermitian relative to ψ_i'. We let $c(X,\pi)$ denote the set of β as ϕ runs through $\pi(X)$. The condition (4.10b) that $\pi(X_1) \otimes \pi(X_2) \subset \pi(X_1 \otimes X_2)$ becomes $c(X_1,\pi) \otimes c(X_2,\pi) \subset c(X_1 \otimes X_2,\pi)$.

Lemma 4.15. There exists a $b \in K$ with the following properties:

(a) $b_X \in c(X,\pi)$ for all irreducible X;

(b) $\sigma = \sigma_K \circ \underline{\mathrm{ad}}\, b$, where σ_K denotes complex conjugation on G relative to K;

(c) $\varepsilon^{-1}c = \sigma b.b = b^2$.

Proof: Let $a = \varepsilon c^{-1} \in G(\mathbb{C})$. When X is irreducible, (4.14.1) applied twice shows that $\beta^2 gx = \sigma^2(g)\beta^2 x = cgc^{-1}\beta^2 x$ for $\beta \in c(X,\pi)$, $g \in K$, and $x \in X$; therefore $(c^{-1}\beta^2)gx = g(c^{-1}\beta^2)x$, and so $c^{-1}\beta^2$ acts as a scalar on X. Hence $a\beta^2 = \varepsilon\, c^{-1}\beta^2$ also acts as a scalar. Moreover, $\beta^2 a = \beta\beta^*$ (by 4.14.2) and so $\mathrm{Tr}_X(a\beta^2) = \mathrm{Tr}_X(\beta^2 a) > 0$; we conclude that $a_X\beta^2 \in \mathbb{R}_{>0}$. It follows that there is a unique $\beta \in c(X,\pi)$ such that $a_X = \beta^{-2}$, $\beta g_X = \sigma(g)_X\beta$ $(g \in K)$, and $\beta^* = \beta^{-1}$ (so β is unitary).

For an arbitrary X we write $X = \oplus W_i \otimes X_i$ as before, and set $\beta = \oplus\, \mathrm{id} \otimes \beta_i$, where β_i is the canonical element of

$c(X_i,\pi)$ just defined. We still have $a_X = \beta^{-2}$, $\beta g_X = \sigma(g)_X\beta$ $(g \in K)$, and $\beta \in c(X,\pi)$. Moreover, these conditions characterize β : if $\beta' \in c(X,\pi)$ has the same properties, then $\beta' = \Sigma\, \gamma_i \otimes \beta_i$ (this expresses that $\beta' g_X = \sigma(g)_X\beta'$, $g \in K$) with $\gamma_i^2 = 1$ (as $\beta'^2 = a_X^{-1}$) and γ_i positive-definite and Hermitian; hence $\gamma_i = 1$.

The conditions are compatible with tensor products, and so the canonical β are compatible with tensor products: they therefore define an element $b \in G(\mathbb{C})$. As b is unitary on all irreducible representations, it lies in K. The equations $\beta^2 = a_X^{-1}$ show that $b^2 = a^{-1} = \varepsilon^{-1}c$. Finally, $\beta g_X = \sigma(g)_X\beta$ implies that $\sigma(g) = \underline{ad}\, b(g)$ for all $g \in K$; therefore $\sigma\circ\underline{ad}\, b^{-1}$ fixes K and, as it has order 2, it must equal σ_K.

<u>Theorem</u> 4.16. Let $\underline{C}$ be an algebraic Tannakian category over $\mathbb{R}$ and let $G = \underline{Aut}^{\otimes}(\omega)$ where ω is a fibre functor on $\underline{C}$ with values in $\mathbb{C}$; let π be a polarization on $\underline{C}$ with parity ε. For any compact real-form K of G, the pair (σ_K,ε) satisfies (4.2.1), and the equivalence $\underline{C}' \xrightarrow{\sim} \underline{Rep}_{\mathbb{C}}(G)$ defined by ω carries the descent datum on $\underline{C}'$ defined by $\underline{C}$ into that on $\underline{Rep}_{\mathbb{C}}(G)$ defined by (σ_K,ε) : $\omega(\bar{X}) = \overline{\omega(X)}$, $\omega(\mu_X) = \mu_{\omega(X)}$. For any simple X in $\underline{C}'$, $\{\omega(\phi)' \mid \phi \in \pi(X)\}$ is the set of K-invariant positive-definite Hermitian forms on $\omega(X)$.

<u>Proof</u>: Let $(\underline{C},\omega)$ correspond to a triple (G,σ_1,c_1) (see (4.3a)), and let $b \in K$ be the element constructed in the lemma. Then $\sigma_1 = \sigma_K\circ\underline{ad}\, b$ and $c = \varepsilon.\sigma b.b = \sigma b.\varepsilon.b$. Therefore (σ_K,ε) has the same property as (σ_1,c_1) (see (4.3b)), which proves the first assertion. The second assertion follows from the fact that $b \in c(\omega(X),\pi)$ for any simple X.

Classification of polarized Tannakian categories

Theorem 4.17 (a) The category $\underline{C}$ is polarizable if and only if its band is defined by a compact real algebraic group K.

(b) Let K be a compact real algebraic group, and let $\varepsilon \in Z(\mathbb{R})$ where Z is the centre of K ; there exists a Tannakian category $\underline{C}$ over $\mathbb{R}$ whose gerb is bound by the band $B(K)$ of K and a polarization π on $\underline{C}$ with parity ε.

(c) Let $(\underline{C}_1,\pi_1)$ and $(\underline{C}_2,\pi_2)$ be polarized algebraic Tannakian categories over $\mathbb{R}$, and let $B \tilde{\to} B_1$ and $B \tilde{\to} B_2$ be the identifications of the bands of $\underline{C}_1$ and $\underline{C}_2$ with a given band B. If $\varepsilon(\pi_1) = \varepsilon(\pi_2)$ in $Z(B)(\mathbb{R})$ then there is a tensor equivalence $\underline{C}_1 \tilde{\to} \underline{C}_2$ respecting the polarizations and the actions of B (i.e., such that $FIB(\underline{C}_2) \tilde{\to} FIB(\underline{C}_1)$ is a B-equivalence), and this equivalence is unique up to isomorphism.

Proof: We have already seen that if $\underline{C}$ is polarizable, then $\underline{C}'$ is semisimple, and so, for any fibre functor ω with values in $\mathbb{C}$, (the identity component of) $G = \underline{Aut}^{\otimes}(\omega)$ is reductive, and hence has a compact real form K. This proves half of (a). Part (b) is proved in the first lemma below, and the sufficiency in (a) follows from (b) and the second lemma below. Part (c) is essentially proved by (4.16).

Lemma 4.18. Let K be a compact real algebraic group and let $G = K_{\mathbb{C}}$; let $\sigma(g) = \sigma'(\bar{g})$ where σ' is a Cartan involution for K, and let $\varepsilon \in Z(\mathbb{R})$ where Z is the centre of K. Then (σ,ε) satisfies (4.2.1) and the Tannakian category $\underline{C}$ defined by (G,σ,ε) has a polarization with parity ε.

Proof: Since $\sigma^2 = \mathrm{id}$ and σ fixes all elements of K, (4.2.1) is obvious. There exists a polarization π on $\underline{C}$ such that, for all simple X, $\{\phi' \mid \phi \in \pi(X)\}$ is the set of positive-definite K-invariant Hermitian forms on X. (In the notation of (4.15), b=1.) This polarization has parity ε.

Let $\underline{C}$ correspond to $(\underline{C}', X \mapsto \bar{X}, \mu)$; for any $z \in Z(\mathbb{R})$, where Z is the centre of the band B of $\underline{C}$, $(\underline{C}', X \mapsto \bar{X}, \mu \circ z)$ defines a new Tannakian category ${}^{z}\underline{C}$ over $\mathbb{R}$.

Lemma 4.19. Every Tannakian category over $\mathbb{R}$ whose gerb is bound by B is of the form ${}^{z}\underline{C}$ for some $z \in Z(\mathbb{R})$; there is a tensor equivalence ${}^{z}\underline{C} \xrightarrow{\sim} {}^{z'}\underline{C}$ respecting the action of B if and only if $z'z^{-1} \in Z(\mathbb{R})^2$.

Proof: Let ω be a fibre functor on $\underline{C}$, and let $(\underline{C},\omega)$ correspond to (G,σ,c); we can assume that a second category $\underline{C}_1$ corresponds to (G,σ_1,c_1). Let γ and γ_1 be the functors $V \mapsto \bar{V}$ defined by (σ,c) and (σ_1,c_1) respectively. Then $\gamma_1^{-1} \circ \gamma$ defines a tensor automorphism of ω, and so corresponds to an element $m \in G(\mathbb{C})$. We have $\sigma = \sigma_1 \circ \underline{\mathrm{ad}}(m)$, and so we can modify (σ_1,c_1) in order to get $\sigma_1 = \sigma$. Let μ and μ_1 be the functorial isomorphisms $V \to \bar{\bar{V}}$ defined by (σ,c) and (σ,c_1) respectively. Then $\mu_1^{-1} \circ \mu$ defines a tensor automorphism of $\mathrm{id}_{\underline{C}}$, and so $\mu_1^{-1} \circ \mu = z^{-1}$, $z \in Z(\mathbb{R})$. We have $\mu_1 = \mu \circ z$.

The second part of the lemma is obvious.

Remark 4.20. Some of the above results can be given a more cohomological interpretation. Let B be the band defined by a compact real group K , and let Z be the centre of B ; let C be a Tannakian category, whose gerb is B .

(a) As Z is a subgroup of a compact real algebraic group, it is also compact (see (2.33)). It is easy to compute its cohomology; one finds that

$$H^1(\mathbb{R},Z) = {}_2Z(\mathbb{R}) \overset{\mathrm{df}}{=} \ker(2:Z(\mathbb{R}) \to Z(\mathbb{R}))$$

$$H^2(\mathbb{R},Z) = Z(\mathbb{R})/Z(\mathbb{R})^2$$

(b) The general theory shows that there is an isomorphism $H^1(\mathbb{R},Z) \to \mathrm{Aut}_B(\underline{C})$, which can be described explicitly as the map associating to $z \in Z(\mathbb{R})_2$ the automorphism w_z

$$\begin{cases}(X,\ a_X) \longmapsto (X,\ a_X\, z_X)\\ f \longmapsto f\ .\end{cases}$$

(c) The Tannakian categories bound by B , up to B-equivalence, are classified by $H^2(\mathbb{R},B)$, and $H^2(\mathbb{R},B)$ if nonempty is an $H^2(\mathbb{R},Z)$-torsor; the action of $H^2(\mathbb{R},Z) = Z(\mathbb{R})/Z(\mathbb{R})^2$ on the categories is made explicit in (4.19).

(d) Let $\mathrm{Pol}(\underline{C})$ denote the set of polarizations on $\underline{C}$. For $\pi \in \mathrm{Pol}(\underline{C})$ and $z \in Z(\mathbb{R})$ we define $z\pi$ to be the polarization such that

$$\phi(x,y) \in z\pi(X) \iff \phi(x,zy) \in \pi(X) ;$$

it has parity $\varepsilon(z\pi) = z^2\,\varepsilon(\pi)$. The pairing

$$(z,\pi) \mapsto z\pi : Z(\mathbb{R}) \times \mathrm{Pol}(\underline{C}) \to \mathrm{Pol}(\underline{C})$$

makes $\mathrm{Pol}(\underline{C})$ into a $Z(\mathbb{R})$-torsor.

(e) Let $\pi \in \mathrm{Pol}(\underline{C})$ and let $\varepsilon = \varepsilon(\pi)$; then $\underline{C}$ has a polarization with parity $\varepsilon' \in Z(\mathbb{R})$ if and only if $\varepsilon' = \varepsilon z^2$ for some $z \in Z(\mathbb{R})$.

Remark 4.21. In Saavedra [1, V. 1] there is a table of Tannakian categories whose bands are simple, from which it is possible to read off those that are polarizable (loc. cit. V. 2.8.3).

Neutral polarized categories

The above results can be made more explicit when $\underline{C}$ has a fibre functor with values in $\mathbb{R}$.

Let G be an algebraic group over $\mathbb{R}$, and let $C \in G(\mathbb{R})$. A G-invariant sesquilinear form $\psi : V \times V \to \mathbb{C}$ on $V \in \mathrm{ob}(\underline{\mathrm{Rep}}_{\mathbb{C}}(G))$ is said to be a C-polarization if

$$\psi^C(x,y) \overset{\mathrm{df}}{=} \psi(x,Cy)$$

is a positive-definite Hermitian form on V. If every object of $\underline{\mathrm{Rep}}_{\mathbb{C}}(G)$ has a C-polarization then C is called a <u>Hodge element</u>.

<u>Proposition</u> 4.22. Assume that $G(\mathbb{R})$ contains a Hodge element C. There is then a polarization π_C on $\underline{\mathrm{Rep}}_{\mathbb{R}}(G)$ for which the positive forms are exactly the C-polarizations; the parity of π_C is C^2; for any $g \in G(\mathbb{R})$ and $z \in Z(\mathbb{R})$, where Z is the centre of G, $C' = zgCg^{-1}$ is also a Hodge element and $\pi_{C'} = z\pi_C$; every polarization on $\underline{\mathrm{Rep}}_{\mathbb{R}}(G)$ is of the form $\pi_{C'}$ for some Hodge element C'.

<u>Proof</u>: Let ψ be a C-polarization on $V \in ob(\underline{\mathrm{Rep}}_{\mathbb{C}}(G))$; then $\psi(x,y) = \psi(Cx,Cy)$ because ψ is G-invariant, and $\psi(Cx,Cy) = \psi^C(Cx,y) = \overline{\psi^C(y,Cx)} = \overline{\psi(y,C^2x)}$. This shows that ψ has parity C^2. For any V, $\overline{\psi(y,C^2x)} = \psi(x,y) = \psi(gx,gy) = \overline{\psi(gy,C^2gx)} = \overline{\psi(y,g^{-1}C^2gx)}$, $g \in G(\mathbb{R})$, x, $y \in V$; this shows that $C^2 \in Z(\mathbb{R})$. For any $u \in \mathrm{End}(V)$, $u^\psi = u^{\psi^C}$, and so $\mathrm{Tr}(uu^\psi) > 0$ if $u \neq 0$. This shows that ψ is a Weil form with parity C^2. The first assertion of the proposition is now easy to check. The third assertion is straightforward to prove, and the fourth follows from it and (4.19).

<u>Proposition</u> 4.23. The following conditions on G are equivalent:

(a) there is a Hodge element in $G(\mathbb{R})$;

(b) the category $\underline{\mathrm{Rep}}_{\mathbb{R}}(G)$ is polarizable;

(c) G is an inner form of a compact real algebraic group K.

Proof: (a) $\Rightarrow$ (b). This follows from (4.22).

(b) $\Rightarrow$ (c). To say that G is an inner form of K is the same as to say that G and K define the same band; this implication therefore follows from (4.17a).

(c) $\Rightarrow$ (a). Let Z be the centre of K (and therefore also of G) and let $K^{ad} = K/Z$. The assumption says that the isomorphism class of G is in the image of

$$H^1(\mathbb{R}, K^{ad}) \to G^1(\mathbb{R}, \underline{\mathrm{Aut}}(K)) .$$

According to Serre [1, III, Thm 6], the canonical map

$$_2(K^{ad}(\mathbb{R})) = H^1(\mathbb{R}, K^{ad}(\mathbb{R})) \to H^1(\mathbb{R}, K^{ad})$$

is an isomorphism. From the cohomology sequence

$$\begin{array}{ccccccc} K(\mathbb{R}) & \to & K^{ad}(\mathbb{R}) & \to & H^1(\mathbb{R}, Z) & \to & H^1(\mathbb{R}, K) \\ & & & & \| & & \| \\ & & & & _2Z(\mathbb{R}) & \hookrightarrow & _2K(\mathbb{R}) \end{array}$$

we see that $K(\mathbb{R}) \twoheadrightarrow K^{ad}(\mathbb{R})$, and so G is the inner form of K defined by an element $C' \in K(\mathbb{R})$ whose square is in $Z(\mathbb{R})$. Let γ be an isomorphism $K_{\mathbb{C}} \to G_{\mathbb{C}}$ such that $\gamma \circ \underline{ad}C' = \bar{\gamma}$, and let $C = \gamma(C')$; then $\bar{C} = \bar{\gamma}(C') = \gamma(C') = C$ and $\bar{\gamma}^{-1} \circ \underline{ad}(C) = \gamma^{-1}$. This shows that $C \in G(\mathbb{R})$ and that K is the form of G defined by C; the next lemma completes the proof.

Lemma 4.24. An element $C \in G(\mathbb{R})$ such that $C^2 \in Z(\mathbb{R})$ is a Hodge element if and only if the real-form K of G defined by C is a compact real group.

Proof: Identify $K_{\mathbb{C}}$ with $G_{\mathbb{C}}$ and let $\bar{g}$ and g^* respectively be the complex conjugates of $g \in G(\mathbb{C})$ relative to the real structures defined by G and K. Then $g^* = \underline{ad}(C^{-1})(\bar{g}) = C^{1}\bar{g}C$. Let ψ be a sesquilinear form on $V \in ob(\underline{Rep}_{\mathbb{C}}(G))$. Then ψ is a G-invariant if and only if

$$\psi(gx,\bar{g}y) = \psi(x,y) \ , \ g \in G(\mathbb{C}) \ .$$

On the other hand, ψ^C is K-invariant if and only if

$$\psi^C(gx,g^*y) = \psi^C(x,y) \ , \ g \in G(\mathbb{C}) \ .$$

These conditions are equivalent: V has a C-polarization if and only if V has a K-invariant positive-definite Hermitian form. Thus C is a Hodge element if and only if, for every complex representation V of K, the image of K in $\mathrm{Aut}(V)$ is contained in the unitary group of a positive-definite Hermitian form; this last condition is implied by K being compact and implies that K is contained in a compact real group and so is compact (see (2.33)).

Remark 4.25. (a) The centralizer of a Hodge element C of G is a maximal compact subgroup G, and is the only maximal compact subgroup of G containing C; in particular, if G is compact, then

C is a Hodge element if and only if it is in the centre of G (Saavedra [1, 2.7.3.5]).

(b) If C and C' are Hodge elements of G then there exists a $g \in G(\mathbb{R})$ and a unique $z \in Z(\mathbb{R})$ such that $C' = zgCg^{-1}$ (Saavedra [1, 2.7.4]). As $\pi_{C'} = z\pi_C$, this shows that $\pi_{C'} = \pi_C$ if and only if C and C' are conjugate in $G(\mathbb{R})$.

Remark 4.26. It would perhaps have been more natural to express the above results in terms of bilinear forms (see (4.4), (4.9), (4.14)): a G-invariant bilinear form $\phi : V_o \times V_o \to \mathbb{R}$ on $V_o \in \mathrm{ob}(\underline{\mathrm{Rep}}_{\mathbb{R}}(G))$ is a C-polarization if $\phi^C(x,y) \overset{\mathrm{df}}{=} \phi(x,Cy)$ is a positive-definite symmetric form on V_o; C is a Hodge element if every object of $\underline{\mathrm{Rep}}_{\mathbb{R}}(G)$ has a C-polarization; the positive forms for the (bilinear) polarization defined by C are precisely the C-polarizations.

Symmetric polarizations

A polarization is said to be symmetric if its parity is 1.

Let K be a compact real algebraic group. As 1 is a Hodge element (4.24), $\underline{\mathrm{Rep}}_{\mathbb{R}}(K)$ has a symmetric polarization π for which $\pi(X_o)$, $X_o \in \mathrm{ob}(\underline{\mathrm{Rep}}_{\mathbb{R}}(K))$, consists of the K-invariant positive-definite symmetric bilinear forms on X_o (and $\pi(X)$, $X \in \mathrm{ob}(\underline{\mathrm{Rep}}_{\mathbb{C}}(K))$, consists of the K-invariant positive-definite Hermitian forms on X).

Theorem 4.27. Let $\underline{C}$ be an algebraic Tannakian category over $\mathbb{R}$, and let π be a symmetric polarization on $\underline{C}$. Then $\underline{C}$ has a unique (up to isomorphism) fibre functor ω with values in $\mathbb{R}$ transforming positive bilinear forms for π into positive-definite symmetric bilinear forms; ω defines a tensor equivalence $\underline{C} \xrightarrow{\sim} \underline{\mathrm{Rep}}_{\mathbb{R}}(K)$, where $K = \underline{\mathrm{Aut}}^{\otimes}(\omega)$ is a compact real group.

Proof: Let ω_1 be a fibre functor with values in $\mathbb{C}$, and let $G = \underline{\mathrm{Aut}}^{\otimes}(\omega_1)$. Since $\underline{C}$ is polarizable, G has a compact real form K. According to (4.16), $\omega_1' : \underline{C}' \xrightarrow{\sim} \underline{\mathrm{Rep}}_{\mathbb{C}}(G)$ carries the descent datum on $\underline{C}'$ defined by $\underline{C}$ into that on $\underline{\mathrm{Rep}}_{\mathbb{C}}(G)$ defined by $(\sigma_K, 1)$. It therefore defines a tensor equivalence $\omega : \underline{C} \to \underline{\mathrm{Rep}}_{\mathbb{R}}(K)$ transforming π into the polarization on $\underline{\mathrm{Rep}}_{\mathbb{R}}(K)$ defined by the Hodge element 1. The rest of the proof is now obvious.

Remark 4.28. Let π be a polarization on $\underline{C}$. It follows from (4.20d) that $\underline{C}$ has a symmetric polarization if and only if $\varepsilon(\pi) \in Z(\mathbb{R})^2$.

Polarizations with parity ε of order 2

For $u = \pm 1$, define a real u-space to be a complex vector space V together with a semi-linear automorphism σ such that $\sigma^2 = u$. A bilinear form ϕ on a real u-space is u-symmetric if $\phi(x,y) = u\phi(y,x)$; such a form is positive-definite if $\phi(x,\sigma x) > 0$ for all $x \neq 0$. Thus a 1-symmetric form is symmetric, and a (-1)-symmetric form is skew-symmetric.

Let $\underline{V}_o$ be the category whose objects are pairs (V,σ) where $V = V^o \oplus V^1$ is a $\mathbb{Z}/2\mathbb{Z}$ - graded vector space over $\mathbb{C}$ and

$\sigma : V \stackrel{\approx}{\to} V$ is a semi-linear automorphism such that $\sigma^2 x = (-1)^{\deg(x)} x$. With the obvious tensor structure, $\underline{V}_o$ becomes a Tannakian category over $\mathbb{R}$ with $\mathbb{C}$-valued fibre functor $(V,\sigma) \longmapsto V$. There is a polarization $\pi = \pi_{can}$ on $\underline{V}_o$ such that, if V is homogeneous, then $\pi(V,\sigma)$ comprises the $(-1)^{\deg(v)}$-symmetric positive-definite forms on V.

Theorem 4.29. Let $\underline{C}$ be an algebraic Tannakian category over $\mathbb{R}$, and let π be a polarization on $\underline{C}$ with parity ε where $\varepsilon^2=1$, $\varepsilon \neq 1$. There exists a unique (up to isomorphism) exact faithful functor $\omega : \underline{C} \to \underline{V}_o$ such that

(a) ω carries the grading on $\underline{C}$ defined by ε into the grading on $\underline{V}_o$, i.e., $\omega(\varepsilon)$ acts as $(-1)^m$ on $\omega(V)^m$;

(b) ω carries π into π_{can}, i.e., $\phi \in \pi(X)$ if and only if $\omega(\phi) \in \pi_{can}(\omega(X))$.

Proof: Note that $\underline{V}_o$ is defined by the triple $(\mu_2, \sigma_o, \varepsilon_o)$ where σ_o is the unique semi-linear automorphism of μ_2 and ε_o is the unique element of $\mu_2(\mathbb{R})$ of order 2. We can assume (by (4.3)) that $\underline{C}$ corresponds to a triple (G,σ,ε). Let G_o be the subgroup of G generated by ε; then $(G_o, \sigma|G_o, \varepsilon) \approx (\mu_2, \sigma_o, \varepsilon_o)$, and so the inclusion $(G_o, \sigma|G_o, \varepsilon) \hookrightarrow (G,\sigma,\varepsilon)$ induces a functor $\underline{C} \to \underline{V}_o$ having the required properties.

Let $\omega, \omega' : \underline{C} \rightrightarrows \underline{V}_o$ be two functors satisfying (a) and (b). It is clear from (3.2a) that there exists an isomorphism $\lambda : \omega \stackrel{\approx}{\to} \omega'$ from ω to ω' regarded as functors to $\underline{\mathrm{Vec}}_{\mathbb{C}}$. As $\lambda_X : \omega(X) \to \omega'(X)$ commutes with the action of ε, it preserves the gradings; as λ commutes with $\omega(\phi)$, any $\phi \in \pi(X)$, it also commutes with σ; it follows that λ is an isomorphism of ω and ω' as functors to $\underline{V}_o$.

§5. Graded Tannakian categories

Throughout this section, k will be a field of characteristic zero.

Gradings

Let M be set. An M-grading on an object X of an additive category is a decomposition $X = \bigoplus_{m \in M} X^m$; an M-grading on an additive functor $u : \underline{C} \to \underline{C}'$ is an M-grading on each $u(X)$, $X \in ob(\underline{C})$, that depends functorially on X .

Suppose now that M is an abelian group, and let D be the algebraic group of multiplicative type over k whose character group is M (with trivial Galois action; see (2.32)). In the cases of most interest to us, namely $M = \mathbb{Z}$ or $M = \mathbb{Z}/2\mathbb{Z}$, D equals $\mathbb{G}_m$ or μ_2 $(=\mathbb{Z}/2\mathbb{Z})$. An M-grading on a Tannakian category $\underline{C}$ over k can be variously described as follows:

(5.1a) An M-grading, $X = \oplus X^m$, on each object X of $\underline{C}$ that depends functorially on X and is compatible with tensor products in the sense that $(X \otimes Y)^m = \bigoplus_{r+s=m} X^r \otimes Y^s$;

(5.1b) An M-grading on the identity functor $id_{\underline{C}}$ of $\underline{C}$ that is compatible with tensor products;

(5.1c) A homomorphism $D \to \underline{Aut}^{\otimes}(id_{\underline{C}})$

(5.1d) A central homomorphism $D \to G$, $G = \underline{Aut}^{\otimes}(\omega)$, for one (or every) fibre functor ω .

Definitions (a) and (b) are obviously equivalent. By a central homomorphism in (d), we mean a homomorphism from D into the centre of G defined over k ; although G need not be defined over k, its centre is, and equals $\underline{\mathrm{Aut}}^{\otimes}(\mathrm{id}_{\underline{C}})$, whence follows the equivalence of (c) and (d). Finally, a homomorphism $w : D \to \underline{\mathrm{Aut}}^{\otimes}(\mathrm{id}_{\underline{C}})$ corresponds to the family of gradings $X = \oplus X^m$ for which $w(d)$ acts on $X^m \subset X$ as $m(d) \in k$.

Tate triples

A Tate triple $\underline{T}$ over k is a triple $(\underline{C}, w, T)$ comprising a Tannakian category $\underline{C}$ over k, a $\mathbb{Z}$-grading $w: \mathbb{G}_m \to \underline{\mathrm{Aut}}^{\otimes}(\mathrm{id}_{\underline{C}})$ on $\underline{C}$ (called the weight grading), and an invertible object T (called the Tate object) of weight -2. For any $X \in \mathrm{ob}(\underline{C})$ and $n \in \mathbb{Z}$, we write $X(n) = X \otimes T^{\otimes n}$. A fibre functor on $\underline{T}$ with values in R is a fibre functor $\omega : \underline{C} \to \underline{\mathrm{Mod}}_R$ together with an isomorphism $\omega(T) \xrightarrow{\approx} \omega(T^{\otimes 2})$, i.e., the structure of an identity object on $\omega(T)$. If $\underline{T}$ has a fibre functor with values in k, then $\underline{T}$ is said to be neutral. A morphism of Tate triples $(\underline{C}_1, w_1, T_1) \to (\underline{C}_2, w_2, T_2)$ is a tensor functor $\eta : \underline{C}_1 \to \underline{C}_2$ preserving the gradings together with an isomorphism $\eta(T_1) \xrightarrow{\approx} T_2$.

Example 5.2 (a). The triple $(\underline{\mathrm{Hod}}_{\mathbb{R}}, w, \mathbb{R}(1))$ in which $\underline{\mathrm{Hod}}_{\mathbb{R}}$ is the category of real Hodge structures (see (2.31)), w is the weight grading on $\underline{\mathrm{Hod}}_{\mathbb{R}}$, and $\mathbb{R}(1)$ is the unique real Hodge structure with weight -2 and underlying vector space $2\pi i\mathbb{R}$, is a neutral Tate triple.

(b) The category of $\mathbb{Z}$-graded vector spaces over $\mathbb{Q}$, together with the object $T = \mathbb{Q}_B(1)$ (see I.1), forms a Tate triple $\underline{T}_B$; the category of $\mathbb{Z}$-graded vector spaces over $\mathbb{Q}_\ell$, together with the object $T = \mathbb{Q}_\ell(1)$, forms a Tate triple $\underline{T}_\ell$; the category of $\mathbb{Z}$-graded vector spaces over k, together with the object $T = \mathbb{Q}_{DR}(1)$, forms a Tate triple $\underline{T}_{DR}$.

Example 5.3. Let $\underline{V}$ be the category of $\mathbb{Z}$-graded complex vector spaces V with a semi-linear automorphism a such that $a^2v = (-1)^n v$ if $v \in V^n$. With the obvious tensor structure, $\underline{V}$ becomes a Tannakian category over $\mathbb{R}$, and $\omega : (V,a) \longmapsto V$ is a fibre functor with values in $\mathbb{C}$. Clearly $\mathbb{G}_m = \underline{\mathrm{Aut}}^{\otimes}(\omega)$, and $\underline{V}$ corresponds (as in (4.3a)) to the pair $(g \longmapsto \bar{g}, -1)$. Let $w : \mathbb{G}_m \to \mathbb{G}_m$ be the identity map, and let $T = (V,a)$ where V is $\mathbb{C}$ regarded as a homogeneous vector space of weight -2 and a is $z \longmapsto \bar{z}$. Then $(\underline{V},w,T)$ is a (non-neutral) Tate triple over $\mathbb{R}$.

Example 5.4. Let G be an affine group scheme over k and let $w : \mathbb{G}_m \to G$ be a central homomorphism and $t : G \to \mathbb{G}_m$ a homomorphism such that $t \circ w = -2$ $(\overset{\mathrm{df}}{=} s \mapsto s^{-2})$. Let T be the representation of G on k such that g acts as multiplication by $t(g)$. Then $(\underline{\mathrm{Rep}}_k(G),w,T)$ is a neutral Tate triple over k.

The following proposition is obvious.

Proposition 5.5. Let $\underline{T} = (\underline{C},w,T)$ be a Tate triple over k, and let ω be a fibre functor on $\underline{T}$ with values in k. Let $G = \underline{\mathrm{Aut}}^{\otimes}(\omega)$, so that w is a homomorphism $\mathbb{G}_m \to Z(G) \subset G$. There is a homomorphism $t : G \to \mathbb{G}_m$ such that g acts on T as multiplication by $t(g)$, and $t \circ w = -2$. The equivalence $\underline{C} \tilde{\rightarrow} \underline{\mathrm{Rep}}_k(G)$ carries w and T into the weight grading and Tate object defined by t and w .

More generally, a Tate triple $\underline{T}$ defines a band B , a homomorphism $w : \mathbb{G}_m \rightarrow Z$ into the centre Z of B , and a homomorphism $t : B \rightarrow \mathbb{G}_m$ such that $t \circ w = -2$. We say that $\underline{T}$ is __bound__ by (B,w,t).

Let G, w, and t be as in (5.4). Let $G_o = \mathrm{Ker}(t : G \rightarrow \mathbb{G}_m)$, and let $\varepsilon : \mu_2 \rightarrow G_o$ be the restriction of w to μ_2 ; we often identify ε with $\varepsilon(-1) = w(-1) \in Z(G_o)(k)$. Note that ε defines a $\mathbb{Z}/2\mathbb{Z}$-grading on $\underline{C}_o = \underline{\mathrm{Rep}}_k(G_o)$. The inclusion $G_o \hookrightarrow G$ defines a tensor functor $Q : \underline{C} \rightarrow \underline{C}_o$ with the following properties:

(5.6a) if X is homogeneous of weight n , then $Q(X)$ is homogeneous of weight n (mod 2),

(5.6b) $Q(T) = 1$;

(5.6c) if X and Y are homogeneous of the same weight, then

$$\mathrm{Hom}(X,Y) \stackrel{\approx}{\rightarrow} \mathrm{Hom}(Q(X), Q(Y)) ;$$

(5.6d) if X and Y are homogeneous with weights m and n respectively and $Q(X) \approx Q(Y)$, then $m-n$ is an even integer $2k$ and $X(k) \approx Y$;

(5.6e) Q is essentially surjective.

The first four of these statements are obvious. For the last, note that $G = \mathbb{G}_m \times G/\mu_2$, and so we only have to show that any representation of μ_2 extends to a representation of $\mathbb{G}_m$, but this is obvious.

Remark 5.7 (a) The identity component of G_o is reductive if and only if the identity component of G is reductive; if G_o is connected, so also is G, but the converse statement is false (e.g., $G_o = \mu_2$, $G = \mathbb{G}_m$).

(b) It is possible to reconstruct $(\underline{C},w,T)$ from $(\underline{C}_o,\varepsilon)$ — the following diagram makes it clear how to reconstruct (G,w,t) from (G_o,ε):

$$\begin{array}{ccccccccc} 1 & \longrightarrow & \mu_2 & \longrightarrow & \mathbb{G}_m & \xrightarrow{-2} & \mathbb{G}_m & \longrightarrow & 1 \\ & & \downarrow\varepsilon & & \downarrow w & & \| & & \\ 1 & \longrightarrow & G_o & \longrightarrow & G & \xrightarrow{t} & \mathbb{G}_m & \longrightarrow & 1 \end{array}$$

Proposition 5.8. Let $\underline{T} = (\underline{C},w,T)$ be a Tate triple over k with $\underline{C}$ algebraic. There exists a Tannakian category $\underline{C}_o$ over k, an element ε in $\underline{\mathrm{Aut}}^{\otimes}(\mathrm{id}_{\underline{C}_o})$ with $\varepsilon^2 = 1$, and a functor $Q : \underline{C} \to \underline{C}_o$ having the properties (5.6).

Proof: For any fibre functor ω on $\underline{C}$ with values in an algebra R, $\underline{\mathrm{Isom}}(R,\omega(T))$ regarded as a sheaf on spec R is a torsor for $\mathbb{G}_m$. This association gives rise to a morphism of gerbs $\underline{G} \overset{\mathrm{df}}{=} \mathrm{FIB}(\underline{C}) \xrightarrow{t} \mathrm{TORS}(\mathbb{G}_m)$, and we define $\underline{G}_o$ to be the kernel of t; thus $\underline{G}_o$ is the gerb of pairs (Q,ξ) where $Q \in \mathrm{ob}(\underline{G})$ and ξ is an isomorphism $t(Q) \xrightarrow{\approx} \mathbb{G}_{m,X}$, i.e., $\underline{G}_o$ is the gerb of fibre functors on $\underline{T}$. Let $\underline{C}_o$ be the category $\underline{\mathrm{Rep}}_k(G_o)$ which (see (3.14)) is Tannakian. If $Z = \underline{\mathrm{Aut}}^{\otimes}(\mathrm{id}_{\underline{C}})$ and $Z_o = \underline{\mathrm{Aut}}^{\otimes}(\mathrm{id}_{\underline{C}_o})$, then the homomorphism $Z \to \underline{\mathrm{Aut}}(T) = \mathbb{G}_m$, $\alpha \longmapsto \alpha_T$, determined by t has kernel Z_o, and the composite $t \circ w = -2$; we let $\varepsilon = w(-1) \in Z_o$.

There is an obvious (restriction) functor $Q : \underline{C} \to \underline{C}_o$. In showing that Q had the properties (5.6), we can make a finite field extension $k \to k'$. We can therefore assume that $\underline{T}$ is neutral, but this case is covered by (5.5) and (5.6).

Example 5.9. Let $(\underline{V},w,T)$ be the Tate triple defined in (5.3); then $(\underline{V}_o,\varepsilon)$ is the pair defined in the paragraph preceding (4.29).

Example 5.10. Let $\underline{T} = (\underline{C},w,T)$ be a Tate triple over $\mathbb{R}$, and let ω be a fibre functor on $\underline{T}$ with values in $\mathbb{C}$. On combining (4.3) with (5.5) we find that $(\underline{T},\omega)$ corresponds to a quintuple (G,σ,c,w,t) in which

(a) G is an affine group scheme over $\mathbb{C}$;

(b) (σ,c) satisfies (4.2.1);

(c) $w : \mathbb{G}_m \to G$ is a central homomorphism; that the grading is defined over $\mathbb{R}$ means that w is defined over $\mathbb{R}$, i.e., $\sigma(w(g)) = w(\bar{g})$.

(d) $t : G \to \mathbb{G}_m$ is such that $t \circ w = -2$; that T is defined over $\mathbb{R}$ means that $t(\sigma(g)) = \overline{t(g)}$ and there exists $a \in \mathbb{G}_m(\mathbb{C})$ such that $t(c) = \sigma(a)a$.

Let $G_o = \mathrm{Ker}(t)$, and let $m \in G(\mathbb{C})$ be such that $t(m) = a^{-1}$. After replacing (σ,c) with $(\sigma \circ \underline{\mathrm{ad}}\,m, \sigma(m)cm)$ we find that the new c is in G_o. The pair $(\underline{C}_o,\omega|\underline{C}_o)$ corresponds to $(G_o,\sigma|G_o,c)$.

Remark 5.11. As in the neutral case, $\underline{T}$ can be reconstructed from

$(\underline{C}_o, \varepsilon)$. This can be proved by substituting bands for group schemes in the argument used in the neutral case (Saavedra [1, V. 3.14.1]), or by using descent theory to deduce it from the neutral case.

There is a stronger result: $\underline{T} \mapsto (\underline{C}_o, \varepsilon)$ defines an equivalence between the 2-category of Tate triples and that of $\mathbb{Z}/2\mathbb{Z}$-graded Tannakian categories.

Graded polarizations

For the remainder of this section, $\underline{T} = (\underline{C}, w, T)$ will be a Tate triple over $\mathbb{R}$ with $\underline{C}$ algebraic. We use the notations of §4; in particular $\underline{C}' = \underline{C}_{(\mathbb{C})}$. Let U be an invertible object of $\underline{C}'$ that is defined over $\mathbb{R}$, i.e., U is provided with an identification $U \overset{\approx}{\to} \bar{U}$; then in the definitions and results in §4 concerning sesquilinear forms and Weil forms, it is possible to replace $\underline{1}$ with U .

For each $X \in ob(\underline{C}')$ that is homogeneous of degree n , let $\pi(X)$ be an equivalence class of Weil forms $X \otimes \bar{X} \to \underline{1}(-n)$ of parity $(-1)^n$; we say that π is a (graded) <u>polarization</u> on $\underline{T}$ if

(5.12a) for all X , $\bar{\phi} \in \pi(X)$ whenever $\phi \in \pi(\bar{X})$;

(5.12b) for all X and Y that are homogeneous of the same degree, $\phi \oplus \psi \in \pi(X \oplus Y)$ whenever $\phi \in \pi(X)$ and $\psi \in \pi(Y)$;

(5.12c) for all homogeneous X and Y , $\phi \otimes \psi \in \pi(X \otimes Y)$ whenever $\phi \in \pi(X)$ and $\psi \in \pi(Y)$;

(5.12d) the map $T \otimes \bar{T} \to T^{\otimes 2} = \underline{1}(2)$, defined by $T \overset{\approx}{\to} \bar{T}$, is in $\pi(T)$.

Proposition 5.13. Let $(\underline{C}_o, \varepsilon)$ be the pair associated with $\underline{T}$ by (5.8). There is a canonical bijection

$$Q : \mathrm{Pol}(\underline{T}) \to \mathrm{Pol}_{\varepsilon}(\underline{C}_o)$$

from the set of polarizations on $\underline{T}$ to the set of polarizations on $\underline{C}_o$ with parity ε.

Proof: For any $X \in \mathrm{ob}(\underline{C}')$ that is homogeneous degree n, (5.6b) and (5.6c) give an isomorphism

$$Q : \mathrm{Hom}\,(X \otimes \bar{X}, \underline{1}(-n)) \overset{\approx}{\to} \mathrm{Hom}(Q(X) \otimes \overline{Q(X)}, \underline{1}) .$$

We define $Q\pi$ to be the polarization such that, for any homogeneous X, $Q\pi(QX) = \{Q\phi \mid \phi \in \pi(X)\}$. It is clear that $\pi \mapsto Q\pi$ is a bijection.

Corollary 5.14. The Tate triple $\underline{T}$ is polarizable if and only if $\underline{C}_o$ has a polarization π with parity $\varepsilon(\pi) \equiv \varepsilon \pmod{Z_o(\mathbb{R})^2}$.

Proof: See (4.20e).

Corollary 5.15. The map $(z,\pi) \mapsto z\pi : {}_2Z_o(\mathbb{R}) \times \mathrm{Pol}(\underline{T}) \to \mathrm{Pol}(\underline{T})$ (where $\phi(x,y) \in z\pi(X) \iff \phi(x,zy) \in \pi(X)$) makes $\mathrm{Pol}(\underline{T})$ into a pseudo-torsor for ${}_2Z_o(\mathbb{R})$.

Proof: See (4.20d).

Theorem 5.16. Let π be a polarization on $\underline{T}$, and let ω be a fibre functor on $\underline{C}'$ with values in $\mathbb{C}$. Let (G,w,t) correspond to $(\underline{T}_{\mathbb{C}},\omega)$. For any real form K of G such that $K_o = \mathrm{Ker}\,(t)$ is compact, the pair (σ_K,ε) where $\varepsilon = w(-1)$ satisfies (4.2.1), and ω defines an equivalence between $\underline{T}$ and the Tate triple defined by $(G,\sigma_K,\varepsilon,w,t)$. For any simple X in $\underline{C}'$ $\{\omega(\phi)'\,|\,\phi \in \pi(X)\}$ is the set of K_o-invariant positive-definite Hermitian forms on $\omega(X)$.

Proof: See (4.16).

Remark 5.17. From (4.17) one can deduce the following: A triple (B,w,t), where B is an affine algebraic band over $\mathbb{R}$ and $t\circ w = -2$, bounds a polarizable Tate triple if and only if $B_o = \mathrm{Ker}(t{:}B \to \mathbb{G}_m)$ is the band defined by a compact real algebraic group; when this condition holds, the polarizable Tate triple bound by (B,w,t) is unique up to a tensor equivalence preserving the action of B and the polarization, and the equivalence is unique up isomorphism. The Tate triple is neutral if and only if $\varepsilon = w(-1) \in Z_o(\mathbb{R})^2$.

Let (G,w,t) be a triple as in (5.4) defined over $\mathbb{R}$, and let $G_o = \mathrm{Ker}(t)$ and $\varepsilon = w(-1)$. A Hodge element $C \in G_o(\mathbb{R})$ is said to be a Hodge element for (G,w,t) if $C^2 = \varepsilon$. A G-invariant sesquilinear form $\psi : V \times V \to \underline{1}(-n)$ on a homogeneous complex representation V of G of degree n is said to be a C-polarization if

$$\psi^C(x,y) \overset{\mathrm{df}}{=\!=} \psi(x,Cy)$$

is a positive-definite Hermitian form on V. When C is a Hodge element for (G,w,t) there is a polarization π_C on the Tate triple defined by (G,w,t) for which the positive forms are exactly the C-polarizations.

Proposition 5.18. Every polarization on the Tate triple defined by (G,w,t) is of the form π_C for some Hodge element C.
Proof: See (4.22) and (4.23).

Proposition 5.19. Assume that $w(-1)=1$. Then there is a unique (up to isomorphism) fibre functor ω on $\underline{T}$ with values in $\mathbb{R}$ transforming positive bilinear forms for π into positive-definite symmetric bilinear forms.
Proof: See (4.27).

Proposition 5.20. Let $(\underline{V},w,T)$ be the Tate triple defined in (5.3), and let π_{can} be the polarization on $\underline{V}$ such that, if $(V,a) \in ob(\underline{V})$ is homogeneous, then $\pi(V,a)$ comprises the $(-1)^{deg(V)}$-symmetric positive-definite forms on V. If $w(-1) \neq 1$ for $\underline{T}$ and π is a polarization on $\underline{T}$, then there exists a unique (up to isomorphism) exact faithful functor $\omega : \underline{C} \to \underline{V}$ preserving the Tate-triple structures and carrying π into π_{can}.
Proof: Combine (4.29) and (5.9).

Example 5.21. Let $\underline{T}$ be the Tate triple $(\underline{Hod}_{\mathbb{R}},w,\mathbb{R}(1))$ defined in (5.2). A polarization on a real Hodge structure V of weight n is a bilinear form $\phi : V \times V \to \mathbb{R}(-n)$ such that the real-valued form $(x,y) \mapsto (2\pi i)^n \phi(x,Cy)$, where C denotes the element $i \in \mathbb{S}(\mathbb{R}) = \mathbb{C}^\times$, is positive-definite and symmetric. These polarizations are the positive (bilinear) forms for a polarization

π on the Tate triple $\underline{T}$. The functor $\omega : \underline{\mathrm{Hod}}_{\mathbb{R}} \to \underline{V}$ provided by the last proposition is $V \longmapsto (V \otimes \mathbb{C}, v \mapsto C\bar{v})$. (Note that $(\underline{\mathrm{Hod}}_{\mathbb{R}}, w, \mathbb{R}(1))$ is not quite the Tate triple associated, as in (5.4), with $(\mathbb{S}, w, t)$ because we have chosen a different Tate objects; this difference explains the occurrence of $(2\pi i)^n$ in the above formula; π is essentially the polarization defined by the canonical Hodge element C .)

Filtered Tannakian Categories

For this topic we refer the reader to Saavedra [1, IV.2].

§6. Motives for absolute Hodge cycles

Throughout this section, k will denote a field of characteristic zero with algebraic closure $\bar{k}$ and Galois group $\Gamma = \mathrm{Gal}(\bar{k}/k)$. All varieties will be projective and smooth, and, for X a variety (or motive) over k , $\bar{X}$ denotes $X \otimes_k \bar{k}$. We shall freely use the notations and results of Article I; for example, if $k = \mathbb{C}$ then $H_B(X)$ denotes the graded vector space $\oplus H_B^i(X)$.

Complements on absolute Hodge cycles

For X a variety over k , $C_{AH}^p(X)$ denotes the rational vector space of absolute Hodge cycles on X (see I.2). When X has pure dimension n , we write

$$\mathrm{Mor}^p_{AH}(X,Y) = C^{n+p}_{AH}(X \times Y) .$$

Then $\mathrm{Mor}^p_{AH}(X,Y) \subset H^{2n+2p}(X \times Y)(p+n) = \bigoplus_{r+s=2n+2p} H^r(X) \otimes H^s(Y)(p+n)$

$$= \bigoplus_{s=r+2p} H^r(X)^\vee \otimes H^s(Y)(p)$$

$$= \bigoplus_r \mathrm{Hom}(H^r(X), H^{r+2p}(Y)(p))$$

The next proposition is obvious from this and the definition of an absolute Hodge cycle.

<u>Proposition 6.1.</u> An element f of $\mathrm{Mor}^p_{AH}(X,Y)$ gives rise to

(a) for each prime ℓ, a homomorphism $f_\ell : H_\ell(\bar{X}) \to H_\ell(\bar{Y})(p)$ of graded vector spaces (meaning that f_ℓ is a family of homomorphisms $f^r_\ell : H^r_\ell(\bar{X}) \to H^{r+2p}_\ell(\bar{Y})(p))$;

(b) a homomorphism $f_{DR} : H_{DR}(X) \to H_{DR}(Y)(p)$ of graded vector spaces;

(c) for each $\sigma : k \hookrightarrow \mathbb{C}$, a homomorphism $f_\sigma : H_\sigma(X) \to H_\sigma(Y)(p)$ of graded vector spaces.

These maps satisfy the following conditions:

(d) for all $\gamma \in \Gamma$ and primes ℓ, $\gamma(f_\ell) = f_\ell$;

(e) f_{DR} is compatible with the Hodge filtrations on each homogeneous factor;

(f) for each $\sigma : k \hookrightarrow \mathbb{C}$, the maps f_σ, f_ℓ, and f_{DR} correspond under the comparison isomorphisms (I.1).

Conversely, assume that k is embeddable in $\mathbb{C}$; then any family of maps f_ℓ, f_{DR} as in (a), (b) arises from $f \in \mathrm{Mor}^p_{AH}(X,Y)$ provided (f_ℓ) and f_{DR} satisfy (d) and (e) respectively and for every $\sigma : k \hookrightarrow \mathbb{C}$

there exists an f_σ such that (f_ℓ), f_{DR}, and f_σ satisfy condition (f) ; moreover, f is unique.

Similarly, a $\psi \in C_{AH}^{2n-r}(X \times X)$ gives rise to pairings

$$\psi^s : H^s(X) \times H^{2r-s}(x) \to \mathbb{Q}(-r) .$$

Proposition 6.2. On any variety X (of dimension n) there exists a $\psi \in C_{AH}^{2n-r}(X \times X)$ such that, for every $\sigma : k \hookrightarrow \mathbb{C}$,

$$\psi_\sigma^r : H_\sigma^r(X,\mathbb{R}) \times H_\sigma^r(X,\mathbb{R}) \to \mathbb{R}(-r)$$

is a polarization of real Hodge structures (in the sense of (5.21)).

Proof: Choose a projective embedding of X , and let L be a hyperplane section of X . Let ℓ be the class of L in $H^2(X)(1)$, and write ℓ also for the map $H(X) \to H(X)(1)$ sending a class to its cup-product with ℓ . Assume X is connected and define the primitive cohomology of X by

$$H^r(X)_{prim} = \mathrm{Ker}(\ell^{n-r+1} : H^r(X) \to H^{2n-r+2}(X)(n-r+1)) .$$

The hard Lefschetz theorem states that

$$\ell^{n-r} : H^r(X) \to H^{2n-r}(X)(n-r)$$

is an isomorphism for $r \leq n$; it implies that

$$H^r(X) = \bigoplus_{s\geq r-n,\, s\geq o} \ell^s H^{r-2s}(X)(-s)_{prim} .$$

Thus any $x \in H^r(X)$ can be written uniquely, $x = \Sigma \ell^s(x_s)$, with $x_s \in H^{r-2s}(X)(-s)_{prim}$; define

$$*x = \sum (-1)^{(r-2s)(r-2s+1)/2} \ell^{n-r+s} x_s \in H^{2n-r}(X)(n-r) .$$

Then $x \mapsto *x : H^r(X) \to H^{2n-r}(X)(n-r)$ is a well-defined map for each of the three cohomology theories, ℓ-adic, de Rham, and Betti. Proposition 6.1 shows that it is defined by an absolute Hodge cycle (rather, the map $H(X) \to H(X)(n-r)$ that is $x \mapsto *x$ on H^r and zero otherwise is so defined). We take ψ^r to be

$$H^r(X)\otimes H^r(X) \xrightarrow{id\otimes *} H^r(X)\otimes H^{2n-r}(X)(n-r) \to H^{2n}(X)(n-r) \xrightarrow{Tr} \mathbb{Q}(-r) .$$

Clearly it is defined by an absolute Hodge cycle, and the Hodge-Riemann bilinear relations (see Wells [1, 5.3]) show that it defines a polarization on the real Hodge structure $H^r_\sigma(X,\mathbb{R})$ for each $\sigma : k \hookrightarrow \mathbb{C}$.

<u>Proposition</u> 6.3. For any $u \in Mor^o_{AH}(Y,X)$ there exists a unique $u' \in Mor^o_{AH}(X,Y)$ such that

$$\psi_X(uy,x) = \psi_Y(y,u'x) \ , \ x \in H^r(X),\ y \in H^r(Y)$$

where ψ_X and ψ_Y are the forms defined in (6.2); moreover,

$$\mathrm{Tr}(uu') = \mathrm{Tr}(u'u) \in \mathbb{Q}$$

$$\mathrm{Tr}(uu') > 0 \quad \text{if} \quad u \neq 0$$

Proof: The first part is obvious; the last assertion follows from the fact that the ψ_X and ψ_Y are positive forms for a polarization in $\underline{\mathrm{Hod}}_{\mathbb{R}}$.

Note that the proposition shows that $\mathrm{Mor}^{\circ}_{AH}(X,X)$ is a semisimple $\mathbb{Q}$-algebra (see 4.5).

Construction of the category of motives

Let $\underline{V}_k$ be the category of (smooth projective, not necessarily connected) varieties over k . The category $\underline{CV}_k$ is defined to have as objects symbols $h(X)$, one for each $X \in ob(\underline{V}_k)$, and as morphisms $\mathrm{Hom}(h(X),h(Y)) = \mathrm{Mor}^{\circ}_{AH}(X,Y)$. There is a map $\mathrm{Hom}(Y,X) \to \mathrm{Hom}(h(X),h(Y))$ sending a homomorphism to the cohomology class of its graph which makes h into contravariant functor $\underline{V}_k \to \underline{CV}_k$.

Clearly $\underline{CV}_k$ is a $\mathbb{Q}$-linear category, and $h(X \coprod Y) = h(X) \oplus h(Y)$. There is a $\mathbb{Q}$-linear tensor structure on $\underline{CV}_k$ for which $h(X) \otimes h(Y) = h(X \times Y)$, the associativity constraint is induced by $(X \times Y) \times Z \to X \times (Y \times Z)$, the commutativity constraint is induced by $Y \times X \to X \times Y$, and the identity object is h(point).

The false category of effective (or positive) motives $\dot{M}^+_k$ is defined to be the pseudo-abelian (Karoubian) envelope of $\underline{CV}_k$. Thus an object of $\dot{\underline{M}}^+_k$ is a pair (M,p) with $M \in \underline{CV}_k$ and p an idempotent in $\mathrm{End}(M)$, and

$$\mathrm{Hom}((M,p),\ (N,q)) = \{f : M \to N \mid f\circ p = q\circ f\}/\sim \qquad (6.3.1)$$

where $f \sim 0$ if $f\circ p = 0 = q\circ f$. The rule

$$(M,p) \otimes (N,q) = (M\otimes N,\ p\otimes q)$$

defines a $\mathbb{Q}$-linear tensor structure on $\dot{\underline{M}}_k^+$, and
$M \to (M,\mathrm{id}) : \underline{CV}_k \to \dot{\underline{M}}_k^+$ is a fully faithful functor which we use to identify $\underline{CV}_k$ with a subcategory of $\dot{\underline{M}}_k^+$. With this identification, (M,p) becomes the image of $p : M \to M$. The category $\dot{\underline{M}}_k^+$ is pseudo-abelian: any decomposition of id_M into a sum of pairwise orthogonal idempotents

$$\mathrm{id}_M = e_1 + \cdots + e_m$$

corresponds to a decomposition

$$M = M_1 \oplus \cdots \oplus M_m$$

with $e_i | M_i = \mathrm{id}_{M_i}$. The functor $\underline{CV}_k \to \dot{\underline{M}}_k^+$ is universal for functors from $\underline{CV}_k$ into pseudo-abelian categories.

For any $X \in \mathrm{ob}(\underline{V}_k)$, the projection maps $p^r : H(X) \to H^r(X)$ define an element of $\mathrm{Mor}^0_{AH}(X,X) = \mathrm{End}(h(X))$. Corresponding to the decomposition

$$\mathrm{id}_{h(X)} = p^0 + p^1 + p^2 + \cdots$$

there is a decomposition (in $\dot{\underline{M}}_k^+$)

$$h(X) = h^\circ(X) + h^2(X) + h^2(X) + \cdots .$$

This grading of objects of $\underline{CV}_k$ extends in an obvious way to objects of $\dot{\underline{M}}_k^+$, and the Künneth formulas show that these gradings are compatible with tensor products (and therefore satisfy (5.1a)).

Let L be the Lefschetz motive $h^2(\mathbb{P}^1)$. With the notations of (I.1), $H(L) = \mathbb{Q}(-1)$, whence it follows that $\mathrm{Hom}(M,N) \tilde{\to} \mathrm{Hom}(M\otimes L, N\otimes L)$ for any effective motives M and N. This means that $V \mapsto V \otimes L$ is a fully faithful functor and allows us to invert L. The false category $\dot{M}_k$ of <u>motives</u> is defined as follows:

(6.4a) an object of $\dot{\underline{M}}_k$ is a pair (M,m) with $M \in \mathrm{ob}(\dot{\underline{M}}_k^+)$ and $m \in \mathbb{Z}$;

(6.4b) $\mathrm{Hom}((M,m), (N,n)) = \mathrm{Hom}(M \otimes L^{N-m}, N \otimes L^{N-n}), N \geq m,n$ (for different N, these groups are canonically isomorphic);

(6.4c) composition of morphisms is induced by that in $\dot{\underline{M}}_k^+$.

This category of motives is $\mathbb{Q}$-linear and pseudo-abelian and has a tensor structure

$$(M,m) \otimes (N,n) = (M\otimes N, m+n)$$

and grading

$$(M,m)^r = M^{r-2m}$$

We identify $\dot{M}_k^+$ with a subcategory of $\dot{\underline{M}}_k$ by means of $M \mapsto (M,0)$. The Tate motive T is $L^{-1} = (\underline{1},1)$. We abbreviate $M \otimes T^{\otimes m} = (M,m)$ by $M(m)$.

We shall see shortly that $\dot{\underline{M}}_k$ is a rigid abelian tensor category, and $\mathrm{End}(\underline{1}) = \mathbb{Q}$. It is not however a Tannakian category because, for $X \in \mathrm{ob}(\underline{V}_k)$, $\mathrm{rk}(h(X))$ is the Euler-Poincaré characteristic, $\sum (-1)^r \dim H^r(X)$, of X, which is not necessarily positive. To remedy this we modify the commutativity constraint as follows: let

$$\dot{\psi} : M \otimes N \xrightarrow{\approx} N \otimes M,\ \dot{\psi} = \oplus\, \dot{\psi}^{r,s},\ \dot{\psi}^{r,s} : M^r \otimes N^s \xrightarrow{\approx} N^s \otimes M^r$$

by the commutativity constraint on $\dot{\underline{M}}_k$; define a new commutativity constraint by

$$\psi : M \otimes N \xrightarrow{\approx} N \otimes M\ ,\ \psi = \oplus\, \psi^{r,s},\ \psi^{r,s} = (-1)^{rs}\dot{\psi}^{r,s} \tag{6.4.1}$$

Then $\dot{\underline{M}}_k$, with $\dot{\psi}$ replaced by ψ, is the true category $\underline{M}_k$ of

Proposition 6.5. The category $\underline{M}_k$ is a semisimple Tannakian category over $\mathbb{Q}$.

Proof: We first need a lemma.

Lemma 6.6. Let C be a $\mathbb{Q}$-linear pseudo-abelian cateogry, and let $\omega : \underline{C} \to \underline{\mathrm{Vec}}_{\mathbb{Q}}$ be a faithful $\mathbb{Q}$-linear functor. If every indecomposable object ov $\underline{C}$ is simple, then $\underline{C}$ is a semisimple abelian category and ω is exact.

Proof: The existence of ω shows that each object of $\underline{C}$ has finite length and hence is a finite direct sum of simple objects. For any map $f : X \to Y$, $\mathrm{Ker}(f)$ is the largest subobject of X on which f is zero, and $\mathrm{Coker}(f)$ is the largest quotient of

Y such that the composite $X \to Y \to \mathrm{Coker}(f)$ is zero. The rest of the proof is easy.

Proof of (6.5): We can replace $\underline{M}_k$ with the tensor subcategory generated by a finite number of objects, and consequently we can assume that there exists an embedding $\sigma : k \hookrightarrow \mathbb{C}$. The functor $H_\sigma : \underline{M}_k \to \underline{\mathrm{Vec}}_{\mathbb{Q}}$ is faithful and $\mathbb{Q}$-linear. Let M be an indecomposable motive, and let $i : N \hookrightarrow M$ be a nonzero simple subobject of M. Clearly M is homogeneous, and after tensoring it with a power of T we can assume that N and M are effective, and therefore

$$M \oplus M' = h^r(X) \quad \text{with} \quad X \in \underline{V}_k \quad \text{and}$$
$$N \oplus N' = h^r(Y) \quad \text{with} \quad Y \in \underline{V}_k .$$

Let $u : h^r(Y) \to h^r(X)$ be the morphism $\begin{pmatrix} i & 0 \\ 0 & 0 \end{pmatrix}$ and let $u' = \begin{pmatrix} a & b \\ c & d \end{pmatrix}$ be its transpose (see 6.3). As $\mathrm{Tr}(u'u) > 0$, and $\mathrm{Tr}(u'u) = \mathrm{Tr}(ai)$, we see that $ai \neq 0$. It is therefore an automorphism of N, and $(ai)^{-1} a : M \to N$ projects M on N. As M is indecomposable, this shows that $M = N$, and M is simple. The lemma can therefore be applied, and shows that $\underline{M}_k$ is a semisimple $\mathbb{Q}$-linear abelian tensor category. It remains to show that it is rigid. Let X and Y be varieties of pure dimension m and n respectively. Then

$$\begin{aligned}\mathrm{Hom}(h(X),h(Y)) = C^m_{AH}(X\times Y) = C^m_{AH}(Y\times X) &= \mathrm{Hom}(h(Y),h(X)(m-n)) \\ &= \mathrm{Hom}(h(Y)(n),h(X)(m)) .\end{aligned}$$

The functor $h(X) \mapsto h(X)^{\vee} \overset{\text{df}}{=} h(X)(m)$ extends to a fully faithful contravariant functor $M \mapsto M^{\vee} : \underline{M}_k \to \underline{M}_k$, and we set $\underline{\mathrm{Hom}}(M,N) = M^{\vee} \otimes N$. It is straightforward now to check that $\underline{M}_k$ is Tannakian (especially if one applies (1.20)).

The following theorem summarizes what we have (essentially) shown about $\underline{M}_k$.

Theorem 6.7. (a) Let w be the grading on $\underline{M}_k$; then $(\underline{M}_k, w, T)$ is a Tate triple over $\mathbb{Q}$.

(b) There is a contravariant functor $h : \underline{V}_k \to \underline{M}_k$; every effective motive is the image $(h(X),p)$ of an idempotent $p \in \mathrm{End}(h(X))$ for some $X \in \mathrm{ob}(\underline{V}_k)$; every motive is of the form $M(n)$ for some effective M and some $n \in \mathbb{Z}$.

(c) For all varieties X,Y with X of pure dimension m, $C_{AH}^{m+s-r}(X \times Y) = \mathrm{Hom}(h(X)(r), h(Y)(s))$; in particular, $C_{AH}^{m}(X \times Y) = \mathrm{Hom}(h(X), h(Y))$; morphisms of motives can be expressed in terms of absolute Hodge cycles on varieties by means of (6.3.1) and (6.4b).

(d) The constraints on $\underline{M}_k$ have an obvious definition, except that the obvious commutativity constraint has to be modified by (6.4.1).

(e) For varieties X and Y,

$$h(X \amalg Y) = h(X) \oplus h(Y)$$
$$h(X \times Y) = h(X) \otimes h(Y)$$
$$h(X)^{\vee} = h(X)(m), \text{ if } X \text{ is of pure dimension } m.$$

(f) The functors H_{ℓ}, H_{DR}, and H_{σ} define fibre functors on $\underline{M}_k$; these fibre functors define morphisms of Tate triples, $\underline{M}_k \to \underline{T}_{\ell}, T_{DR}, T_B$

(see (5.2b)); in particular $H(T) = \mathbb{Q}(1)$.

(g) When k is embeddable in $\mathbb{C}$, $\mathrm{Hom}(M,N)$ is the vector space of families of maps

$$f_\ell : H_\ell(\bar{M}) \to H_\ell(\bar{N})$$

$$f_{DR} : H_{DR}(M) \to H_{DR}(N)$$

such that f_{DR} preserves the Hodge filtration, $\gamma(f_\ell) = f_\ell$ for all $\gamma \in \Gamma$, and for any $\sigma : k \hookrightarrow \mathbb{C}$ there exists a map $f_\sigma : H_\sigma(M) \to H_\sigma(N)$ agreeing with f_ℓ and f_{DR} under the comparison isomorphisms.

(h) The category $\underline{M}_k$ is semisimple.

(i) There exists a polarization on $\underline{M}_k$ for which $\pi(h^r(X))$ consists of the forms defined in (6.2).

<u>Some calculations</u>

According to (6.7g), to define a map $M \to N$ of motives it suffices to give a procedure for defining a map of cohomology groups $H(M) \to H(N)$ that works (compatibly) for all three theories: Betti, deRham, and ℓ-adic. The map will be an isomorphism if its realization in one theory is an isomorphism.

Let G be a finite group acting on a variety. The group algebra $\mathbb{Q}[G]$ acts on $h(X)$, and we define $h(X)^G$ to be the motive $(h(X),p)$ where p is the idempotent $(\mathrm{ord}\ G)^{-1}\Sigma g$. Note that $H(h(X)^G) = H(X)^G$.

Proposition 6.8. Assume that the finite group G acts freely on X, so that X/G is also smooth; then $h(X/G) = h(X)^G$.

Proof: Since cohomology is functorial, there exists a map $H(X/G) \to H(X)$ whose image lies in $H(X)^G = H(h(X)^G)$. The Hochschild-Serre spectral sequence $H^r(G,H^s(X)) \Rightarrow H^{r+s}(X/G)$ shows that the map $H(X/G) \to H(X)^G$ is an isomorphism for, say the ℓ-adic cohomology, because $H^r(G,V) = 0$, $r>0$, if V is a vector space over a field of characteristic zero.

Remark 6.9. More generally, if $f : Y \to X$ is a map of finite (generic) degree n between connected varieties of the same dimension, then the composite $H(X) \xrightarrow{f^*} H(Y) \xrightarrow{f_*} H(X)$ is multiplication by n; there therefore exist maps $h(X) \to h(Y) \to h(X)$ with composite n, and $h(X)$ is a direct summand of $h(Y)$.

Proposition 6.10. Let E be a vector bundle of rank $m+1$ over a variety X and let $p : \mathbb{P}(E) \to X$ be the associated projective bundle; then $h(\mathbb{P}(E)) = h(X) \oplus h(X)(-1) \oplus \cdots \oplus h(X)(-m)$.

Proof: Let γ be the class in $H^2(\mathbb{P}(E))(1)$ of the canonical line bundle on $\mathbb{P}(E)$, and let $p^* : H(X) \to H(\mathbb{P}(E))$ be the map induced by p. The map

$$(c_o,\dots,c_m) \mapsto \sum p^*(c_i)\,\gamma^i : H(X) \oplus \cdots \oplus H(X)(-m) \to H(\mathbb{P}(E))$$

has the requisite properties.

Proposition 6.11. Let Y be a smooth closed subvariety of codimension c in the variety X, and let X' be the variety obtained from X

by blowing up Y ; then there is an exact sequence

$$0 \to h(Y)(-c) \to h(X) \oplus h(Y')(-1) \to h(X') \to 0$$

where Y' is the inverse image of Y .

<u>Proof</u>: From the Gysin sequences

$$\begin{array}{ccccccc} \cdots \to & H^{r-2c}(Y)(-c) & \to & H^r(X) & \to & H^r(X-Y) & \to \cdots \\ & \downarrow & & \downarrow & & \| & \\ \cdots \to & H^{r-2}(Y')(-1) & \to & H^r(X') & \to & H^r(X'-Y') & \to \cdots \end{array}$$

we obtain a long exact sequence

$$\cdots \to H^{r-2c}(Y)(-c) \to H^r(X) \oplus H^{r-2}(Y')(-1) \to H^r(X') \to \cdots .$$

But Y' is a projective bundle over Y , and so $H^{r-2c}(Y)(-c) \to H^{r-2}(Y')(-1)$ is injective. Therefore there are exact sequences

$$0 \to H^{r-2c}(Y)(-c) \to H^r(X) \oplus H^{r-2}(Y')(-1) \to H^r(X') \to 0 ,$$

which can be rewritten as

$$0 \to H(Y)(-c) \to H(X) \oplus H(Y')(-1) \to H(X') \to 0 .$$

We have constructed a sequence of motives, which is exact because the cohomology functors are faithful and exact.

Corollary 6.12. With the notations of the proposition,

$$h(X') = h(X) \oplus \bigoplus_{r=1}^{c-1} h(Y)(-r) .$$

Proof: (6.10) shows that $h(Y') = \bigoplus_{r=0}^{c-1} h(Y)(r)$.

Proposition 6.13. If X is an abelian variety, then $h(X) = \Lambda(h^1(X))$.

Proof: Cup-product defines a map $\Lambda(H^1(X)) \to H(X)$ which, for the Betti cohomology say, is known to be an isomorphism. (See Mumford [1,I.1].)

Proposition 6.14. If X is a curve with Jacobian J , then

$$h(X) = \underline{1} \oplus h^1(J) \oplus L .$$

Proof: The map $X \to J$ (well-defined up to translation) defines an isomorphism $H^1(J) \to H^1(X)$.

Proposition 6.15. Let X be a unirational variety of dimension $d \leq 3$ over an algebraically closed field; then

(d=1) $h(X) = \underline{1} \oplus L$;

(d=2) $h(X) = 1 \oplus rL \oplus L^2$, some $r \in \mathbb{N}$;

(d=3) $h(X) = \underline{1} \oplus rL \oplus h^1(A)(-1) \oplus rL^2 \oplus L^3$, some $r \in \mathbb{N}$,

where A is an abelian variety.

Proof: We prove the proposition only for $d=3$. According to the resolution theorem of Abhyankar [1], there exist maps

$$\mathbb{P}^3 \xleftarrow{u} X' \xrightarrow{v} X$$

with v surjective of finite degree and u a composite of blowing-ups. We know $h(\mathbb{P}^3) = \underline{1} \oplus L \oplus L^2 \oplus L^3$ (special case of (6.10)). When a point is blown up, a motive $L \oplus L^2$ is added, and when a curve Y is blown up, a motive $L \oplus h^1(Y)(-1) \oplus L^2$ is added. Therefore

$$h(X') = \underline{1} \oplus sL \oplus M(-1) \oplus sL^2 \oplus L^3$$

where M is a sum of motives of the form $h^1(Y)$, Y a curve. A direct summand of such an M is of the form $h^1(A)$ for A an abelian variety (see (6.21) below). As $h(X)$ is a direct summand of $h(X')$ (see (6.9)) and Poincaré duality shows that the multiples of L^2 and L^3 occurring in $h(X)$ are the same as those of L and $\underline{1}$ respectively the proof is complete.

Proposition 6.16. Let X_d^n denote the Fermat hypersurface of dimension n and degree d:

$$T_o^d + T_1^d + \cdots + T_{n+1}^d = 0 .$$

Then

$$h^n(X_d^n) \oplus d\, h^n(\mathbb{P}^n) = h^n(X_d^{n-1} \times X_d^1)^{\mu_d} \oplus (d-1) h^{n-2}(X_d^{n-2})(-1)$$

where μ_d, the group of dth roots of 1, acts on $X_d^{n-1} \times X_d^1$ according to

$$\zeta(t_o:\cdots:t_n;s_o:s_1:s_2) = (t_o:\cdots:t_{n-1}:\zeta t_n;s_o:s_1:\zeta s_2)$$

Proof: See Shioda-Katsura [1, 2.5].

Artin Motives

Let $\underline{V}_k^o$ be the category of zero-dimensional varieties over k, and let $C\underline{V}_k^o$ be the image of $\underline{V}_k^o$ in $\underline{M}_k$. The Tannakian subcategory $\underline{M}_k^o$ of $\underline{M}_k$ generated by the objects of $C\underline{V}_k^o$ is called the category of (E.) Artin motives.

For any X in $ob(\underline{V}_k^o)$, $X(\bar{k})$ is a finite set on which Γ acts continuously. Thus $\mathbb{Q}^{X(\bar{k})}$ is a finite-dimensional continuous representation of Γ. If we regard Γ, in the obvious way, as a (constant, pro-finite) affine group scheme over k, $\mathbb{Q}^{X(\bar{k})} \in \underline{Rep}_{\mathbb{Q}}(\Gamma)$. For $X, Y \in ob(\underline{V}_k^o)$,

$$\mathrm{Hom}(h(X),h(Y)) \overset{df}{=} C^o_{AH}(X \times Y) = (\mathbb{Q}^{X(\bar{k})\times Y(\bar{k})})^{\Gamma} = \mathrm{Hom}_{\Gamma}(\mathbb{Q}^{X(\bar{k})}, \mathbb{Q}^{Y(\bar{k})}).$$

Thus $h(X) \mapsto \mathbb{Q}^{X(\bar{k})} : C\underline{V}_k^o \to \underline{Rep}_{\mathbb{Q}}(\Gamma)$ is fully faithful, and Grothendieck's formulation of Galois theory shows that it is essentially surjective. Therefore $C\underline{V}_k^o$ is abelian and $\underline{M}_k^o = C\underline{V}_k^o$. We have shown:

Proposition 6.17. The category of Artin motives $\underline{M}_k^o = C\underline{V}_k^o$; the functor $h(X) \mapsto \mathbb{Q}^{X(\bar{k})}$ defines an equivalence of tensor categories $\underline{M}_k^o \xrightarrow{\approx} \underline{Rep}_{\mathbb{Q}}(\Gamma)$.

Remark 6.18. Let M be an Artin motive, and regard M as an object of $\underline{Rep}_{\mathbb{Q}}(\Gamma)$. Then

$H_\sigma(M) = M$ (underlying vector space) for any $\sigma : k \hookrightarrow \mathbb{C}$;

$H_\ell(\bar{M}) = M \otimes_{\mathbb{Q}} \mathbb{Q}_\ell$, as a Γ-module;

$H_{DR}(M) = (M \otimes_{\mathbb{Q}} \bar{k})^\Gamma$.

Note that, if $M = h(X)$ where $X = \operatorname{spec} A$, then

$$H_{DR}(M) = (\mathbb{Q}^{X(\bar{k})} \otimes_{\mathbb{Q}} \bar{k})^\Gamma = (A \otimes_k \bar{k})^\Gamma = A .$$

Remark 6.19. The proposition shows that $\underline{M}_k^o$ is equivalent to the category of sheaves of finite-dimensional $\mathbb{Q}$ vector spaces on the étale site $\operatorname{spec}(k)_{et}$.

Effective motives of degree 1

A $\mathbb{Q}$-rational Hodge structure is a finite-dimensional vector space V over $\mathbb{Q}$ together with a real Hodge structure on $V \otimes \mathbb{R}$ whose weight filtration is defined over $\mathbb{Q}$. Let $\underline{Hod}_{\mathbb{Q}}$ be the category of $\mathbb{Q}$-rational Hodge structures. A polarization on an object V of $\underline{Hod}_{\mathbb{Q}}$ is bilinear pairing $\psi : V \times V \to \mathbb{Q}(-n)$ such that $\psi \otimes \mathbb{R}$ is a polarization on the real Hodge structure $V \otimes \mathbb{R}$.

Let $\underline{Isab}_k$ be the category of abelian varieties up to isogeny over k . The following theorem summarizes part of the analytic theory of abelian varieties.

Theorem 6.20. (Riemann) The functor $H_B^1 : \underline{Isab}_{\mathbb{C}} \to \underline{Hod}_{\mathbb{Q}}$ is fully faithful; the essential image consists of polarizable Hodge structures of weight 1.

Let $\underline{M}_k^{+1}$ be the pseudo-abelian subcategory of $\underline{M}_k$ generated by motives of the form $h^1(X)$ for X a geometrically connected curve; according to (6.14), $\underline{M}_k^{+1}$ can also be described as the category generated by motives of the form $h^1(J)$ for J a Jacobian.

<u>Proposition</u> 6.21. (a) The functor $h^1 : \underline{\text{Isab}}_k \to \underline{M}_k$ factors through $\underline{M}_k^{+1}$ and defines an equivalence of categories, $\underline{\text{Isab}}_k \xrightarrow{\sim} \underline{M}_k^{+1}$.

(b) The functor $H^1 : \underline{M}_{\mathbb{C}}^{+1} \to \underline{\text{Hod}}_{\mathbb{Q}}$ is fully faithful; its essential image consists of polarizable Hodge structures of weight 1.

<u>Proof</u>: Every object of $\underline{\text{Isab}}_k$ is a direct summand of a Jacobian, which shows that h^1 factors through $\underline{M}_k^{+1}$. Assume, for simplicity, that k is algebraically closed. Then, for any $A,B \in \text{ob}(\underline{\text{Isab}}_k)$,

$$\text{Hom}(B,A) \subset \text{Hom}(h^1(A), h^1(B)) \subset \text{Hom}(H_\sigma(A), H_\sigma(B)) ,$$

and (6.20) shows that $\text{Hom}(B,A) = \text{Hom}(H_\sigma(A),H_\sigma(B))$. Thus h^1 is fully faithful and (as $\underline{\text{Isab}}_k$ is abelian) essentially surjective. This proves (a), and (b) follows from (a) and (6.20).

<u>The motivic Galois group</u>

Let k be a field that is embeddable in $\mathbb{C}$. For any $\sigma : k \hookrightarrow \mathbb{C}$, we define $G(\sigma) = \underline{\text{Aut}}^{\otimes}(H_B)$. Thus $G(\sigma)$ is an affine group scheme over $\mathbb{Q}$, and H_B defines an equivalence of tensor categories $\underline{M}_k \xrightarrow{\sim} \underline{\text{Rep}}_{\mathbb{Q}}(G(\sigma))$. Because $G(\sigma)$ plays the same role for $\underline{M}_k$ as $\Gamma = \text{Gal}(\bar{k}/k)$ plays for $\underline{M}_k^{o}$, it is called the <u>motivic Galois group</u>.

<u>Proposition</u> 6.22. (a) If k is algebraically closed, then $G(\sigma)$ is a connected pro-reductive affine group scheme over $\mathbb{Q}$.
(b) Let $k \subset k'$ be algebraically closed fields, let $\sigma' : k' \hookrightarrow \mathbb{C}$, and let $\sigma = \sigma'|k$. The homomorphism $G(\sigma') \to G(\sigma)$ induced by $\underline{M}_k \to \underline{M}_{k'}$ is faithfully flat; if k has infinite transcendence degree over $\mathbb{Q}$, then $G(\sigma') \to G(\sigma)$ is an isomorphism.

<u>Proof</u>: (a) Let $X \in ob(\underline{M}_k)$, and let $\underline{C}_X$ be the abelian tensor subcategory of $\underline{M}_k$ generated by X, $X^\vee$, T, and $T^\vee$. According to (I 3.4), $G_X \overset{df}{=} \underline{Aut}^{\otimes}(H_\sigma|\underline{C}_X)$ is the smallest subgroup of $Aut(H_\sigma(X)) \times \mathbb{G}_m$ such that $(G_X)_{\mathbb{C}}$ contains the image of the homomorphism $\mu : \mathbb{G}_{m\mathbb{C}} \to Aut(H_\sigma(X,\mathbb{C})) \times \mathbb{G}_{m\mathbb{C}}$ defined by the Hodge structure on $H_\sigma(X)$. As $Im(\mu)$ is connected, so also is G_X. As $\underline{C}_X$ is semisimple (see (6.5)), G_X is a reductive group (2.23). Therefore $G = \varprojlim G_X$ is connected and pro-reductive.
(b) According to (I 2.9), $\underline{M}_k \to \underline{M}_{k'}$ is fully faithful, and so (2.29) shows that $G(\sigma') \twoheadrightarrow G(\sigma)$. When k has infinite transcendence degree over $\mathbb{Q}$, $\underline{M}_k \to \underline{M}_{k'}$ is essentially surjective, and so $G(\sigma') \overset{\approx}{\to} G(\sigma)$.

Now let k be arbitrary, and fix an embedding $\sigma : \bar{k} \hookrightarrow \mathbb{C}$. The inclusion $\underline{M}^o_k \to \underline{M}_k$ defines a homomorphism $\pi : G(\sigma) \to \Gamma$ because $\Gamma = \underline{Aut}^{\otimes}(H_\sigma|\underline{M}^o_k)$ (see (6.17)), and the functor $\underline{M}_k \to \underline{M}_{\bar{k}}$ defines a homomorphism $i : G^o(\sigma) \to G(\sigma)$ where $G^o(\sigma) \overset{df}{=} \underline{Aut}^{\otimes}(H_\sigma|\underline{M}_{\bar{k}})$.

<u>Proposition</u> 6.23. (a) The sequence

$$1 \to G^o(\sigma) \overset{i}{\to} G(\sigma) \overset{\pi}{\to} \Gamma \to 1$$

is exact.
(b) The identity component of $G(\sigma)$ is $G^o(\sigma)$.

(c) For any $\tau \in \Gamma$, $\pi^{-1}(\tau) = \underline{\mathrm{Hom}}^{\otimes}(H_\sigma, H_{\tau\sigma})$, regarding H_σ and $H_{\tau\sigma}$ as functors on $\underline{M}_{\bar{k}}$.

(d) For any prime ℓ , there is a canonical continuous homomorphism $sp_\ell : \Gamma \to G(\sigma)(\mathbb{Q}_\ell)$ such that $\pi \circ sp_\ell = id$.

Proof: (a) As $\underline{M}_k^o \to \underline{M}_k$ is fully faithful, π is surjective (2.29). To show that i is injective, it suffices to show that every motive $h(X)$, $X \in \underline{V}_{\bar{k}}$, is a subquotient of a motive $h(\bar{X}')$ for some $X' \in \underline{V}_k$; but X has a model X_o over a finite extension k' of k , and we can take $X' = \mathrm{Res}_{k'/k} X_o$. The exactness of $G(\sigma)$ is a special case of (c) .

(b) This is an immediate consequence of (6.22a) and (a).

(c) Let $M, N \in \mathrm{ob}(\underline{M}_k)$. Then $\mathrm{Hom}(\bar{M},\bar{N}) \in \mathrm{ob}(\underline{\mathrm{Rep}}_{\mathbb{Q}}(\Gamma))$, and so we can regard it as an Artin motive over k. There is a canonical map of motives $\mathrm{Hom}(\bar{M},\bar{N}) \hookrightarrow \underline{\mathrm{Hom}}(M,N)$ giving rise to

$$H_\sigma(\mathrm{Hom}(\bar{M},\bar{N})) = \mathrm{Hom}(\bar{M},\bar{N}) \xrightarrow{H_\sigma} \mathrm{Hom}(H_\sigma(\bar{M}),H_\sigma(\bar{N})) = H_\sigma(\underline{\mathrm{Hom}}(M,N)) .$$

Let $\tau \in \Gamma$; then $H_\sigma(\bar{M}) = H_\sigma(M) = H_{\tau\sigma}(M) = H_{\tau\sigma}(\bar{M})$ and, for $f \in \mathrm{Hom}(\bar{M},\bar{N})$, $H_\sigma(f) = H_{\tau\sigma}(\tau f)$.

Let $g \in G(R)$; for any $f : M \to N$ in $\underline{M}_k$, there is a commutative diagram

$$\begin{array}{ccc} H_\sigma(M,R) & \xrightarrow{g_M} & H_\sigma(M,R) \\ \downarrow{\scriptstyle H_\sigma(f)} & & \downarrow{\scriptstyle H_\sigma(f)} \\ H_\sigma(N,R) & \xrightarrow{g_N} & H_\sigma(N,R) \end{array}$$

Let $\tau = \pi(g)$, so that g acts on $\mathrm{Hom}(\bar{M},\bar{N}) \subset \underline{\mathrm{Hom}}(M,N)$ as τ . Then for any $f : \bar{M} \to \bar{N}$ in $\underline{M}_{\bar{k}}$

$$\begin{array}{ccccc} H_\sigma(\bar{M},R) & \xrightarrow{g_M} & H_\sigma(\bar{M},R) & = & H_{\tau\sigma}(\bar{M},R) \\ \downarrow H_\sigma(f) & & \downarrow H_\sigma(\tau^{-1}f) & & \downarrow H_{\tau\sigma}(f) \\ H_\sigma(\bar{N},R) & \xrightarrow{g_N} & H_\sigma(\bar{N},R) & = & H_{\tau\sigma}(\bar{N},R) \end{array}$$

commutes. The diagram shows that $g_M : H_\sigma(\bar{M},R) \to H_{\tau\sigma}(\bar{M},R)$ depends only on M as an object of $\underline{M}_{\bar{k}}$. We observed in the proof of (a) above that $\underline{M}_{\bar{k}}$ is generated by motives of the form $\bar{M}$, $M \in \underline{M}_k$. Thus g defines an element of $\underline{\mathrm{Hom}}^{\otimes}(H_\sigma, H_{\tau\sigma})(R)$, where H_σ and $H_{\tau\sigma}$ are to be regarded as functors on $\underline{M}_{\bar{k}}$. We have defined a map $\pi^{-1}(\tau) \to \underline{\mathrm{Hom}}^{\otimes}(H_\sigma, H_{\tau\sigma})$, and it is easy to see that it is surjective.

(d) After (c), we have to find a canonical element of $\mathrm{Hom}^{\otimes}(H_\ell(\sigma M), H_\ell(\tau\sigma M))$ depending functorially on $M \in \underline{M}_{\bar{k}}$. Extend τ to an automorphism $\bar{\tau}$ of $\mathbb{C}$. For any variety X over $\bar{k}$, there is a $\bar{\tau}^{-1}$-linear isomorphism $\sigma X \leftarrow \tau\sigma X$ which induces an isomorphism $\tau : H_\ell(\sigma X) \xrightarrow{\approx} H_\ell(\tau\sigma X)$.

The "espoir" (Deligne [2, 0.10) that every Hodge cycle is absolutely Hodge has a particularly elegant formulation in terms of motives.

Conjecture 6.24. For any algebraically closed field k and embedding $\sigma : k \hookrightarrow \mathbb{C}$, the functor $H_\sigma : \underline{M}_k \to \underline{\mathrm{Hod}}_{\mathbb{Q}}$ is fully faithful.

The functor is obviously faithful. There is no description, not even conjectural, for the essential image of H_σ.

Motives of abelian varieties

Let $\underline{M}_k^{av}$ be the Tannakian subcategory of $\underline{M}_k$ generated by motives of abelian varieties and Artin motives. The main theorem of I has the following restatement.

Theorem 6.25. For any algebraically closed field k and embedding $\sigma : k \hookrightarrow \mathbb{C}$, the functor $H_\sigma : \underline{M}_k^{av} \to \underline{\mathrm{Hod}}_{\mathbb{Q}}$ is fully faithful.

Proposition 6.26. The motive $h(X) \in ob(\underline{M}_k^{av})$ if

(a) X is a curve;

(b) X is a unirational variety of dimension ≤ 3 ;

(c) X is a Fermat hypersurface;

(d) X is a K3-surface

Before proving this, we note the following consequence.

Corollary 6.27. Every Hodge cycle on a variety that is a product of abelian varieties, zero-dimensional varieties, and varieties of type (a), (b), (c) and (d), is absolutely Hodge.

Proof of 6.26. Cases (a) and (b) follow immediately from (6.14) and (6.15), and (c) follows by induction (on n) from (6.16). In fact one does not need the full strength of (6.16). There is a rational map

$$X_d^r \times X_d^s \dashrightarrow X_d^{r+s}$$

$(x_o:\cdots:x_{r+1}), (y_o:\cdots:y_{s+1}) \longmapsto (x_o y_{s+1}:\cdots:x_r y_{s+1}:\varepsilon x_{r+1} y_o:\cdots:\varepsilon x_{r+1} y_s)$
where ε is a primitive $2m^{th}$ root of 1 . The map is not defined on the subvariety $Y : x_{r+1} = y_{s+1} = 0$. On blowing up $X_d^r \times X_d^s$

along the nonsingular centre Y , one obtains maps

$$\begin{array}{ccc} Z_d^{r,s} & & \\ \downarrow & \searrow & \\ X_d^r \times X_d^s & \dashrightarrow & X_d^{r+s} \end{array} .$$

By induction, we can assume that the motives of X_d^r, X_d^s, and $Y(=X_d^{r-1} \times X_d^{s-1})$ are in $\underline{M}_k^{av}$. Corollary (6.12) now shows that $h(Z_d^{r,s}) \in ob(\underline{M}_k^{av})$ and (6.9) that $h(X_d^{r+s}) \in ob(\underline{M}_k^{av})$.

For (d), we first note that the proposition is obvious if X is a Kummer surface, for then $X = \tilde{A}/\langle\sigma\rangle$ where $\tilde{A}$ is an abelian variety A with its 16 points of order ≤ 2 blown up and σ induces $a \longmapsto -a$ on A .

Next consider an arbitrary K3-surface X , and fix a projective embedding of X . Then

$$h(X) = h(\mathbb{P}^2) \oplus h^2(X)_{prim}$$

and so it suffices to show that $h^2(X)_{prim}$ is in $\underline{M}^{av}$. We can assume $k = \mathbb{C}$. It is known (Kuga-Shimura [1], Deligne [1, 6.5]) that there is a smooth connected variety S over $\mathbb{C}$ and families $f : Y \to S$, $a : A \to S$ of polarized K3-surfaces and abelian varieties respectively parametrized by S having the following properties:

(a) for some $o \in S$, $Y_o \overset{df}{=} f^{-1}(o)$ is X together with its given polarization;

(b) for some $1 \in S$, Y_1 is a polarized Kummer surface;

(c) there is an inclusion $u : R^2 f_* \mathbb{Q}(1)_{prim} \hookrightarrow \underline{End}(R^1 a_* \mathbb{Q})$

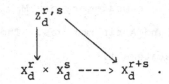

compatible with the Hodge filtrations.

The map $u_o : H_B^2(X)(1)_{prim} \hookrightarrow End(H^1(A_o,\mathbb{Q}))$ is therefore defined by a Hodge cycle, and it remains to show that it is defined by an absolutely Hodge cycle. But the initial remark shows that u_1, being a Hodge cycle on a product of Kummer and abelian surfaces, is absolutely Hodge, and principle B (I2.12) completes the proof.

Motives of abelian varieties of potential CM-type

An abelian variety A over k is said to be of potential CM-type if it becomes of CM-type over an extension of k. Let A be such an abelian variety defined over $\mathbb{Q}$, and let MT(A) be the Mumford-Tate group of $A_{\mathbb{C}}$ (see I.5). Since $A_{\mathbb{C}}$ is of CM-type, MT(A) is a torus, and we let $L \subset \mathbb{C}$ be a finite Galois extension of $\mathbb{Q}$ splitting MT(A). Let $\underline{M}_{\mathbb{Q}}^{A,L}$ be the Tannakian subcategory of $\underline{M}_{\mathbb{Q}}$ generated by A, the Tate motive, and the Artin motives split by L^{ab}, and let G^A be affine group scheme associated with this Tannakian category and the fibre functor H_B.

Proposition 6.28. There is an exact sequence of affine group schemes

$$1 \to MT(A) \xrightarrow{i} G^A \xrightarrow{\pi} Gal(L^{ab}/\mathbb{Q}) \to 1 .$$

Proof: Let $\underline{M}_{\mathbb{C}}^{A}$ be the image of $\underline{M}_{\mathbb{Q}}^{A,L}$ in $\underline{M}_{\mathbb{C}}$; then MT(A) is the affine group scheme associated with $\underline{M}_{\mathbb{C}}^{A}$, and so the above sequence is a subsequence of the sequence in (6.23a).

Remark 6.29. If we identify MT(A) with the subgroup of $Aut(H_B^1(A))$, then (as in 6.23c) $\pi^{-1}(\tau)$ becomes identified with the MT(A)-torsor whose R-points, for any $\mathbb{Q}$-algebra R, are the

R-linear isomorphisms $a : H^1(A_{\mathbb{C}},R) \to H^1(\tau A_{\mathbb{C}},R)$ such that $a(s) = \tau s$ for all (absolute) Hodge cycles on $A_{\bar{\mathbb{Q}}}$. We can also identify $MT(A)$ with a subgroup of $Aut(H_1^B(A))$ and then it becomes more natural to identify $\pi^{-1}(\tau)$ with the torsor of R-linear isomorphisms $a^\vee : H_1(A_{\mathbb{C}},R) \to H_1(\tau A_{\mathbb{C}},R)$ preserving Hodge cycles.

On passing to the inverse limit over all A and L, we obtain an exact sequence

$$1 \to S^o \to S \to Gal(\bar{\mathbb{Q}}/\mathbb{Q}) \to 1$$

with S^o and S respectively the connected Serre group and the Serre group. This sequence plays an important role in the next three articles.

Appendix: Terminology from non-abelian cohomology

We review some definitions from Giraud [1].

Fibred categories

Let $\alpha : \underline{F} \to \underline{A}$ be a functor. For any object U of $\underline{A}$ we write $\underline{F}_U$ for the category whose objects are those F in $\underline{F}$ such that $\alpha(\underline{F}) = U$ and whose morphisms are those f such that $\alpha(f) = id_U$. For any morphism $a : \alpha(F_1) \to \alpha(F_2)$, we write $Hom_a(F_1,F_2)$ for the set of $f : F_1 \to F_2$ such that $\alpha(f) = a$. A morphism $f : F_1 \to F_2$ in $\underline{F}$ is said to be cartesian, and F_1 is said to be the inverse image $\alpha(f)^*F_2$ of F_2 relative to $\alpha(f)$, if, for any $F' \in \underline{F}_{\alpha(F_1)}$ and $h \in Hom_{\alpha(f)}(F',F_2)$, there is a unique $g \in \mathrm{Hom}_{id}(F',F_1)$ such that $fg=h$:

$$\begin{array}{ccc} F' & & \\ {\scriptstyle g}\downarrow & \searrow^{h} & \\ F_1 & \xrightarrow{\quad f \quad} & F_2 \end{array}$$

$$\alpha(F_1) \xrightarrow{\alpha(f)} \alpha(F_2) \ .$$

We say that $\alpha : \underline{F} \to \underline{A}$ is a <u>fibered category</u> if

(a) for any morphism $a : U_1 \to U_2$ in $\underline{A}$ and $F_2 \in \mathrm{ob}(\underline{F}_{U_2})$, the inverse image $a^*(F_2)$ of F_2 exists;

(b) the composite of two cartesian morphisms is cartesian. (Existence and transitivity of inverse images.) Then a^* can be made into a functor $\underline{F}_{U_2} \to \underline{F}_{U_1}$, and $(ab)^*$ is canonically isomorphic to b^*a^*.

Let $\alpha : \underline{F} \to \underline{A}$ and $\alpha' : \underline{F}' \to \underline{A}$ be fibred categories over $\underline{A}$, and let $\beta : \underline{F} \to \underline{F}'$ be a functor such that $\alpha' \circ \beta = \alpha$; one says β is <u>cartesian</u> if it maps cartesian morphisms to cartesian morphisms.

<u>Stacks</u> (Champs)

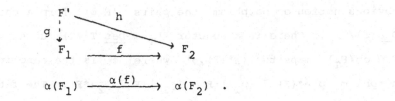

Let $\alpha : \underline{F} \to \underline{\mathrm{Aff}}_S$ be a fibred category where $\underline{\mathrm{Aff}}_S$ is the category of affine schemes over $S = \mathrm{spec}\ R$. We endow $\underline{\mathrm{Aff}}_S$ with the f.p.q.c. topology. Let $a : T' \to T$ be a faithfully flat map of affine S-schemes and let $F \in \mathrm{ob}(\underline{F}_{T'})$; a <u>descent datum</u> on F relative to a is an isomorphism $\phi : p_1^*(F) \to p_2^*(F)$ such that $p_{31}^*(\phi) = p_{32}^*(\phi)\ p_{21}^*(\phi)$ where p_1 and p_2 are the projections $T'' = T' \times_T T' \rightrightarrows T'$ and the p_{ij} are the projections $T''' = T' \times_T T' \times_T T' \rightrightarrows T' \times_T T'$. With an obvious notion of morphism, the pairs (F,ϕ) form a category $\underline{Des}(T'/T)$. There is a functor $\underline{F}_T \to \underline{Des}(T'/T)$ under which $F \in ob(\underline{F}_T)$ maps to $(a^*(F),\phi)$ where ϕ is the canonical morphism $p_1^* a^*(F) \approx (p_1 a)^* F = (p_2 a)^* F \approx p_2^* a^* F$. The fibred category $\alpha : \underline{F} \to \underline{Aff}_S$ is a <u>stack</u> if, for all faithfully flat maps $a : T' \to T$, $\underline{F}_T \to \underline{Des}(T'/T)$ is an equivalence of categories.

For example, let $\alpha : MOD \to \underline{Aff}_S$ be the fibred category such that MOD_T is the category of finitely presented $\Gamma(T,O_T)$-modules; descent theory shows that this is a stack (Waterhouse [1,17.2], Bourbaki [2,I.3.6]). Similarly, there is a stack $PROJ \to \underline{Aff}_S$ for which $PROJ_T$ is the category of finitely generated projective $\Gamma(T,O_T)$-modules (ibid.) and a stack $AFF \to \underline{Aff}_S$ for which $AFF_T = \underline{Aff}_T$.

<u>Gerbs</u> (Gerbes)

A stack $\underline{G} \to \underline{Aff}_S$ is a <u>gerb</u> if

(a) each fibre $\underline{G}_T$ is a groupoid (i.e., all morphisms in $\underline{G}_T$ are isomorphisms);

(b) there is a faithfully flat map $T \to S$ such that $\underline{G}_T$ is nonempty;

(c) any two objects of a fibre $\underline{G}_T$ are locally isomorphic (i.e., their inverse images relative to some faithfully flat map $T' \to T$ are isomorphic).

By a <u>morphism of gerbs</u> over $\underline{Aff}_S$ we mean a cartesian functor. A gerb $\underline{G} \to \underline{Aff}_S$ is said to be <u>neutral</u> (or trivial) if $\underline{G}_S$ is nonempty.

Let F be a sheaf of groups on S for the f.p.q.c. topology. The fibred category $\mathrm{TORS}(F) \to \underline{\mathrm{Aff}}_S$ for which $\mathrm{TORS}(F)_T$ is the category $\underline{\mathrm{Tors}}_T(F)$ of right F-torsors on T is a neutral gerb. Conversely, let $\underline{G}$ be a neutral gerb and let $Q \in \mathrm{ob}(\underline{G}_S)$; then $F = \underline{\mathrm{Aut}}(Q)$ is a sheaf of groups on $\underline{\mathrm{Aff}}_S$ and $\underline{G} \to \mathrm{TORS}(F)$, $P \mapsto \underline{\mathrm{Isom}}_T(a^*Q,P)$ (for $a : T \to S$) is an equivalence of gerbs.

Bands (Liens)

Let F and G be sheaves of groups for the f.p.q.c. topology on S, and let G^{ad} be the quotient sheaf G/Z where Z is the centre of G. The action of G^{ad} on G by conjugation induces an action of G^{ad} on the sheaf $\underline{\mathrm{Isom}}(F,G)$ and we set $\mathrm{Isex}(F,G) = \Gamma(S,G^{ad}\backslash\underline{\mathrm{Isom}}(F,G))$. As G^{ad} acts faithfully on $\underline{\mathrm{Isom}}(F,G)$,

$$\mathrm{Isex}(F,G) = \varinjlim \mathrm{Ker}(G^{ad}(T)\backslash\mathrm{Isom}(F|T,G|T) \rightrightarrows G^{ad}(T\times T)\backslash\mathrm{Isom}(F|T\times T,G|T\times T))$$

where the limit is over all $T \to S$ faithfully flat and affine.

A band B on S is defined by a triple (S',G,ϕ) where S' is an affine S-scheme, faithfully flat over S, G is a sheaf of groups on S', and $\phi \in \mathrm{Isex}(p_1^*G, p_2^*G)$ is such that $p_{31}^*(\phi) = p_{32}^*(\phi)p_{21}^*(\phi)$. (As before, the p_i and p_{ij} are the various projection maps $S'' \rightrightarrows S'$ and $S''' \rightrightarrows S''$). If T is also a faithfully flat affine S-scheme, and $a : T \to S'$ is an S-morphism, then we do not distinguish between the bands defined by (S',G,ϕ) and $(T,a^*(G),(a\times a)^*(\phi))$. Let B_1 and B_2 be the bands defined by (S',G_1,ϕ_1) and (S',G_2,ϕ_2); an isomorphism $B_1 \xrightarrow{\approx} B_2$ is an element $\psi \in \mathrm{Isex}(G_1,G_2)$ such that $p_2^*(\psi)\circ\phi_1 = \phi_2\circ p_1^*(\psi)$.

If G is a sheaf of groups on S , we write $B(G)$ for the band defined by (S,G,id). One shows that $\mathrm{Isom}(B(G_1),B(G_2)) = \mathrm{Isex}(G_1,G_2)$. Thus $B(G_1)$ and $B(G_2)$ are isomorphic if and only if G_2 is an inner form of G_1, i.e. G_2 becomes isomorphic to G_1 on some faithfully flat S-scheme T , and the class of G_2 in $H^1(S,\underline{\mathrm{Aut}}(G_1))$ comes from $H^1(S,G_1^{ad})$. When G_2 is commutative, then $\mathrm{Isom}(B(G_1),B(G_2)) = \mathrm{Isex}(G_1,G_2) = \mathrm{Isom}(G_1,G_2)$, and we usually do not distinguish $B(G_2)$ from G_2 .

The centre $Z(B)$ of the band B defined by (S',G,ϕ) is defined by $(S',Z,\phi|p_1^*Z)$ where Z is the centre of G . The above remark shows that $\phi|p_1^*Z$ lifts to an element $\phi_1 \in \mathrm{Isom}(p_1^*Z, p_2^*Z)$, and one checks immediately that $p_{31}^*(\phi_1) = p_{32}^*(\phi_1)\, p_{21}^*(\phi_1)$. Thus $(S',Z,\phi|p_1^*Z)$ arises from a sheaf of groups on S , which we identify with $Z(B)$.

Let $\underline{G}$ be a gerb on $\underline{\mathrm{Aff}}_S$. By definition, there exists an object $Q \in \underline{G}_{S'}$, for some $S' \to S$ faithfully flat and affine. Let $G = \underline{\mathrm{Aut}}(Q)$; it is a sheaf of groups on S' . Again by definition, p_1^*Q and p_2^*Q are locally isomorphic on S'', and the locally-defined isomorphisms determine an element $\phi \in \mathrm{Isex}(p_1^*(G),p_2^*(G))$. The triple (S',G,ϕ) defines a band B which is uniquely determined up to a unique isomorphism. This band B is called the band associated with the gerb $\underline{G}$, and $\underline{G}$ is said to be bound by B . For example, the gerb $\mathrm{TORS}(G)$ is bound by $B(G)$.

A band B is said to be affine (or algebraic) if it can be defined by a triple (S',G,ϕ) with G an affine (or algebraic) group scheme over S'. A gerb is said to be affine (or algebraic) if it is bound by an affine (or algebraic) band.

Cohomology

Let B be a band. Two gerbs $\underline{G}_1$ and $\underline{G}_2$ bound by B are said to be B-equivalent if there is an isomorphism $m : \underline{G}_1 \to \underline{G}_2$ with the following property: for some triple (S',G,ϕ) defining B there is an object $Q \in \underline{G}_{1S'}$ such that the automorphism $G \approx \underline{Aut}(Q) \overset{\approx}{\to} \underline{Aut}(m(Q)) \approx G$ defined by m is equal to id in $Isex(G,G)$. The cohomology set $H^2(S,B)$ is defined to be the set of B-equivalence classes of gerbs bound by B . If Z is the centre of B , then $H^2(S,Z)$ is equal to the cohomology group of Z in the usual sense of the f.p.q.c. topology on S , and either $H^2(S,B)$ is empty or $H^2(S,Z)$ acts simply transitively on it (Giraud [1, IV. 3.3.3]).

Proposition: Let S = spec k , k a field, and let $\underline{G}$ be an affine algebraic gerb on S ; then there is a finite field extension k' of k such that $G_{S'}$, S' = spec k' , is nonempty.

Proof: By assumption, the band B of $\underline{G}$ is defined by a triple (S',G,ϕ) with G of finite type over S' . Let S' = spec R'; R' can be replaced by a finitely generated subalgebra, and then by a quotient modulo a maximal ideal, and so we may suppose S' = speck' where k' is a finite field extension of k . We shall show that the gerbs $\underline{G}$ and TORS(G) become B-equivalent over some finite field extension of k' . The statement preceding the proposition shows we have to prove that an element of $H^2(S',Z)$, Z the centre of B , is killed by a finite field extension of k' . But this assertion is obvious for elements of $H^1(S',Z)$ and is easy to prove for elements of the Čech groups $\check{H}^r(S',Z)$, and so the exact sequence

$$0 \to \check{H}^2(S',Z) \to H^2(S',Z) \to \check{H}^1(S',\underline{H}^1(Z))$$

completes the proof. (See Saavedra [1, III 3.1] for more details.)

<u>Note</u>: (Added July, 1981): It seems likely that the final question in (3.5) can be shown to have a positive answer when k has characteristic zero. In particular this would show that any rigid abelian tensor category $\underline{C}$ with $End(\underline{1}) = k$ having a fibre functor with values in some extension of k is Tannakian, provided k is a field of characteristic zero.

REFERENCES

Abhyankar, S.
1. Resolution of Singularities of Embedded Algebraic Surfaces, Academic Press, 1966.

Bourbaki, N.
1. Algèbre; Modules et Anneaux Semi-Simples. Hermann, Paris (1958).
2. Algèbre Commutative; Modules Plats, Localisation. Hermann, Paris (1961).

Deligne, P.
1. La conjecture de Weil pour les surfaces K3, Invent. Math. 15 (1972) 206-222.
2. Valeurs de fonctions L et périodes d'integrales. Proc. Symp. Pure Math., A.M.S., 33 (1979) part 2, 313-346.

Giraud, J.
1. Cohomologie Non Abélienne, Springer, Heidelberg, 1971.

Hochschild, G.
1. The Structure of Lie Groups, Holden-Day, San Francisco, 1965.

Humphries, J.
1. Introduction to Lie Algebras and Representation Theory, Springer, Heidelberg, 1972.

Kuga, M. and Satake, I.
1. Abelian varieties attached to polarized K3-surfaces, Math. Ann. 169 (1967) 239-242.

MacLane, S.
1. Natural associativity and commutativity. Rice University Studies 69 (1963) 28-46.
2. Categories for the Working Mathematician. Springer, Heidelberg, 1972.

Mumford, D.
1. Abelian Varieties, Oxford U.P., Oxford, 1970.

Nori, M.
1. On the representations of the fundamental group. Compositio Math. 33 (1976) 29-41.

Saavedra Rivano, N.
1. Catégories Tannakiennes, Lecture Notes in Math 265, Springer, Heidelberg, 1972.

REFERENCES

Serre, J.-P.
1. Cohomologie Galoisienne, Lecture Notes in Math 5, Springer, Heidelberg, 1964.
2. Groupes algébriques associés aux modules de Hodge-Tate, (Journées de Géométrie Algébrique de Rennes), Astérisque 65 (1979) 155-187.

Springer, T.
1. Reductive groups, Proc. Symp. Pure Math., A.M.S., 33 (1979) part 1, 3-27.

Waterhouse, W.
1. Introduction to Affine Group Schemes, Springer, Heidelberg, 1979.

Wells, R.
1. Differential Analysis on Complex Manifolds. Prentice-Hall, Englewood Cliffs, 1973.

III. LANGLANDS'S CONSTRUCTION OF THE TANIYAMA GROUP

J. S. Milne and K.-y. Shih

Introduction: In this article we give a detailed description of Langlands's construction of his Taniyama group. The first section reviews the definition and properties of the Serre group, and the following section discusses extensions of Galois groups by the Serre group. The construction itself is carried out in the third section, which also contains additional material required for V.

We mention that in [1] Langlands is using the opposite sign convention for the reciprocity law in class field theory from us and hence the opposite notion of the Weil group (although his statement at the bottom of p. 224 is misleading on this point). Thus, there are many sign differences between his article and ours.

Notation: Vector spaces are finite-dimensional, number fields are of finite degree over $\mathbb{Q}$ (and usually contained in $\mathbb{C}$), and $\bar{\mathbb{Q}}$ is the algebraic closure of $\mathbb{Q}$ in $\mathbb{C}$. For L a number field, $L^{ab} \subset \bar{\mathbb{Q}}$ denotes its abelian closure. For the Weil group, we follow the notations of Tate [2]. In particular, for a topological group Γ, Γ^c denotes the closure of the commutator subgroup of Γ and $\Gamma^{ab} = \Gamma/\Gamma^c$.

§1. The Serre group.

Let $L \subset \mathbb{C}$ be a finite extension of $\mathbb{Q}$, let Γ be the set of embeddings of L into $\mathbb{C}$, and write $L^{\times}$ for $\mathrm{Res}_{L/\mathbb{Q}}\mathbb{G}_m$. Any $\rho \in \mathrm{Gal}(\overline{\mathbb{Q}}/\mathbb{Q})$ defines an element $[\rho]$ of Γ, which may be regarded as a character of $L^{\times}$. Then Γ is a basis for $X^*(L^{\times})$. An element σ of $\mathrm{Gal}(\overline{\mathbb{Q}}/\mathbb{Q})$ acts on $X^*(L^{\times})$ by $\sigma(\Sigma b_{\rho}[\rho]) = \Sigma b_{\rho}[\sigma\rho] = \Sigma b_{\sigma^{-1}\rho}[\rho]$. The quotient of $L^{\times}$ by the Zariski closure of any sufficiently small arithmetic subgroup has character group $X^*(L^{\times}) \cap (Y^{o} \oplus Y^{-})$ where

$$Y^{o} = \{\chi \in X^*(L^{\times}) \otimes \mathbb{Q} \mid \sigma\chi = \chi \text{ , all } \sigma \in \mathrm{Gal}(\overline{\mathbb{Q}}/\mathbb{Q})\}$$

$$Y^{-} = \{\chi \in X^*(L^{\times}) \otimes \mathbb{Q} \mid c\chi = -\chi, \text{ all } c \text{ of the form } c = \sigma\iota\sigma^{-1}\}$$

(Serre [1, II-31, Cor.1]). Thus this quotient is independent of the arithmetic subgroup; it is called the <u>Serre group</u> S^{L} of L (or, sometimes, the <u>connected</u> Serre group). One checks easily that $X^*(S^{L})$ is the subgroup of $X^*(L^{\times})$ of χ satisfying

(1.1) $$(\sigma-1)(\iota+1)\chi = 0 = (\iota+1)(\sigma-1)\chi \text{ , all } \sigma \in \mathrm{Gal}(\overline{\mathbb{Q}}/\mathbb{Q}) .$$

There is a canonical homomorphism $h = h^{L}\colon \mathbb{S} \to S^{L}_{\mathbb{R}}$ and hence corresponding homomorphisms $w_h\colon \mathbb{G}_m \to S^{L}_{\mathbb{R}}$ and $\mu = \mu^{L}\colon \mathbb{G}_m \to S^{L}_{\mathbb{C}}$. They determine the following maps on the character groups:

$$X^*(h) = (\Sigma b_{\rho}[\rho] \mapsto (b_1, b_{\iota}) \colon X^*(S^{L}) \longrightarrow X^*(\mathbb{S}) = \mathbb{Z} \oplus \mathbb{Z})$$

$$X^*(w_h) = (\Sigma b_{\rho}[\rho] \mapsto -b_1 - b_{\iota})$$

$$X^*(\mu) = (\Sigma b_{\rho}[\rho] \mapsto b_1)$$

Note that w_h is defined over $\mathbb{Q}$. The pair (S^L, μ^L) is universal: for any $\mathbb{Q}$-rational torus T that is split over L and cocharacter μ of T satisfying (1.1) there is a unique $\mathbb{Q}$-rational homomorphism $S^L \xrightarrow{\rho_\mu} T$ such that $\rho_\mu \circ \mu^L = \mu$. In particular there are no nontrivial automorphisms of (S^L, μ^L).

For $\mathbb{C} \supset L' \supset L \supset \mathbb{Q}$ and L' of finite degree over $\mathbb{Q}$, the norm map induces a homomorphism $S^{L'} \to S^L$ sending $h^{L'}$ to h^L. The (connected) Serre group S is defined to be the pro-algebraic group $\varprojlim S^L$. There is a canonical homomorphism $h = h_{can} = \varprojlim h^L : \mathbb{S} \to S_{\mathbb{R}}$ and corresponding cocharacter $\mu = \mu_{can} : \mathbb{G}_m \to S_{\mathbb{C}}$. For any L, S^L is the largest quotient of S that splits over L.

We review the properties of S that we shall need to use.

(1.2). The topology induced on $S^L(\mathbb{Q})$ by the embedding $S^L(\mathbb{Q}) \hookrightarrow S^L(\mathbb{A}^f)$ is the discrete topology; thus $S^L(\mathbb{Q})$ is closed in $S^L(\mathbb{A}^f)$. This is a consequence of Chevalley's theorem, which says that any arithmetic subgroup of the $\mathbb{Q}$-rational points of a torus is open relative to the adèlic topology, because the subgroup $\{1\}$ of $S^L(\mathbb{Q})$ is arithmetic.

(1.3). Make $\mathrm{Gal}(\overline{\mathbb{Q}}/\mathbb{Q})$ act on the group Λ of locally constant functions $\mathrm{Gal}(\overline{\mathbb{Q}}/\mathbb{Q}) \to \mathbb{Z}$ by transport of structure: thus $(\sigma\lambda)(\rho) = \lambda(\sigma^{-1}\rho)$. The map $X^*(S^L) \to \Lambda$ that sends $\chi = \Sigma b_\rho[\rho]$ to the function $\rho \mapsto b_\rho$ identifies $X^*(S^L)$ with the subset Λ^L of Λ comprising those functions that are constant on left cosets of $\mathrm{Gal}(\overline{\mathbb{Q}}/L)$ in $\mathrm{Gal}(\overline{\mathbb{Q}}/\mathbb{Q})$ and satisfy (1.1). On passing to the limit over L, we find that $X^*(S)$ becomes identified with the subgroup of Λ of functions satisfying (1.1).

(1.4). Let $\mathbb{Q}^{cm}$ be the union of all subfields of $\bar{\mathbb{Q}}$ of CM-type; it is the largest subfield on which ι and σ commute for all $\sigma \in Gal(\bar{\mathbb{Q}}/\mathbb{Q})$. The condition (1.1) is equivalent to the following conditions:

(1.1') λ is fixed by $Gal(\bar{\mathbb{Q}}/\mathbb{Q}^{cm})$ and $\lambda(\iota\sigma) + \ (\sigma)$ is independent of σ.

In particular, for a given L, $\Lambda^L \ \Lambda^F$ where $F = L \ \mathbb{Q}^{cm}$ is the maximal CM-subfield of L (or is $\mathbb{Q}$). Since obviously $\Lambda^L \ \Lambda^F$, they must be equal: $S^L \xrightarrow{\approx} S^F$.

(1.5). (Deligne) Let F be a CM-field with maximal real subfield F_0. There is an exact commutative diagram (of algebraic groups)

$$\begin{array}{ccccccccc}
 & & & & 1 & & 1 & & \\
 & & & & \uparrow & & \uparrow & & \\
1 & \longrightarrow & Ker & \longrightarrow & F^{\times}/F_0^{\times} & \xrightarrow{\approx} & S^F/hw(\mathbb{Q}^{\times}) & & \\
 & & \uparrow & & \uparrow & & \uparrow & & \\
 & & \approx & & F^{\times} & & S^F & \longrightarrow & 1 \\
 & & \uparrow & & \uparrow & & \uparrow hw & & \\
1 & \longrightarrow & Ker & \longrightarrow & F_0^{\times} & \xrightarrow{norm} & \mathbb{Q}^{\times} & \longrightarrow & 1 \\
 & & & & \uparrow & & \uparrow & & \\
 & & & & 1 & & 1 & &
\end{array}$$

To prove this it suffices to show that the square at bottom-right commutes, and the top horizontal arrow is injective, but both of these are easily seen on the character groups. Thus there is an exact sequence

$$1 \longrightarrow F_0^\times \longrightarrow F^\times \times \mathbb{Q}^\times \longrightarrow S^F \longrightarrow 1 .$$

We can deduce that, for any field $k \supset \mathbb{Q}$, there is an injection $H^1(k,S^F) \hookrightarrow Br(F_0 \otimes k)$ where Br denotes the Brauer group. It follows that, when k is a number field, the Hasse principle holds for $H^1(k,S^F)$: the map $H^1(k,S^F) \to \oplus H^1(k_v,S^F)$ is injective. The remark (1.4) shows that this is also true without assuming F to be a CM-field.

(1.6) Let $\lambda \in X^*(S)$ and let T_λ be the $\mathbb{Q}$-rational torus such that $X^*(T_\lambda)$ is the $Gal(\overline{\mathbb{Q}}/\mathbb{Q})$-submodule of $X^*(S)$ generated by λ. Thus T_λ is a quotient of S and h_{can} defines a homomorphism $h: \mathbb{S} \to T_\lambda$. For any $\mathbb{Q}$-rational representation of T_λ, $T_\lambda \hookrightarrow GL(V)$, (V,h) is a $\mathbb{Q}$-rational Hodge structure with weight $n = -(\lambda(1) + \lambda(\iota))$ and Mumford-Tate group $MT(V,h) = T_\lambda$ (See II). The condition (1.1') shows that ι acts as -1 on $Ker(\lambda' \mapsto \lambda'(1) + \lambda'(\iota): X^*(T_\lambda) \to \mathbb{Z})$; thus $(T_\lambda/w_h(\mathbb{G}_m))(\mathbb{R})$ is compact, and (V,h) is polarizable (Deligne [1,2.8]). It follows easily that $S = \varprojlim MT(V,h)$ where the limit is over the $\mathbb{Q}$-rational polarizable Hodge structures (V,h) of CM-type. In other words, S is the group associated with the Tannakian category of Hodge structures of this type.

(1.7) (Serre). It is an easy combinatorial exercise to show that $X^*(S)$ is generated by functions λ such that $\lambda(\sigma)$ is 0 or 1 and $\lambda(\sigma) + \lambda(\iota\sigma) = 1$. If λ is of this type then, for any representation $T_\lambda \hookrightarrow GL(V)$ of T_λ, (V,h) is a $\mathbb{Q}$-rational polarizable Hodge structure of CM-type and weight -1; it therefore corresponds to an abelian variety. Thus $S = \varprojlim MT(A)$ where the limit is over abelian varieties (over $\mathbb{C}$) of CM-type. In other words, the Tannakian category of $\mathbb{Q}$-rational polarizable Hodge structures of CM-type is generated by those arising from abelian varieties.

(1.8) If L is Galois over $\mathbb{Q}$, then $\mathrm{Gal}(L/\mathbb{Q})$ acts on $L^\times = \mathrm{Res}_{L/\mathbb{Q}}\mathbb{G}_m$ and this action induces an action on the quotient S^L. Thus there is an action of $\mathrm{Gal}(\overline{\mathbb{Q}}/\mathbb{Q})$ on the $\mathbb{Q}$-rational pro-algebraic group S. It is important to distinguish carefully between the two natural actions of $\mathrm{Gal}(\overline{\mathbb{Q}}/\mathbb{Q})$ on $S(\overline{\mathbb{Q}})$, the first of which arises from the (algebraic) action of $\mathrm{Gal}(\overline{\mathbb{Q}}/\mathbb{Q})$ on S and the second from the (Galois) action of $\mathrm{Gal}(\overline{\mathbb{Q}}/\mathbb{Q})$ on $\overline{\mathbb{Q}}$. See Langlands [1, p.220].

2. Extensions of $\mathrm{Gal}(\overline{\mathbb{Q}}/\mathbb{Q})$ by S.

By an extension of $\mathrm{Gal}(\overline{\mathbb{Q}}/\mathbb{Q})$ by S we shall mean a projective system

$$\begin{array}{ccccccccc}
1 & \longrightarrow & S^{L'} & \longrightarrow & \underset{\sim}{T}^{L'} & \longrightarrow & \mathrm{Gal}(L'^{ab}/\mathbb{Q}) & \longrightarrow & 1 \\
 & & \downarrow{\scriptstyle N_{L'/L}} & & \downarrow & & \downarrow{\scriptstyle \mathrm{can}} & & \\
1 & \longrightarrow & S^{L} & \longrightarrow & \underset{\sim}{T}^{L} & \longrightarrow & \mathrm{Gal}(L^{ab}/\mathbb{Q}) & \longrightarrow & 1
\end{array} \qquad (L \subset L')$$

of extensions of $\mathbb{Q}$-rational pro-algebraic groups; the indexing set is all finite Galois extensions of $\mathbb{Q}$ contained in $\overline{\mathbb{Q}}$. The group $\mathrm{Gal}(L^{ab}/\mathbb{Q})$ is to be regarded as a pro-system of finite constant algebraic groups in the obvious way, and the action of $\mathrm{Gal}(L^{ab}/\mathbb{Q})$ on S^L determined by the extension is to be the algebraic action described in (1.8). On passing to the limit we obtain an extension

$$1 \longrightarrow S \longrightarrow \underset{\sim}{T} \longrightarrow \mathrm{Gal}(\overline{\mathbb{Q}}/\mathbb{Q}) \longrightarrow 1 .$$

We shall always assume there to be a splitting of the extension over $\mathbb{A}^f$, i.e., a compatible family of continuous homomorphic sections $sp^L\colon \mathrm{Gal}(L^{ab}/\mathbb{Q}) \to \underset{\sim}{T}^L(\mathbb{A}^f)$. In the limit this defines a continuous homomorphism $sp\colon \mathrm{Gal}(\overline{\mathbb{Q}}/\mathbb{Q}) \to \underset{\sim}{T}(\mathbb{A}^f)$.

Fix an L. The general theory of affine group schemes (Demazure-Gabriel [1,V.2]) shows that, for some finite quotient $\mathfrak{g}'$ of $\mathfrak{g} = \mathrm{Gal}(L^{ab}/\mathbb{Q})$, $\underset{\sim}{T}^L$ will be the pull-back of an extension of $\mathfrak{g}'$ by S^L:

$$\begin{array}{ccccccccc} 1 & \longrightarrow & S^L & \longrightarrow & \underset{\sim}{T}^L & \longrightarrow & \mathfrak{g} & \longrightarrow & 1 \\ & & \| & & \downarrow & & \downarrow & & \\ 1 & \longrightarrow & S^L & \longrightarrow & \underset{\sim}{T}' & \longrightarrow & \mathfrak{g}' & \longrightarrow & 1 . \end{array}$$

Since S^L splits over L, Hilbert's theorem 90 shows that $H^1(L,S^L) = 0$, and so $\underset{\sim}{T}'(L) \twoheadrightarrow \mathfrak{g}'$ is surjective. Thus we can choose a section $a'\colon \mathfrak{g}' \to \underset{\sim}{T}'_L$, which will automatically be a morphism of algebraic varieties. On pulling back to $\mathfrak{g}$, we get a section $a = a^L$: $\mathrm{Gal}(L^{ab}/\mathbb{Q}) \to \underset{\sim}{T}^L_L$ which is a morphism of pro-algebraic varieties. The choice of such an a gives us the following data.

(2.1). A 2-cocycle (d_{τ_1,τ_2}) for $\mathrm{Gal}(L^{ab}/\mathbb{Q})$ with values in the algebraic group S^L_L, defined by $d_{\tau_1,\tau_2} = a(\tau_1)a(\tau_2)a(\tau_1\tau_2)^{-1}$.

(2.2). A family of 1-cocycles $c(\tau) \in Z^1(L/\mathbb{Q}, S^L(L))$, one for each $\tau \in \mathrm{Gal}(L^{ab}/\mathbb{Q})$, defined by $c_\sigma(\tau)a(\tau) = \sigma a(\tau)$. ($\mathrm{Gal}(L/\mathbb{Q})$ acts on $S^L(L)$ through its action on the field L.)

(2.3). A continuous map b: $\mathrm{Gal}(L^{ab}/\mathbb{Q}) \to S^L(\mathbb{A}^f_L)$ defined by $b(\tau)\mathrm{sp}^L(\tau) = a(\tau)$.

These satisfy the following relations:

(2.4). $d_{\tau_1,\tau_2} \cdot c_\sigma(\tau_1) \cdot \tau_1(c_\sigma(\tau_2)) = \sigma d_{\tau_1,\tau_2} \cdot c_\sigma(\tau_1\tau_2)$,

(2.5). $d_{\tau_1,\tau_2} = b(\tau_1) \cdot \tau_1 b(\tau_2) \cdot b(\tau_1\tau_2)^{-1}$,

(2.6). $c_\sigma(\tau) = b(\tau)^{-1} \cdot \sigma(b(\tau))$

for $\tau_1, \tau_2, \tau \in \mathrm{Gal}(L^{ab}/\mathbb{Q})$ and $\sigma \in \mathrm{Gal}(L/\mathbb{Q})$. (We have used the convention that $\tau \in \mathrm{Gal}(L^{ab}/\mathbb{Q})$ acts on $S^L(L)$ through its action on S^L, and $\sigma \in \mathrm{Gal}(L/\mathbb{Q})$ acts on $S^L(L)$ through its action on the field L.) In fact, the first relation is a consequence of the other two.

Note that b determines (d_{τ_1,τ_2}) and the $(c_\sigma(\tau))$, and that the image $\bar{b}(\tau)$ of $b(\tau)$ in $S^L(\mathbb{A}^f_L)/S^L(L)$ is uniquely determined by the extension and sp^L (independently of the choice of a) .

<u>Proposition 2.7.</u> A mapping $\bar{b}$: $\mathrm{Gal}(L^{ab}/\mathbb{Q}) \to S^L(\mathbb{A}_L^f)/S^L(L)$ arises (as above) from an extension of S^L by $\mathrm{Gal}(L^{ab}/\mathbb{Q})$ and a splitting if and only if it satisfies the following conditions:

(a) $\sigma(\bar{b}(\tau)) = \bar{b}(\tau)$, all $\tau \in \mathrm{Gal}(L^{ab}/\mathbb{Q})$, $\sigma \in \mathrm{Gal}(L/\mathbb{Q})$;

(b) $\bar{b}(\tau_1\tau_2) = \bar{b}(\tau_1)\cdot\tau_1\bar{b}(\tau_2)$, all $\tau_1, \tau_2 \in \mathrm{Gal}(L^{ab}/\mathbb{Q})$;

(c) $\bar{b}$ lifts to a continuous map b: $\mathrm{Gal}(L^{ab}/\mathbb{Q}) \to S^L(\mathbb{A}_L^f)$ such that the map $(\tau_1,\tau_2) \mapsto d_{\tau_1,\tau_2} \overset{df}{=} b(\tau_1)\cdot\tau_1 b(\tau_2)\cdot b(\tau_1\tau_2)^{-1}$ is locally constant. Moreover, the extension (together with the splitting) is determined by $\bar{b}$ up to isomorphism.

<u>Proof.</u> We shall only show how to construct the extension from $\bar{b}$, the rest being easy. Choose a lifting b of $\bar{b}$ as in (c) . The family d_{τ_1,τ_2} is a 2-cocycle which takes values in the algebraic group S_L^L . It therefore defines an extension

$$1 \longrightarrow S_L^L \longrightarrow \underset{\sim}{T}_L^L \longrightarrow \mathrm{Gal}(L^{ab}/\mathbb{Q}) \longrightarrow 1$$

of pro-algebraic groups over L together with a section a: $\mathrm{Gal}(L^{ab}/\mathbb{Q}) \to \underset{\sim}{T}_L^L$ that is a morphism of pro-varieties. Define $\underset{\sim}{T}^L$ to be the pro-algebraic group scheme over $\mathbb{Q}$ such that $\underset{\sim}{T}^L(\bar{\mathbb{Q}}) = \underset{\sim}{T}_L^L(\bar{\mathbb{Q}})$ with $\mathrm{Gal}(\bar{\mathbb{Q}}/\mathbb{Q})$ acting by the formula:

$\sigma(s\cdot a(\tau)) = c_\sigma(\tau)\cdot \sigma s\cdot a(\tau)$, $\sigma \in \mathrm{Gal}(\overline{\mathbb{Q}}/\mathbb{Q})$, $s \in S^L(\overline{\mathbb{Q}})$

$\tau \in \mathrm{Gal}(L^{ab}/\mathbb{Q})$, $c_\sigma(\tau) \overset{\mathrm{df}}{=} b(\tau)^{-1}\cdot \sigma b(\tau) \in S^L(L)$. There is an exact sequence

$$1 \longrightarrow S^L \longrightarrow \underset{\sim}{T}^L \longrightarrow \mathrm{Gal}(L^{ab}/\mathbb{Q}) \longrightarrow 1 .$$

For each $\tau \in \mathrm{Gal}(L^{ab}/\mathbb{Q}$, $b(\tau)^{-1}a(\tau) \in S^L(\mathbb{A}^f_L)^{\mathrm{Gal}(L/\mathbb{Q})} = S^L(\mathbb{A}^f)$, and $\tau \longmapsto sp(\tau) \overset{\mathrm{df}}{=} b(\tau)^{-1}a(\tau)$ is a homomorphism. As b is continuous, so also is sp .

<u>Corollary 2.8</u>. To define an extension of $\mathrm{Gal}(\overline{\mathbb{Q}}/\mathbb{Q})$ by S (together with a splitting over $\mathbb{A}^f$) it suffices to give maps $\bar{b}^L$: $\mathrm{Gal}(L^{ab}/\mathbb{Q}) \to S^L(\mathbb{A}^f_L)/S^L(L)$ satisfying the conditions of (2.7) and such that, whenever $L \subset L'$,

$$\begin{array}{ccccc}
\mathrm{Gal}(L'^{ab}/\mathbb{Q}) & & \xrightarrow{\quad \bar{b}^{L'} \quad} & & S^{L'}(\mathbb{A}^f_{L'})/S^{L'}(L') \\
\Big\downarrow \text{can} & & & & \Big\downarrow N_{L'/L} \\
\mathrm{Gal}(L^{ab}/\mathbb{Q}) & \xrightarrow{\bar{b}^L} & S^L(\mathbb{A}^f_L)/S^L(L) & \hookrightarrow & S^L(\mathbb{A}^f_{L'})/S^L(L')
\end{array}$$

commutes.

<u>Remark 2.9</u>. Let $\underset{\sim}{T}$ be an extension of $\mathrm{Gal}(\overline{\mathbb{Q}}/\mathbb{Q})$ by S . For any $\tau \in \mathrm{Gal}(\overline{\mathbb{Q}}/\mathbb{Q})$, multiplication in $\underset{\sim}{T}$ makes $\pi^{-1}(\tau)$ into a torsor for S , and $sp(\tau)$ is a point of the torsor with values in $\mathbb{A}^f$ (i.e. a trivialization of the torsor over $\mathbb{A}^f$) . In the above we have implictly regarded $\pi^{-1}(\tau)$ as a left torsor, because that is the convention of Langlands [1]. It is however

both more convenient and more conventional to regard $\pi^{-1}(\tau)$ as a right S-torsor. With this point of view it is natural to associate with $\underset{\sim}{T}$ cocycles $(\gamma_\sigma(\tau))$ and a map β defined as follows: let L be a finite Galois extension of $\mathbb{Q}$ and choose a section $\tau \mapsto a(\tau)$ to $\underset{\sim}{T}^L \to \mathrm{Gal}(L^{ab}/\mathbb{Q})$ that is a morphism of pro-algebraic varieties; then

$$\sigma a(\tau) = a(\tau)\gamma_\sigma(\tau), \quad \text{for } \sigma \in \mathrm{Gal}(L/\mathbb{Q}),\ \tau \in \mathrm{Gal}(L^{ab}/\mathbb{Q}), \text{ and}$$

$$sp(\tau)\beta(\tau) = a(\tau) \quad \text{for } \tau \in \mathrm{Gal}(L^{ab}/\mathbb{Q}) .$$

The following relations hold:

$$\gamma_\sigma(\tau) = \beta(\tau)^{-1} \cdot \sigma(\beta(\tau)) ,$$

$$\overline{\beta}(\tau_1\tau_2) = \tau_2^{-1}\overline{\beta}(\tau_1) \cdot \overline{\beta}(\tau_2) .$$

The new objects are related to the old as follows:

$$\gamma_\sigma(\tau) = \tau^{-1}c_\sigma(\tau) ,$$

$$\beta(\tau) = \tau^{-1}b(\tau) .$$

Define $c'(\tau)$ and $b'(\tau)$ by the formulas (2.2) and (2.3) but with $a(\tau)$ replaced by the section $\tau \mapsto a'(\tau) = a(\tau^{-1})^{-1}$. Then

$$\gamma_\sigma(\tau) = c'_\sigma(\tau^{-1})^{-1} ,$$

$$\beta(\tau) = b'(\tau^{-1})^{-1} .$$

In particular, we see that $\gamma(\tau)$ and $c(\tau^{-1})^{-1}$ are cohomologous and $\overline{\beta}(\tau) = \overline{b}(\tau^{-1})^{-1}$.

Example 2.10. In the preceding discussion there is no need to take the base field to be $\mathbb{Q}$. We shall use this method to construct for any number field $L \subset \overline{\mathbb{Q}}$, a canonical extension

$$1 \longrightarrow S^L \longrightarrow (\underset{\sim}{T}^L)^{ab} \xrightarrow{\pi} \mathrm{Gal}(L^{ab}/L) \longrightarrow 1$$

of pro-algebraic groups over $\mathbb{Q}$, together with a splitting over $\mathbb{A}^f$. According to (2.7), such an extension corresponds to a map $\overline{b}$: $\mathrm{Gal}(L^{ab}/L) \to S^L(\mathbb{A}_L^f)/S^L(L)$ satisfying conditions similar to (a), (b), and (c) of that proposition. In fact we shall define a map $\overline{b}$: $\mathrm{Gal}(L^{ab}/L) \to S^L(\mathbb{A}^f)/S^L(\mathbb{Q}) \subset S^L(\mathbb{A}_L^f)/S^L(L)$ and so (a) will be obvious (and the cocycles $c(\tau)$ trivial). Note that $\mathrm{Gal}(L^{ab}/L)$ acts trivially on S^L and so (b) requires that $\overline{b}$ be a homomorphism.

The canonical element $\mu^L \in X_*(S^L)$ is defined over L, and so gives rise to a homomorphism of algebraic groups,

$$NR:\ L^\times \xrightarrow{\mathrm{Res}_{L/\mathbb{Q}}(\mu^L)} \mathrm{Res}_{L/\mathbb{Q}}\, S_L^L \xrightarrow{N_{L/\mathbb{Q}}} S^L .$$

Consider

$$\begin{array}{lccc} NR(\mathbb{A}): & \mathbb{A}_L^\times & \longrightarrow & S^L(\mathbb{A}) \\ & \cup & & \cup \\ NR(L): & L^\times & \longrightarrow & S^L(\mathbb{Q}) \end{array}$$

The reciprocity morphism (Deligne [2,2.2.3])

$$r_L = r_L(S^L, h^L): \quad \mathrm{Gal}(L^{ab}/L) \longrightarrow S^L(\mathbb{A}^f)/S^L(\mathbb{Q})$$

is defined to be the reciprocal of the composite of the following maps: the reciprocity law isomorphism $\mathrm{Gal}(L^{ab}/L) \overset{\approx}{\longleftrightarrow} \pi_0(\mathbb{A}_L^\times/L^\times)$, the map $\pi_0(\mathbb{A}_L^\times/L^\times) \to \pi_0(S^L(\mathbb{A})/S^L(\mathbb{Q}))$ defined by NR, and the projection $\pi_0(S^L(\mathbb{A})/S^L(\mathbb{Q})) \to S^L(\mathbb{A}^f)/S^L(\mathbb{Q}))$. We define $\bar{b}(\tau) = r_L(\tau)^{-1}$. It satisfies (a) and (b) of (2.7).

According to (1.2), $S^L(\mathbb{Q})$ is a discrete subgroup of $S^L(\mathbb{A}^f)$, and hence of $S^L(\mathbb{A})$. Thus there is an open subgroup U of $\mathbb{A}_L^\times$ such that $NR: \mathbb{A}_L^\times \to S^L(\mathbb{A})$ is 1 on $U \cap L^\times$. If $F \supset L$ corresponds to $U \subset \mathbb{A}_L^\times$, then there is a commutative diagram

$$\begin{array}{ccccccc}
1 \longrightarrow & \mathrm{Gal}(L^{ab}/F) & \longrightarrow & \mathrm{Gal}(L^{ab}/L) & \longrightarrow & \mathrm{Gal}(F/L) \longrightarrow 1 \\
& \downarrow b & \swarrow b \text{ (dashed)} & \downarrow \bar{b} & & \\
& S^L(\mathbb{A}^f) & \longrightarrow & S^L(\mathbb{A}^f)/S(L) & &
\end{array}$$

in which $b^{-1}: \mathrm{Gal}(L^{ab}/F) \to S^L(\mathbb{A})$ is induced by $NR: U/U \cap L^\times \to S^L(\mathbb{A})$. It is easy to extend b to a continuous map $\mathrm{Gal}(L^{ab}/L) \to S^L(\mathbb{A}^f)$ lifting $\bar{b}$: choose a set S' of representatives for $\mathrm{Gal}(F/L)$ in $\mathrm{Gal}(L^{ab}/L)$, choose an element $b(s) \in S^L(\mathbb{A}^f)$ mapping to $\bar{b}(s)$ for each $s \in S'$, and define $b(sg) = b(s)b(g)$ for $s \in S'$, $g \in \mathrm{Gal}(L^{ab}/F)$. This map b satisfies (c) of (2.7) because, when restricted to $\mathrm{Gal}(L^{ab}/F)$, it is a homomorphism.

Remark 2.11. The extension constructed in (2.10) is, up to sign, that defined by Serre [1]. For a sufficiently large modulus $\mathfrak{m}$ the group $T_{\mathfrak{m}} = T/\bar{E}_{\mathfrak{m}}$ of (ib.,p II-8) is the Serre group S^L, and $C_{\mathfrak{m}} = \mathrm{Gal}(L_{\mathfrak{m}}/L)$ for some $L_{\mathfrak{m}} \subset L^{ab}$. Thus the sequence (ib.,p II-9) can be written

$$1 \longrightarrow S^L \longrightarrow S_{\mathfrak{m}} \longrightarrow \mathrm{Gal}(L_{\mathfrak{m}}/L) \longrightarrow 1 .$$

On passing to the limit over increasing $\mathfrak{m}$, this becomes

$$1 \longrightarrow S^L \longrightarrow (\underset{\sim}{T}^L)^{ab} \longrightarrow \mathrm{Gal}(L^{ab}/L) \longrightarrow 1 .$$

The splitting (over $\mathbb{Q}_\ell$) is defined in (ib., 2.3).

§3. The Taniyama group.

We denote the Weil group of a local or global field L by W_L. Let v denote the prime induced on $\bar{\mathbb{Q}}$, or a subfield L of $\bar{\mathbb{Q}}$, by the fixed inclusion $\bar{\mathbb{Q}} \hookrightarrow \mathbb{C}$, and let L_v denote the closure of L in $\bar{\mathbb{Q}}_v = \mathbb{C}$. According to Tate [2] there is a homomorphism $i_v\colon W_{\mathbb{Q}_v} \to W_{\mathbb{Q}}$ such that the diagrams

$$\begin{array}{ccc} L_v & \xrightarrow[\approx]{r_v} & W_{L_v}^{ab} \\ \downarrow{\scriptstyle \mathrm{can}} & & \downarrow{\scriptstyle i_v^{ab}} \\ C_L & \xrightarrow[\approx]{r} & W_L^{ab} \end{array} \qquad\qquad \begin{array}{ccc} W_{\mathbb{Q}_v} & \xrightarrow{\phi_v} & \mathrm{Gal}(\bar{\mathbb{Q}}_v/\mathbb{Q}_v) \\ \downarrow{\scriptstyle i_v} & & \cap \\ W_{\mathbb{Q}} & \xrightarrow{\phi} & \mathrm{Gal}(\bar{\mathbb{Q}}/\mathbb{Q}) \end{array}$$

commute for all number fields L contained in $\bar{\mathbb{Q}}$. The constructions that follow will be independent of the choice of

i_v, but we shall ignore this question by fixing an i_v. If $L \subset \overline{\mathbb{Q}}$ is a finite Galois extension of $\mathbb{Q}$ then i_v induces a map from $W_{L_v/\mathbb{Q}_v} \overset{\text{df}}{=} W_{\mathbb{Q}_v}/W^c_{L_v}$ to $W_{L/\mathbb{Q}} \overset{\text{df}}{=} W_{\mathbb{Q}}/W^c_L$ which makes

$$\begin{array}{ccccccccc} 1 & \longrightarrow & L_v^\times & \longrightarrow & W_{L_v/\mathbb{Q}_v} & \longrightarrow & \mathrm{Gal}(L_v/\mathbb{Q}_v) & \longrightarrow & 1 \\ & & \downarrow & & \downarrow i_v & & \downarrow & & \\ 1 & \longrightarrow & C_L & \longrightarrow & W_{L/\mathbb{Q}} & \longrightarrow & \mathrm{Gal}(L/\mathbb{Q}) & \longrightarrow & 1 \end{array} \tag{3.1}$$

commute.

We note that there is a commutative diagram

$$\begin{array}{ccccccccc} 1 & \longrightarrow & C_L & \longrightarrow & W_{L/\mathbb{Q}} & \longrightarrow & \mathrm{Gal}(L/\mathbb{Q}) & \longrightarrow & 1 \\ & & \downarrow & & \downarrow & & \| & & \\ 1 & \rightarrow & \mathrm{Gal}(L^{ab}/L) & \rightarrow & \mathrm{Gal}(L^{ab}/\mathbb{Q}) & \longrightarrow & \mathrm{Gal}(L/\mathbb{Q}) & \longrightarrow & 1 \end{array} \tag{3.2}$$

in which the vertical arrows are surjective.

Let T be a torus over $\mathbb{Q}$; by analogy with $T(L) = X_*(T) \otimes L^\times$, $T(\mathbb{A}^f) = X_*(T) \otimes \mathbb{A}^f$ etc., we shall write $T(C_L)$ for $X_*(T) \otimes C_L$. If $\mu \in X_*(T)$ and a belongs to a $\mathbb{Q}$-algebra R (or C_L) then we write a^μ for $\mu \otimes a \in T(R)$.

Fix such a torus T and an element $\mu \in X_*(T)$, and let $L \subset \overline{\mathbb{Q}}$ be a number field splitting T. For each $\tau \in \mathrm{Gal}(L^{ab}/\mathbb{Q})$ that satisfies

$$(1 + \iota)(\tau^{-1} - 1)\mu = 0 \tag{3.3}$$

and lifting $\tilde{\tau}$ of τ to $W_{L/\mathbb{Q}}$ (using the map in (3.2)) we shall define an element $b_0(\tilde{\tau}, \mu) \in T(C_L)/T(L_\infty^\times)$, where $L_\infty = L \otimes_{\mathbb{Q}} \mathbb{R}$.

Choose a section $\sigma \longmapsto w_\sigma$ to $W_{L/\mathbb{Q}} \longrightarrow \mathrm{Gal}(L/\mathbb{Q})$ such that:

(3.4a) $w_1 = 1$;

(3.4b) $w_\iota \in W_{L_v/\mathbb{Q}_v} \subset W_{L/\mathbb{Q}}$;

(3.4c) for some choice of H containing 1 and such that $Gal(L/\mathbb{Q}) = H \cup H\iota$ (disjoint union), $w_{\sigma\iota} = w_\sigma w_\iota$ for all $\sigma \in H$.

Of course, the last two conditions are trivial if $L \subset \mathbb{R}$.

Corresponding to w there is a 2-cocycle $(a_{\sigma,\tau})$, defined by $w_\sigma w_\tau = a_{\sigma,\tau} w_{\sigma\tau}$. Let $\tau \in Gal(L^{ab}/\mathbb{Q})$ satisfy (3.3) and let $\tilde{\tau} \in W_{L/\mathbb{Q}}$ map to it. Choose $c_{\sigma,\tilde{\tau}} \in C_L$ to satisfy $w_\sigma \tilde{\tau} = c_{\sigma,\tilde{\tau}} w_{\sigma\tau}$, and define

$$b_0(\tilde{\tau},\mu) = \prod_{\sigma \in Gal(L/\mathbb{Q})} c_{\sigma,\tilde{\tau}}^{\sigma\mu} \in T(C_L)/T(L_\infty).$$

Lemma 3.5. The element $b_0(\tilde{\tau},\mu)$ is independent of the choice of the section w; it is fixed by $Gal(L/\mathbb{Q})$.

Proof. (Langlands [1. p. 221; p. 223].) Suppose $\sigma \mapsto w'_\sigma = e_\sigma w_\sigma$, $e_\sigma \in C_L$, is another section. We use $'$ to denote objects defined using this section. It is easy to see that

$$c'_{\sigma,\tilde{\tau}} = e_{\sigma\tau}^{-1} e_\sigma c_{\sigma,\tilde{\tau}} \quad \text{for all } \sigma \in Gal(L/\mathbb{Q}).$$

Therefore

$$b_0(\tilde{\tau},\mu)' = \{\prod_{\sigma \in Gal(L/\mathbb{Q})} (e_{\sigma\tau}^{\sigma\mu})^{-1} e_\sigma^{\sigma\mu}\} \, b_0(\tilde{\tau},\mu).$$

We have to show that the product in $\{ \}$ is congruent to 1 modulo $L_\infty^\times$. Consider $\sigma \in Gal(L_v/\mathbb{Q}_v)$. Because of (3.4b), we have $e_\sigma \in L_v^\times$ and hence $\rho(e_\sigma) \in L_\infty^\times$ for all $\rho \in \mathrm{Gal}(L/\mathbb{Q})$. As

$$a'_{\rho,\sigma} = e_\rho \rho(e_\sigma)\, e_{\rho\sigma}^{-1}\, a_{\rho,\sigma} ,$$

and both $a_{\rho,\sigma}$ and $a'_{\rho,\sigma}$ belong to $L_\infty^\times$, we have $e_\rho \equiv e_{\rho\sigma}$ (mod $L_\infty^\times$) . Thus

$$\prod_{\sigma \in \mathrm{Gal}(L/\mathbb{Q})} (e_{\sigma\tau}^{\sigma\mu})^{-1}\, e_\sigma^{\sigma\mu} = \prod_\sigma (e_\sigma^{\sigma\tau^{-1}\mu})\, e_\sigma^{\sigma\mu}$$

$$= \prod_\sigma e_\sigma^{\sigma(1-\tau^{-1})\mu}$$

$$= \prod_{\eta \in H} \prod_{\sigma \in \mathrm{Gal}(L_v/\mathbb{Q}_v)} e_{\eta\sigma}^{\eta\sigma(1-\tau^{-1})\mu}$$

is congruent modulo $L_\infty^\times$ to

$$\prod_{\eta \in H} \prod_{\sigma \in \mathrm{Gal}(L_v/\mathbb{Q}_v)} e_\eta^{\eta\sigma(1-\tau^{-1})\mu} ,$$

which is 1 , because in view of (3.3) ,

$$\sum_{\sigma \in \mathrm{Gal}(L_v/\mathbb{Q}_v)} \sigma(1-\tau^{-1})\mu = 0 .$$

Next we show that $b_0(\tilde{\tau},\mu)$ is fixed by $\mathrm{Gal}(L/\mathbb{Q})$. We have

$$\rho(c_{\sigma,\tilde{\tau}}) = a_{\rho,\sigma}\, a^{-1}_{\rho,\sigma\tau}\, c_{\rho\sigma,\tilde{\tau}} \quad \text{for all} \quad \rho,\sigma \in \mathrm{Gal}(L/\mathbb{Q}) \ ,$$

and hence

$$\rho(b_o(\tilde{\tau},\mu)) = \{\prod_{\sigma} (a_{\rho,\sigma}\, a^{-1}_{\rho,\sigma\tau})^{\rho\sigma\mu}\}\, b_o(\tilde{\tau},\mu) \ .$$

We can write the product in { } as

$$\prod_{\sigma} (a_{\rho,\sigma}{}^{\rho\sigma\mu})(a^{-1}_{\rho,\sigma}{}^{\rho\sigma\tau^{-1}\mu}) = \prod_{\sigma} a_{\rho,\sigma}{}^{\rho\sigma(1-\tau^{-1})\mu}$$

$$= \prod_{\eta \in H} \ \prod_{\sigma \in \mathrm{Gal}(L_v/\mathbb{Q}_v)} a_{\rho,\eta\sigma}{}^{\rho\eta\sigma(1-\tau^{-1})\mu} \ .$$

In view of (3.4c) we have $a_{\rho,\eta\sigma} \equiv a_{\rho,\eta} \pmod{L_\infty^\times}$ for all $\eta \in H$ and $\sigma \in \mathrm{Gal}(L_v/\mathbb{Q}_v)$. Hence the above product is congruent modulo $L_\infty^\times$ to

$$\prod_{\eta \in H} \ \prod_{\sigma \in \mathrm{Gal}(L_v/\mathbb{Q}_v)} a_{\rho,\eta}{}^{\rho\eta\sigma(1-\tau^{-1})\mu} \ ,$$

which is 1 because of (3.3) .

On tensoring

$$\begin{array}{ccccccccc} 1 & \longrightarrow & L^{\times} & \longrightarrow & \mathbb{A}_L^{f\times} & \longrightarrow & \mathbb{A}_L^{f\times}/L^{\times} & \longrightarrow & 1 \\ & & {\scriptstyle -1}\downarrow & & \downarrow & & \downarrow{\scriptstyle\approx} & & \\ 1 & \longrightarrow & L_\infty^{\times} & \longrightarrow & C_L & \longrightarrow & C_L/L_\infty^{\times} & \longrightarrow & 1 \end{array} \qquad (3.6a)$$

with $X_*(T)$ we obtain an exact commutative diagram

$$\begin{array}{ccccccccc} 1 & \longrightarrow & T(L) & \longrightarrow & T(\mathbb{A}_L^{f}) & \longrightarrow & T(\mathbb{A}_L^{f})/T(L) & \longrightarrow & 1 \\ & & {\scriptstyle -1}\downarrow & & \downarrow & & \downarrow{\scriptstyle\approx} & & \\ 1 & \longrightarrow & T(L_\infty) & \longrightarrow & T(C_L) & \longrightarrow & T(C_L)/T(L_\infty) & \longrightarrow & 1 \end{array} \qquad (3.6b)$$

(The -1 reminds us that the map is the reciprocal of the obvious inclusion.) We define $\bar{b}(\tilde{\tau},\mu)$ to be the element of $T(\mathbb{A}_L^f)/T(L)$ corresponding to $b_0(\tilde{\tau},\mu)$. Lemma 3.5 shows that it lies in $(T(\mathbb{A}_L^f)/T(L))^{\mathrm{Gal}(L/\mathbb{Q})}$ and hence gives rise to an element $c(\tilde{\tau},\mu) \in H^1(L/\mathbb{Q}, T(L))$ through the boundary map in the exact sequence

$$1 \to T(\mathbb{Q}) \longrightarrow T(\mathbb{A}^f) \longrightarrow (T(\mathbb{A}_L^f)/T(L))^{\mathrm{Gal}(L/\mathbb{Q})} \longrightarrow H^1(L/\mathbb{Q}, T(L)).$$

<u>Lemma 3.7.</u> The cohomology class $c(\tilde{\tau},\mu)$ depends only on the image of $\tilde{\tau}$ in $\mathrm{Gal}(L/\mathbb{Q})$.

<u>Proof</u>. Suppose $\tilde{\tau}'$ and $\tilde{\tau}$ have the same image in $\mathrm{Gal}(L/\mathbb{Q})$; then $\tilde{\tau}' = u\tilde{\tau}$ with $u \in C_L$, and $c_{\sigma,\tilde{\tau}'} = \sigma(u)c_{\sigma,\tilde{\tau}}$. Thus $b_0(\tilde{\tau},\mu)$ is multiplied by $\Pi\sigma(u)^{\sigma\mu} = NR(u)$, where NR is the map of algebraic groups $L^{\times} \xrightarrow{\mathrm{Res}(\mu)} \mathrm{Res}_{L/\mathbb{Q}} T_L \xrightarrow{N_{L/\mathbb{Q}}} T$. Choose an element $\tilde{u} \in \mathbb{A}_L^f$ such that $\tilde{u}$ and u represent the same element in $C_L/L_\infty^{\times}$. Then $NR(\tilde{u}) \in T(\mathbb{A}^f)$ has the same image as $NR(u)$ in $T(C_L)/T(L_\infty^{\times})$, and we see that $\bar{b}(\tilde{\tau}',\mu) = \overline{NR}(\tilde{u})\ \bar{b}(\tilde{\tau},\mu)$ where $\overline{NR}(\tilde{u})$ denotes the image of $NR(\tilde{u})$ in $T(\mathbb{A}^f)/T(\mathbb{Q}) \subset T(\mathbb{A}_L^f)/T(L)$. Hence $c(\tilde{\tau},\mu) = c(\tilde{\tau}',\mu)$.

Thus we can write $c(\tau,\mu)$ for $c(\tilde{\tau},\mu)$ where $\tau \in \mathrm{Gal}(L^{ab}/\mathbb{Q})$ (or even $\mathrm{Gal}(L/\mathbb{Q})$).

<u>Lemma 3.8</u>. Up to multiplication by an element of the closure $T(\mathbb{Q})^{\wedge}$ of $T(\mathbb{Q})$ in $T(\mathbb{A}^f)$, $\bar{b}(\tilde{\tau},\mu)$ depends only on τ (and not $\tilde{\tau}$).

<u>Proof</u>. From (3.2) we see that $\tilde{\tau}$ can be multiplied only by an element u of the identity component of C_L. An argument as in the proof of (3.7) shows that multiplying $\tilde{\tau}$ by u corresponds to multiplying $\bar{b}(\tilde{\tau},\mu)$ by $\overline{NR}(\tilde{u})$, where $\tilde{u}$ is a lifting of u to $\mathbb{A}_L^f$. But $\tilde{u}$ is in the closure of $L^{\times} \subset (\mathbb{A}_L^f)^{\times}$, and so $\overline{NR}(\tilde{u})$ is in the closure of $T(\mathbb{Q})$.

Thus, for any $\tau \in \mathrm{Gal}(L^{ab}/\mathbb{Q})$ satisfying (3.3), there is a well-defined element $\bar{b}(\tau,\mu) \in T(\mathbb{A}_L^f)/T(L)\ T(\mathbb{Q})^{\wedge}$.

Example 3.9. For any T and μ, $\bar{b}(\iota,\mu)$ is defined; we show that it is 1. We can take $\tilde{\iota} = w_\iota$. If $\sigma \in H$ (see 3.4), then $w_\sigma \tilde{\iota} = w_\sigma w_\iota = w_{\sigma\iota}$, and $c_{\sigma,\tilde{\iota}} = 1$; moreover $w_{\sigma\iota}\tilde{\iota} = w_\sigma w_\iota w_\iota = w_\sigma a_{\iota,\iota} = \sigma(a_{\iota,\iota}) w_\sigma$ and $c_{\sigma\iota,\tilde{\iota}} = \sigma(a_{\iota,\iota}) \in L_\infty^\times$. Clearly $b_0(\tilde{\iota},\mu) = 1$.

Proposition 3.10. Let $h : \mathbb{S} \to T_{\mathbb{R}}$ be a homomorphism and $\mu = \mu_h$ be the corresponding cocharacter. Assume that μ is defined over $E \subset L$. Then $\bar{b}(\tau,\mu)$ is defined for all $\tau \in \mathrm{Gal}(L^{ab}/E)$ and there is a commutative diagram

$$\begin{array}{ccc}
\mathrm{Gal}(L^{ab}/E) & \xrightarrow{\bar{b}(-,\mu)} & T(\mathbb{A}_L^f)/T(L)\,T(\mathbb{Q})^{\wedge} \\
\big\downarrow \mathrm{rest} & & \big\uparrow \\
\mathrm{Gal}(E^{ab}/E) & \xrightarrow{r_E(T,h)^{-1}} & T(\mathbb{A}^f)/T(\mathbb{Q})^{\wedge}
\end{array}$$

in which $r_E(T,h)$ is the reciprocity morphism (Deligne [2,2.2.3]). In particular, $c(\tau,\mu)$ is trivial.

Proof. Let $\tau \in \mathrm{Gal}(L^{ab}/E)$. Then τ fixes μ, and so (3.3) is satisfied and $\bar{b}(\tau,\mu)$ is defined. We may choose the section w to $W_{L/\mathbb{Q}} \to \mathrm{Gal}(L/\mathbb{Q})$ in such a way that $w_\tau = \tilde{\tau}$ maps to τ in $\mathrm{Gal}(L^{ab}/\mathbb{Q})$. Then $c_{\sigma,\tilde{\tau}} = a_{\sigma,\tau}$. Let R be a set of representatives for $\mathrm{Gal}(L/\mathbb{Q})/\mathrm{Gal}(L/E)$. We have

$$b_0(\tilde{\tau},\mu) = \prod_{\rho \in R} \prod_{\sigma \in \mathrm{Gal}(L/E)} a_{\rho\sigma,\tau}^{\rho\mu} \quad (\text{since } \sigma\mu = \mu)$$

$$= \prod_{\rho \in R} \Big(\prod_{\sigma}(\rho a_{\sigma,\iota} \cdot a_{\rho,\sigma\tau} \cdot a_{\rho,\sigma}^{-1})\Big)^{c\mu}$$

$$= \prod_{\rho \in R} (\rho a)^{\rho\mu}, \quad \text{where } a = \prod_{\sigma} a_{\sigma,\tau}.$$

To evaluate a, we use the commutative diagram (Tate[2, W_3])

$$\begin{array}{ccc} C_E & \xrightarrow{r_E} & W_E^{ab} \\ \downarrow & & \downarrow t \\ C_L & \xrightarrow{r_L} & W_L^{ab} \end{array}$$

where t is the transfer map arising from the inclusion $W_L \hookrightarrow W_E$. Clearly $r_L(a) = \prod r_L(a_{\sigma,\tau}) = t(\tilde{\gamma} W_E^c)$. Thus a is an element of C_E that maps to $\tau|E^{ab}$ in $\mathrm{Gal}(E^{ab}/E)$. Let $\tilde{a} \in \mathbb{A}_E^f$ represent the same element in $C_E/E_\infty^\times$ as a. Then $\bar{b}(\tau,\mu)$ is the image of $\dot{a}$ under $\mathbb{A}_E^{f\times} \xrightarrow{NR} T(\mathbb{A}^f)/T(\mathbb{Q})^\wedge$, and this equals $r_E(T,h)(\tau|E^{ab})^{-1}$.

We now apply the above theory to construct the Taniyama group of a finite Galois extension L of $\mathbb{Q}$, $L \subset \bar{\mathbb{Q}}$. To do so, we take the torus T to be S^L and μ to be the canonical co-character of S^L (see §1). Since $S^L(\mathbb{Q})$ is closed in $S^L(\mathbb{A}^f)$ the above constructions give a map $\mathrm{Gal}(L^{ab}/\mathbb{Q}) \longrightarrow (S^L(\mathbb{A}_L^f)/S^L(L))^{\mathrm{Gal}(L/\mathbb{Q})}$ which we denote by $\bar{b}$ (or $\bar{b}^L$).

Proposition 3.11. The map $\bar{b}$ satisfies the conditions of (2.7) and so defines an extension

$$1 \longrightarrow S^L \longrightarrow \underset{\sim}{T}^L \longrightarrow \mathrm{Gal}(L^{ab}/\mathbb{Q}) \longrightarrow 1$$

together with a continuous splitting over $\mathbb{A}^f$.

Proof. We have already observed (Lemma 3.5) that $\bar{b}(\tau)$ is fixed by $\mathrm{Gal}(L/\mathbb{Q})$ for all τ. To show $\bar{b}$ satisfies (2.7b), let $\tau = \tau_1\tau_2$ and lift τ_1 and τ_2 to elements $\tilde{\tau}_1$ and $\tilde{\tau}_2$ of $W_{L/\mathbb{Q}}$. We take $\tilde{\tau}_1\tilde{\tau}_2$ to be the lift of $\tau = \tau_1\tau_2$. Then we have

$$c_{\sigma,\tilde{\tau}} = c_{\sigma,\tilde{\tau}_1}\, c_{\sigma\tau_1,\tilde{\tau}_2} \quad \text{for all} \quad \sigma \in \mathrm{Gal}(L/\mathbb{Q}) .$$

Hence

$$b_o(\tilde{\tau},\mu) = \prod_{\sigma} (c_{\sigma,\tilde{\tau}_1})^{\sigma\mu} \cdot \prod_{\sigma} (c_{\sigma\tau_1,\tilde{\tau}_2})^{\sigma\mu} .$$

The first factor is $b_o(\tilde{\tau}_1,\mu)$, and the second one is $\prod_{\sigma} (c_{\sigma,\tilde{\tau}_2})^{\sigma\tau_1^{-1}\mu}$, which is $\tau_1(b_o(\tilde{\tau}_2,\mu))$ (recall that the action of τ_1 on S^L is the 'algebraic' one, see §1.8). Thus $\bar{b}(\tau_1\tau_2) = \bar{b}(\tau_1)\,\tau_1(\bar{b}(\tau_2))$. To prove (2.7c), consider the diagram

$$\begin{array}{ccc} \mathrm{Gal}(L^{ab}/\mathbb{Q}) & \xrightarrow{\ \bar{b}\ } & S^L(\mathbb{A}_L^f)/S^L(L) \\ \uparrow & & \uparrow \\ \mathrm{Gal}(L^{ab}/L) & \xrightarrow{\ b\ } & S^L(\mathbb{A}^f) \end{array}$$

where b is the map defined in (2.10). The diagram commutes because of (3.10). It is easy to extend b to a continuous map $\mathrm{Gal}(L^{ab}/\mathbb{Q}) \to S^L(\mathbb{A}^f)$ lifting $\bar{b}$ (see the proof of 2.10). Then b satisfies (2.7c) because its restriction to $\mathrm{Gal}(L^{ab}/F)$ is a homomorphism, where F is the finite extension of L defined in the proof of (2.10).

The extension, together with the splitting, is the Taniyama group of L. The next lemma implies that the Taniyama groups for varying L form a projective system: we have an extension of $\mathrm{Gal}(\bar{\mathbb{Q}}/\mathbb{Q})$ by S in the sense of §2.

<u>Lemma 3.12</u>. If $L' \supset L$ then

$$\begin{array}{ccccc}
\mathrm{Gal}(L'^{ab}/\mathbb{Q}) & & \xrightarrow{\quad\quad\bar{b}^{L'}\quad\quad} & & S^{L'}(\mathbb{A}^f_{L'})/S^{L'}(L') \\
\Big\downarrow \text{rest.} & & & & \Big\downarrow N_{L'/L} \\
\mathrm{Gal}(L^{ab}/\mathbb{Q}) & \xrightarrow{\bar{b}^{L}} & S^{L}(\mathbb{A}^f_{L})/S^{L}(L) & \hookrightarrow & S^{L}(\mathbb{A}^f_{L'})/S^{L}(L')
\end{array}$$

commutes.

<u>Proof</u>. We discuss the case $\mathrm{Gal}(L'/L) \cap \mathrm{Gal}(L'_v/\mathbb{Q}_v) = \{1\}$ first. Let R be a set of representatives for the coset space

$$\mathrm{Gal}(L'/L)\backslash \mathrm{Gal}(L'/\mathbb{Q})/\mathrm{Gal}(L'_v/\mathbb{Q}_v) .$$

We choose R such that $1 \in R$. For elements ξ in $\mathrm{Gal}(L'/L) \cup R \cup \mathrm{Gal}(L'_v/\mathbb{Q}_v)$, choose $w'_\xi \in W_{L'/\mathbb{Q}}$ lifting ξ; we choose $w'_1 = 1$ and for $\rho \in \mathrm{Gal}(L'_v/\mathbb{Q}_v)$, choose w'_ρ to be in $W_{L'_v/\mathbb{Q}_v}$. Write an element σ of $\mathrm{Gal}(L'/\mathbb{Q})$ uniquely as $\sigma = \zeta\eta\rho$ with $\zeta \in \mathrm{Gal}(L'/L)$, $\eta \in R$ and $\rho \in \mathrm{Gal}(L'_v/\mathbb{Q}_v)$, and put

$$w'_\sigma = w'_\zeta \, w'_\eta \, w'_\rho \quad .$$

Then $\sigma \mapsto w'_\sigma$ is a section of $W_{L'/\mathbb{Q}} \to \mathrm{Gal}(L'/\mathbb{Q})$ satisfying (3.4). We choose a section $\sigma \mapsto w_\sigma$ of $W_{L/\mathbb{Q}} \to \mathrm{Gal}(L/\mathbb{Q})$ as follows: for $\sigma \in \mathrm{Gal}(L/\mathbb{Q})$, σ extends to a unique $\eta\rho$ in $\mathrm{Gal}(L'/\mathbb{Q})$ with $\eta \in R$ and $\rho \in \mathrm{Gal}(L'_v/\mathbb{Q}_v)$; we take w_σ to be the image of $w'_{\eta\rho} = w'_\eta \, w'_\rho$ in $W_{L/\mathbb{Q}}$.

Let τ be an element of $\mathrm{Gal}(L'^{ab}/\mathbb{Q})$. We lift $\tau|L'$ to $\tilde{\tau}'$ in $W_{L'/\mathbb{Q}}$, and let $\tilde{\tau}$ be the image of $\tilde{\tau}'$ in $W_{L/\mathbb{Q}}$. Suppose $\sigma \in \mathrm{Gal}(L/\mathbb{Q})$ lifts to $\eta\rho \in \mathrm{Gal}(L'/\mathbb{Q})$ and $\sigma\tau \in \mathrm{Gal}(L/\mathbb{Q})$ lifts to $\eta'\rho' \in \mathrm{Gal}(L'/\mathbb{Q})$. Then

$$w'_\eta \, w'_\rho \, \tilde{\tau}' = d' \, w'_{\eta'} \, w'_{\rho'} \quad ,$$

with $d' \in W_{L'/L} \subset W_{L'/\mathbb{Q}}$. This shows that under the homomorphism $W_{L'/\mathbb{Q}} \to W_{L/\mathbb{Q}}$, the image d of d' belongs to $W_{L/L} = W_L^{ab}$ and is the image of $c_\sigma(\tilde{\tau})$ under the isomorphism $r_L : C_L \to W_L^{ab}$. On the other hand, for $\zeta \in \mathrm{Gal}(L'/L)$, there is a unique $\zeta' \in \mathrm{Gal}(L'/L)$ such that $\zeta\eta\rho\tau = \zeta'\eta'\rho'$. By definition

$$w'_\zeta \, w'_\eta \, w'_\rho \, \tilde{\tau}' = c_{\zeta\eta\rho}(\tilde{\tau}') \, w'_{\zeta'} \, w'_{\eta'} \, w'_{\rho'} \quad .$$

It follows that

$$w'_\zeta \, d' = c_{\zeta\eta\rho}(\tilde{\tau}') \, w'_{\zeta'} \tag{3.13}$$

This is an equation in $W_{L'/L}$. Let $t : W_L^{ab} \cong (W_{L'/L})^{ab} \to W_{L'}^{ab} = (W_{L'/L'})^{ab}$ be the transfer homomorphism arising from $W_{L'/L'} \hookrightarrow W_{L'/L}$. We have an exact sequence

$$\begin{array}{ccccccccc} 1 & \longrightarrow & C_{L'} & \longrightarrow & W_{L'/L} & \longrightarrow & \mathrm{Gal}(L'/L) & \longrightarrow & 1 \\ & & \downarrow \approx & & & & & & \\ & & W_{L'/L'} & & & & & & \end{array}$$

and $\zeta \mapsto w'_\zeta$ is a section of $W_{L'/L} \to \mathrm{Gal}(L'/L)$; thus (3.13) shows that $t(d) = r_{L'}\left(\prod_{\zeta \in \mathrm{Gal}(L'/L)} c_{\zeta\eta\rho}(\tilde{\tau}')\right)$. Since

$$\begin{array}{ccc} C_{L'} & \xrightarrow{r_{L'}} & W_{L'}^{ab} \\ \uparrow & & \uparrow t \\ C_L & \xrightarrow{r_L} & W_L^{ab} \end{array}$$

commutes (Tate [2, §1, W_3]) and $r_L(c_\sigma(\tilde{\tau})) = d$, $c_\sigma(\tilde{\tau})$ regarded as an element of $C_{L'}$ is $\prod_{\zeta \in \mathrm{Gal}(L'/L)} c_{\zeta\eta\rho}(\tilde{\tau}')$. Now under $N_{L'/L} : X_*(S^{L'}) \to X_*(S^L)$, $\zeta\eta\rho\mu$ maps to $\sigma\mu$ for all $\zeta \in \mathrm{Gal}(L'/L)$. Therefore

$$b_o(\tilde{\tau}',\mu) = \prod_{\zeta,\eta,\rho} c_{\zeta\eta\rho}(\tilde{\tau}')^{\zeta\eta\rho\mu} \in C_{L'} \otimes X_*(S^{L'})$$

maps to

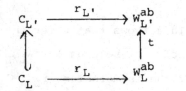

$$\prod_{\substack{\eta,\rho \\ \eta\rho\to\sigma}} (\prod_{\zeta} c_{\zeta\eta\rho}(\tilde{\tau}'))^{\sigma\mu} \in C_{L'} \otimes X_*(S^L) .$$

But

$$\prod_{\substack{\eta,\rho \\ \eta\rho\to\sigma}} (\prod_{\zeta} c_{\zeta\eta\rho}(\tilde{\tau}'))^{\sigma\mu} = \prod_{\sigma} (c_\sigma(\tilde{\tau}))^{\sigma\mu} = b_o(\tilde{\tau},\mu) \in C_L \otimes X_* (S^L) .$$

Hence the diagram in the Lemma commutes.

Now suppose $Gal(L'/L) \cap Gal(L'_v/\mathbb{Q}_v) \neq \{1\}$. This happens only if $L_v = \mathbb{R}$. Thus in this case $X_*(S^L) \simeq \mathbb{Z}$ and the Galois group acts on it trivially. Let $\sigma \mapsto w_\sigma$ be an arbitrary section of $W_{L/\mathbb{Q}} \to Gal(L/\mathbb{Q})$, not necessarily satisfying (3.4). Define $c_\sigma \in C_L$ by $w_\sigma\tilde{\tau} = c_\sigma w_{\sigma\tau}$. Then $\prod_\sigma c_\sigma^{\sigma\mu} = \prod_\sigma c_\sigma^\mu \in C_L$ is independent of the choice of the section $\sigma \mapsto w_\sigma$, for in replacing w_σ by $e_\sigma w_\sigma$, $e_\sigma \in C_L$, $\prod_\sigma c_\sigma^\mu$ is multiplied by the factor $\prod_\sigma (e_\sigma e_{\sigma\tau}^{-1})^\mu$, which is 1. In particular, $b_o(\tilde{\tau},\mu) = \prod_\sigma c_\sigma^{\sigma\mu}$. Similarly, let $\rho \mapsto w'_\rho$ be an arbitrary section of $W_{L'/\mathbb{Q}} \to Gal(L'/\mathbb{Q})$, and define $c'_\rho \in C_L$, by $w'_\rho \tilde{\tau}' = c'_\rho w'_{\rho\tau}$. Then the image of $\prod_\rho (c'_\rho)^{\rho\mu}$ in $S^L(C_{L'})$ is independent of the choice of $\rho \mapsto w'_\rho$; in particular, it is the image of $b_o(\tilde{\tau},\mu)$. For our purpose, we choose $\sigma \mapsto w_\sigma$ and $\rho \mapsto w_\rho$ as follows. Let R be a set of representatives for the coset space $Gal(L'/L)\backslash Gal(L'/\mathbb{Q})$. Fix $w'_\xi \in W_{L'/\mathbb{Q}}$ projecting to ξ for each ξ in $Gal(L'/L) \cup R$. For $\rho \in Gal(L'/\mathbb{Q})$, write $\rho = \zeta\eta$ with $\zeta \in Gal(L'/L)$ and $\eta \in R$, then put $w'_\rho = w'_\zeta w'_\eta$. For $\sigma \in Gal(L/\mathbb{Q})$, let η be the unique element of R extending σ, and let w_σ be the image of w'_η in $W_{L/\mathbb{Q}}$. As before, we have $c_\sigma = \prod_{\zeta \in \mathrm{Gal}(L'/L)} c'_{\zeta\eta}$ if $\eta \in R$ maps to σ in $\mathrm{Gal}(L/\mathbb{Q})$. It follows that the image of $b_o(\tilde{\tau}',\mu)$ is $b_o(\tilde{\tau},\mu)$.

<u>Proposition 3.14</u>. Let T be a torus over $\mathbb{Q}$, let $\mu \in X_*(T)$, and let τ be an automorphism of $\mathbb{C}$. Assume (3.3) holds, so that $c(\tau,\mu) \in H^1(L/\mathbb{Q}, T(L))$ is defined for L a sufficiently larger number field. The image of $c(\tau,\mu)$ in $H^1(L_v/\mathbb{Q}_v, T(L_v))$ is represented by $\mu(-1)/\tau^{-1}\mu(-1) \in \mathrm{Ker}(1+\iota : T(\mathbb{C}) \to T(\mathbb{C}))$.

<u>Proof</u>. The image of $c(\tau,\mu)$ in $H^1(\mathbb{C}/\mathbb{R}, T(\mathbb{C}))$ is the cup-product of the local fundamental class in $H^2(\mathbb{C}/\mathbb{R}, \mathbb{C}^\times)$ with the element of $H^{-1}(\mathbb{C}/\mathbb{R}, X_*(T))$ represented by $(1-\tau^{-1})\mu$. (See Langlands [1, p. 225]). Thus the proposition is a consequence of the following easy lemma.

<u>Lemma 3.15</u>. For any torus T over $\mathbb{R}$, the map $H^{-1}(\mathbb{C}/\mathbb{R}, X_*(T)) \to H^1(\mathbb{C}/\mathbb{R}, T)$ induced by cupping with the fundamental class in $H^2(\mathbb{C}/\mathbb{R}, \mathbb{C}^\times)$ sends the class represented by $\chi \in X_*(T)$ to the class represented by $\chi(-1)$.

<u>Remark 3.16</u>. Thus $c(\tau,\mu)$ has the following property: For any finite prime p of $\mathbb{Q}$ and extension of p to L, $c(\tau,\mu)$ has image 1 in $H^1(L_{\mathfrak{P}}/\mathbb{Q}_p, T(L_{\mathfrak{P}}))$, and the image of $c(\tau,\mu)$

in $H^1(\mathbb{C}/\mathbb{R}, T(\mathbb{C}))$ is represented by $(1-\tau^{-1})\,\mu(-1)$. When $T = S^L$, (2.5) shows that this property determines $c(\tau,\mu)$ uniquely. On the other hand, it is not difficult to construct directly a cohomology class having the property. Consider the exact commutative diagram

$$\begin{array}{ccccc}
H^1(L/\mathbb{Q},T(L)) & \longrightarrow & \oplus_p\, H^1(L_{\mathfrak{P}}/\mathbb{Q}_p,T(L_{\mathfrak{P}})) & \longrightarrow & H^1(L/\mathbb{Q},T(C_L)) \\
 & & \uparrow \wr & & \uparrow \wr \\
 & & \oplus_p\, H^{-1}(L_{\mathfrak{P}}/\mathbb{Q}_p,X_*(T)) & \longrightarrow & H^{-1}(L/\mathbb{Q},X_*(T))
\end{array}$$

in which the vertical maps are the Tate-Nakayama isomorphisms (Tate [1]). For a finite group G and G-module M, $H^{-1}(G,M) = (\mathrm{Ker}\, N : M \to M)/\Sigma(\sigma-1)M$. Thus (3.3) shows that $(1-\tau^{-1})\mu$ defines an element $\alpha_\infty \in H^{-1}(\mathbb{C}/\mathbb{R}, X_*(T))$, and we let $\alpha = (\alpha_p) \in \oplus_p H^{-1}(L_{\mathfrak{P}}/\mathbb{Q}_p,X_*(T))$ with $\alpha_p = 0$ for $p \neq \infty$ and α_∞ the element just defined. Note that the image of α_∞ in $H^1(\mathbb{C}/\mathbb{R},T(\mathbb{C}))$ is represented by $(1-\tau^{-1})\mu(-1)$. The image of α in $H^{-1}(L/\mathbb{Q},X_*(T))$ is represented by $(1-\tau^{-1})\mu$, and is therefore zero. It follows that the image of α in $\oplus_p H^1(L_{\mathfrak{P}}/\mathbb{Q}_p, T(L_{\mathfrak{P}}))$ arises from an element of $H^1(L/\mathbb{Q},T(L))$, and this is the class sought.

The next property of the Taniyama group will be needed in showing that the zeta function of an abelian variety of potential CM-type is the L-series of a representation of the Weil group.

<u>Proposition 3.17</u>. For any finite Galois extension L of $\mathbb{Q}$ that is not totally real, there is a homomorphism $\phi : W_{L/\mathbb{Q}} \to \underset{\sim}{T}^L(\mathbb{C})$ making

$$\begin{array}{ccccccccc} & & & & & & W_{L/\mathbb{Q}} & & \\ & & & & & \overset{\phi}{\swarrow} & \downarrow & & \\ 1 & \longrightarrow & S^L(\mathbb{C}) & \longrightarrow & \underset{\sim}{T}^L(\mathbb{C}) & \longrightarrow & \mathrm{Gal}(L^{ab}/\mathbb{Q}) & \longrightarrow & 1 \end{array}$$

commute. If ϕ' is a second such homomorphism then $\phi' = \phi \cdot \alpha$ with α a 1-cocycle for $W_{L/\mathbb{Q}}$ with values in $S^L(\mathbb{C})$.

<u>Proof</u>. We have to show that the 2-cocycle (d_{τ_1,τ_2}) defining the extension (see 2.1) becomes trivial when inflated to $H^2(W_{L/\mathbb{Q}}, S^L(\mathbb{C}))$. Choose a section $\sigma \mapsto w_\sigma$ to $W_{L/\mathbb{Q}} \to \mathrm{Gal}(L/\mathbb{Q})$ as in (3.4) and a map $b : \mathrm{Gal}(L^{ab}/\mathbb{Q}) \to S^L(\mathbb{A}_L^f)$ lifting the map $\bar{b}$ defined above and satisfying (2.7c). For $w \in W_{L/\mathbb{Q}}$ mapping to $\tau \in \mathrm{Gal}(L^{ab}/\mathbb{Q})$ define $c_{\sigma,w} \in C_L$ by the condition $w_\sigma w = c_{\sigma,w}\, w_{\sigma\tau}$ and set

$$b_o(w) = \prod_{\sigma \in \mathrm{Gal}(L/\mathbb{Q})} c_{\sigma,w}^{\sigma\mu} \in S^L(C_L) .$$

A calculation as in the proof of (3.11) shows that $b_o(w_1w_2) = b_o(w_1) \cdot \tau_1(b_o(w_2))$, where τ_1 as the image of w_1 in $\mathrm{Gal}(L^{ab}/\mathbb{Q})$. Choose a mapping $b : W_{L/\mathbb{Q}} \to S^L(\mathbb{A}_L)$ making

$$\begin{array}{ccc} W_{L/\mathbb{Q}} & \overset{b}{\dashrightarrow} & S^L(\mathbb{A}_L) \\ & \searrow^{b_o} & \downarrow b \\ & & S^L(C_L) \end{array} \quad \text{and} \quad \begin{array}{ccc} W_{L/\mathbb{Q}} & \twoheadrightarrow & \mathrm{Gal}(L^{ab}/\mathbb{Q}) \\ \downarrow b & & \downarrow \\ S^L(\mathbb{A}_L) & \twoheadrightarrow & S^L(\mathbb{A}_L^f) \end{array}$$

commute. Then $b(w_1) \cdot \tau_1 b(w_2) \cdot b(w_1 w_2)^{-1}$ lies in $S^L(L) \subset S^L(\mathbb{A}_L)$, and projects onto d_{τ_1,τ_2} in $S^L(\mathbb{A}_L^f)$. It is therefore equal to d_{τ_1,τ_2}. Let v be an infinite prime of L such that $L_v = \mathbb{C}$, and let $b_v(w) \in S^L(L_v) = S^L(\mathbb{C})$ be the component of $b(w)$ at v. Then $w \mapsto b_v(w)$ is a 1-cochain whose coboundary is (d_{τ_1,τ_2}).

<u>Remark 3.18</u>. In V we shall need to use the following notations. For any $\mathbb{Q}$-rational torus T, split by L, and cocharacter μ satisfying (3.3) relative to $\tau \in \mathrm{Gal}(L^{ab}/\mathbb{Q})$ we have defined an element $\overline{b}(\tau,\mu) \in T(\mathbb{A}_L^f)/T(L)\, T(\mathbb{Q})^\wedge$. It is natural also to define $\overline{\beta}(\tau,\mu) = \overline{b}(\tau^{-1},\mu)^{-1}$ and $\gamma(\tau,\mu) = c(\tau^{-1},\mu)^{-1}$ (c.f. 2.9). If μ satisfies the stronger condition (1.1) then there is a unique homomorphism $\rho_\mu : S^L \to T$ such that $\rho_\mu \circ \mu^L = \mu$, and we have $\overline{\beta}(\tau,\mu) = \rho_\mu(\overline{\beta}(\tau))$ and $\gamma(\tau,\mu) = \rho_\mu(\gamma(\tau))$.

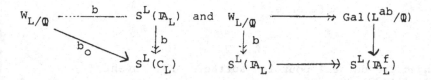

References

Deligne, P.
1. La conjecture de Weil pour les surfaces K3, Invent. Math. 15 (1972) 206-226.
2. Variétés de Shimura : interpretation modulaire, et techniques de construction de modèles canoniques. Proc. Symp. Pure Math., A.M.S., 33 (1979) part 2, 247-290.

Demazure, M. et Gabriel, P.
1. Groupes Algébriques, Tome I: Géométrie algébriques, generalités, groupes commutatifs. Masson, Paris, 1979.

Langlands, R.
1. Automorphic representations, Shimura varieties, and motives. Ein Märchen. Proc. Symp. Pure Math., A.M.S., 33 (1979), part 2, 205-246.

Serre, J.-P.
1. Abelian ℓ-adic representations and elliptic curves. Benjamin, New York, 1968.

Tate, J.
1. The cohomology groups of tori in finite Galois extensions of number fields, Nagoya Math. J. 27 (1966) 709-719.
2. Number theoretic background, Proc. Symp. Pure Math. A.M.S., 33 (1979) part 2, 3-26.

IV. MOTIFS ET GROUPES DE TANIYAMA

par P. Deligne*

... j'ai fait quelques progrès quant à la relation de ton groupe de Taniyama T avec les motifs. Je m'intéresserai aux structures suivantes dont on dispose sur T.

(1) Un morphisme surjectif $T \to \mathrm{Gal}(\overline{\mathbb{Q}}/\mathbb{Q})$ (où $\overline{\mathbb{Q}}$ désigne la clôture algébrique de $\mathbb{Q}$ dans $\mathbb{C}$). On notera T° son noyau.

(2) Un morphisme μ, défini sur $\mathbb{C}$, de $\mathbb{G}_m$ dans T°. Le système (T°,μ) n'a pas d'automorphisme non trivial. C'est le groupe de Serre connexe (ce vol., III, §1)

(3) Une section continue $T(\mathbb{A}^f) \xleftarrow{\pi} \mathrm{Gal}(\overline{\mathbb{Q}},\mathbb{Q})$.

Prendre garde que π n'est pas défini par un morphisme de schémas en groupes (cf. C) : les homomorphismes continus $\pi_\ell : \mathrm{Gal}(\overline{\mathbb{Q}}/\mathbb{Q}) \to T(\mathbb{Q}_\ell)$ déduits de π ont une image Zariski-dense. Le système (T,π) n'a pas d'automorphisme non trivial.

Je ne considérerai des motifs que sur des corps de caractéristique 0. Il ne s'agira pas des motifs de Grothendieck, tels qu'il les définissait en termes de cycles algébriques, mais des <u>motifs de Hodge absolus</u>, définis de même en termes de cycles de Hodge absolus. La catégorie des motifs sur un corps k est semi-simple et tannakienne. De même pour toute sous-catégorie, stable par produits tensoriels, dualité, sommes et facteurs directs.

Pour F un corps de caractérisque 0, je noterai $(CM)_F$ la catégorie tannakienne de motifs sur F engendrée par les motifs d'Artin, i.e. par les H^0 de variétés algébriques finies (la terminologie est celle de [1] §6) et par les variétés abéliennes sur F, potentiellement de type CM (i.e., qui deviennent de type CM sur une extension de F). La catégorie de motifs qui m'intéresse est $(CM)_{\mathbb{Q}}$. Cette catégorie, munie du foncteur fibre H_B (cohomologie rationnelle, sur $\mathbb{C}$) donne lieu à un groupe M, muni de structures (1')(2')(3') parallèles à (1)(2)(3). Le morphisme (1') : $M \to \mathrm{Gal}(\overline{\mathbb{Q}}/\mathbb{Q})$ provient de ce que la restriction de H_B à la catégorie des motifs d'Artin identifie cette dernière à celle des représentations rationnelles de $\mathrm{Gal}(\overline{\mathbb{Q}}/\mathbb{Q})$.

Pour chaque représentation V de M, correspondant à un motif V_1, on a

---

*) Cet article reprend et complète une lettre à Langlands datée du 10 avril 1979, où était obtenu un résultat partiel.

$V = H_B(V_1)$, et $V \otimes \mathbb{C}$ est muni d'une bigraduation de Hodge, et en particulier de l'action de $\mathbb{G}_m$ donnée $z * v^{p,q} = z^{-p} . v^{p,q}$ pour $v^{p,q}$ de type (p,q) . Cette action, étant compatible au produit tensoriel, provient d'un morphisme μ , défini sur $\mathbb{C}$, de $\mathbb{G}_m$ dans M . C'est la structure (2') . De même, $V \otimes \mathbb{A}^f$ est la $\mathbb{A}^f$-cohomologie de V_1 , et, en tant que tel, est muni d'une action de $Gal(\overline{\mathbb{Q}}/\mathbb{Q})$. Ces actions fournissent (3').

<u>Théorème</u>. T , <u>muni de</u> (1)(2)(3) , <u>est isomorphe à</u> M , <u>muni de</u> (1')(2')(3') .

<u>Remarque 1</u>. T , muni des structures (1)(2)(3) , n'a pas d'automorphisme autre que l'identité. L'isomorphisme dont l'existence est garantie par le théorème est donc unique.

<u>Remarque 2</u>. Il résulte du théorème que la fonction L d'une variété abélienne sur $\mathbb{Q}$ potentiellement de type CM est une fonction L de Weil, définie par une représentation complexe du groupe de Weil. H. Yoshida a récemment donné de ce théorème une démonstration directe, dont le point essentiel est l'observation suivante : supposons qu'une $\overline{\mathbb{Q}}_\ell$ - représentation V de $Gal(\overline{\mathbb{Q}}/\mathbb{Q})$, restreinte à un sous-groupe d'indice fini $Gal(\overline{\mathbb{Q}}/F)$, soit somme de représentations de dimension 1 définies par des caractères de Hecke algébriques. Alors, V est la somme directe de représentations induites $\mathrm{Ind}_{Gal(\overline{\mathbb{Q}}/F_i)}^{Gal(\overline{\mathbb{Q}}/\mathbb{Q})}(\chi_i \otimes V_i)$ avec χ_i caractère de Hecke algébrique de F_i et V_i représentation d'un quotient fini de $Gal(\overline{\mathbb{Q}}/F_i)$.

Ma démonstration donne l'information additionnelle que la représentation complexe du groupe de Weil requise se déduit par le procédé de (III § 3.17) d'une représentation définie sur $\mathbb{Q}$ du groupe de Taniyama.

<u>Remarque 3</u>. Soient K et E des corps de nombres. <u>Un motif potentiellement de type CM sur</u> K , <u>à coefficients dans</u> E est un objet H de $(CM)_K$, muni de $E \to End(H)$. Choisissons une clôture algébrique $\overline{K}$ de K . Pour chaque place λ de E , H définit une représentation λ-adique de $Gal(\overline{K}/K)$. Il résulte du théorème que ces représentations forment un système strictement compatibles (au sens de [3] 1.13). Je vois ce résultat comme un premier pas dans l'étude des représentations ℓ-adiques mystérieuses introduites par G. Shimura ([4] § 11), qui sont des représentations ℓ-adiques attachées à des motifs dans la $\otimes$-catégorie engendrée par toutes les variétés abéliennes.

Prenons pour simplifier $K \subset \overline{\mathbb{Q}}$ et $\overline{K} = \overline{\mathbb{Q}}$. Choisissons un plongement complexe σ de E ; de H se déduit alors une fonction L . C'est une fonction L de Weil attachée à une représentation linéaire définie sur E du groupe de Taniyama $_K T$ image inverse de $Gal(\overline{\mathbb{Q}}/K)$ dans T .

Remarque 4. Implicite dans le théorème est une règle disant comment $\mathrm{Aut}(\mathbb{C})$ agit sur l'ensemble des classes d'isomorphie de variétés abéliennes de type CM sur $\mathbb{C}$ munies de polarisations et de structures de niveau. En particulier, pour K un corps CM, d'anneau d'entiers $\mathcal{O}$, et Σ un isomorphisme $K \otimes \mathbb{R} \xrightarrow{\sim} \mathbb{C}^g$, on dispose de la variété abélienne $\mathbb{C}^g/\Sigma(\mathcal{O})$, à multiplication par $\mathcal{O}$, et munie d'une structure de niveau ∞ $\alpha : \hat{T} \xrightarrow{\sim} \hat{\mathcal{O}}$; il faut déterminer ses conjugués. Essentiellement, pour $\sigma \in \mathrm{Aut}(\mathbb{C})$, il faut déterminer la classe d'isomorphie du $\mathcal{O}$-module inversible $L_\sigma := H_1(\sigma(\mathbb{C}^g/\Sigma(\mathcal{O})))$, muni de la structure, déduite de celle de niveau, $\alpha_\sigma : L_\sigma^\wedge \xrightarrow{\sim} \mathcal{O}^\wedge$. J. Tate a su rendre plus explicite la règle donnant $(L_\sigma, \alpha_\sigma)$.

Remarque 5. Je montrerai que la classe d'isomorphie de M, muni de (1') (2') (3'), est uniquement déterminée par les propriétés de M exprimées par les lemmes 1 à 5 qui suivent. Le théorème résulte de ce que le groupe T a ces mêmes propriétés.

Pour K une extension finie de $\mathbb{Q}$, $R_{K/\mathbb{Q}}(\mathbb{G}_m)$ est le groupe multiplicatif de la $\mathbb{Q}$-algèbre K, vu comme groupe algébrique sur $\mathbb{Q}$. Pour F une extension algébriquement close de $\mathbb{Q}$, le groupe $R_{K/\mathbb{Q}}(\mathbb{G}_m) \otimes F$ qui s'en déduit par extension des scalaires à F s'identifie à $\mathbb{G}_m^{\mathrm{Hom}(K,F)}$: c'est le groupe multiplicatif de la F-algèbre $K \otimes_{\mathbb{Q}} F \xrightarrow{\sim} F^{\mathrm{Hom}(K,F)}$, vu comme groupe algébrique sur F. Pour K muni d'un plongement ι dans F, on dispose donc, après extension des scalaires à F, de morphismes $\mathbb{G}_m \xrightarrow{\varepsilon} R_{K/\mathbb{Q}}(\mathbb{G}_m) \xrightarrow{\varepsilon'} \mathbb{G}_m$: l'injection et la projection d'indice ι. Ces morphismes proviennent par extension des scalaires de K à F de morphismes définis sur K.

Je dirai "K est CM" pour "K est une extension quadratique totalement imaginaire d'un corps totalement réel". Pour indiquer que ce corps totalement réel s'appelle L, je dirai "(K,L) est CM". Pour (K,L) CM, posons

$$^{K}S = R_{K/\mathbb{Q}}(\mathbb{G}_m)/\mathrm{Ker}(N_{L/\mathbb{Q}} : R_{L/\mathbb{Q}}(\mathbb{G}_m) \to \mathbb{G}_m) .$$

Pour $K \subset \overline{\mathbb{Q}}$, on définit un morphisme μ, défini sur $\mathbb{C}$, de $\mathbb{G}_m$ dans ^{K}S comme étant le composé

$$\mu : \mathbb{G}_m \xrightarrow{\varepsilon^{-1}} R_{K/\mathbb{Q}}(\mathbb{G}_m) \longrightarrow {}^{K}S .$$

Le groupe de Serre connexe S est la limite projective des ^{K}S, pour (K,L) CM, avec $K \subset \overline{\mathbb{Q}}$, les morphismes de transitions étant déduits par passage au quotient des morphismes surjectifs $N_{K'/K} : R_{K'/\mathbb{Q}}(\mathbb{G}_m) \longrightarrow R_{K/\mathbb{Q}}(\mathbb{G}_m)$. Il est muni de $\mu : \mathbb{G}_m \to S$, défini sur $\mathbb{C}$, limite projective des morphismes μ construits plus haut.

Soit M° le noyau de $(1'){:}M \to \mathrm{Gal}(\overline{\mathbb{Q}}/\mathbb{Q})$. Le morphisme (2'), défini sur

$\mathbb{C}$: $\mu : \mathbb{G}_m \to M$ se factorise par M^o.

Lemme 1. (M^o,μ) est isomorphe à (S,μ).

Le groupe de Serre connexe, muni de μ, n'a pas d'automorphisme non trivial. Le lemme 1 détermine donc (M^o,μ) à isomorphisme unique près. Il est démontré dans la section A. Puisque M^o est abélien, l'action par automorphismes intérieurs de M sur M^o se factorise par (1') : $M \to \mathrm{Gal}(\overline{\mathbb{Q}}/\mathbb{Q})$: le groupe de Galois $\mathrm{Gal}(\overline{\mathbb{Q}}/\mathbb{Q})$ agit sur le schéma en groupes M^o.

Lemme 2. L'action de $\mathrm{Gal}(\overline{\mathbb{Q}}/\mathbb{Q})$ sur $M^o = S$ (lemme 1) est celle décrite dans la section B.

Pour $K \subset \overline{\mathbb{Q}}$, CM, et galoisien sur $\mathbb{Q}$, il résulte des lemmes 1 et 2 que le noyau de la projection de $M^o = S$ sur ${}^KS$ est stable sous $\mathrm{Gal}(\overline{\mathbb{Q}}/\mathbb{Q})$. Ceci permet de passer au quotient et de définir ${}^KM$ comme le quotient de M obtenu en remplaçant M^o par ${}^KS$.

Définissons $w: \mathbb{G}_m \to M$ comme étant l'inverse du produit de μ et de son complexe conjugué $\overline{\mu}$. Définissons $h : \mathbb{C}^* \to M(\mathbb{R})$ par $h(z) = \mu(z).\overline{\mu(z)}$. Pour toute représentation linéaire (V,ρ) de M, correspondant à un motif V_1, $\rho\circ w$ et $\rho\circ h$ s'expriment en terme de la décomposition de Hodge de $V \otimes \mathbb{C} = H_B(V_1) \otimes \mathbb{C}$, et la déterminent : sur un vecteur de type (p,q), $w(z)$ agit par multiplication par z^{p+q}, et $h(z)$ par multiplication par $\overline{z}^{-p}z^{-q}$.

Lemme 3. Quel que soit $K \subset \overline{\mathbb{Q}}$ galoisien sur $\mathbb{Q}$ et CM, $\mathrm{int}(h(i))$ est une involution de Cartan du groupe (pro-)algébrique réel $[{}^KM/w(\mathbb{G}_m)]_{\mathbb{R}}$.

Rappelons qu'une involution ε d'un groupe algébrique réel H définit une nouvelle forme réelle de H, tordue de H par ε, de conjugaison complexe le produit de l'ancienne par ε. Elle est dite de Cartan si cette tordue ${}^{\varepsilon}H$ est compacte, i.e. si ${}^{\varepsilon}H(\mathbb{R})$ est compact et que chaque composante connexe complexe admet un point réel.

Le lemme 3 est prouvé dans la section C. Pour $\sigma \in \mathrm{Gal}(\overline{\mathbb{Q}}/\mathbb{Q})$, soit P_σ l'image inverse de σ dans M. C'est un torseur sous S, pour l'action de S par translations à gauche. Soit ${}^KP_\sigma$ le ${}^KS$-torseur qui s'en déduit; c'est l'image inverse de σ dans ${}^KM$. Je prouverai dans la section C que les lemmes 1 à 3 suffisent à déterminer sa classe d'isomorphie.

Pour $K \subset \overline{\mathbb{Q}}$, on désignera par un indice K à gauche une image inverse de $\mathrm{Gal}(\overline{\mathbb{Q}}/K) \subset \mathrm{Gal}(\overline{\mathbb{Q}}/\mathbb{Q})$. Par exemple : ${}_KM$ est l'image inverse de $\mathrm{Gal}(\overline{\mathbb{Q}}/K)$ dans M.

Ce schéma est une extension de $\mathrm{Gal}(\overline{\mathbb{Q}}/K)$ par S. La structure (3') induit une section continue $\pi : \mathrm{Gal}(\overline{\mathbb{Q}}/K) \to {}_K M(\mathbb{A}^f)$, de composantes $\pi_\ell : \mathrm{Gal}(\overline{\mathbb{Q}}/K) \to {}_K M(\mathbb{Q}_\ell)$. Notons encore H_B le foncteur fibre de $(CM)_K$: cohomologie rationnelle, après extension des scalaires à $\mathbb{C}$, par le plongement donné de K dans $\mathbb{C}$. Le schéma en groupes ${}_K M$ est le schéma en groupes des automorphismes de ce foncteur fibre. On a un diagramme commutatif

$$\begin{array}{ccccc} (CM)_{\mathbb{Q}} & \xrightarrow{\text{extension des scalaires}} & (CM)_K & \xrightarrow{H_B} & \text{vectoriels sur } \mathbb{Q} \\ H_B \downarrow \wr & & H_B \downarrow \wr & \nearrow \text{oubli} & \\ (\text{représentations de } M) & \xrightarrow{\text{restriction à } {}_K M} & (\text{représentations de } {}_K M) & & \end{array}$$

Le morphisme défini sur $\mathbb{C}$, μ, de $\mathbb{G}_m$ dans ${}_K M$, et la section continue π, proviennent respectivement de ce que pour V_1 dans $(CM)_K$, $H_B(V_1) \otimes \mathbb{C}$ est muni d'une structure de Hodge, et $H_B(V_1) \otimes \mathbb{A}^f$ d'une action de $\mathrm{Gal}(\overline{\mathbb{Q}}/K)$.

Soit E une extension finie de $\mathbb{Q}$. Les variétés abéliennes A sur K, à isogénie près, munies de $E \to \mathrm{End}(A)$, telles que $H_B(A)$ soit de dimension 1 sur E (type CM), correspondant, par le foncteur H_B, à la donnée de (a) un espace vectoriel H de dimension 1 sur E et (b) une représentation E-linéaire de ${}_K M$ sur H, telle que la structure de Hodge qui s'en déduit sur $H \otimes_{\mathbb{Q}} \mathbb{C}$ soit de type $\{(0,1),(1,0)\}$. Ceci s'exprime en terme de μ (cf. le laius qui précède le lemme 3). Se donner une représentation linéaire de ${}_K M$ sur H revient à se donner un morphisme défini sur E de ${}_K M$ dans $\mathbb{G}_m$, soit encore un morphisme $\rho : {}_K M \to R_{E/\mathbb{Q}}(\mathbb{G}_m)$. Nous dirons que ρ est de type $\{(0,1),(1,0)\}$ si la structure de Hodge de $H \otimes_{\mathbb{Q}} \mathbb{C}$ l'est.

Pour K non-nécessairement de type CM, on peut encore définir ${}^K S$ comme le quotient de $R_{K/\mathbb{Q}}(\mathbb{G}_m)$ par l'adhérence de Zariski des sous-groupes d'indice fini assez petits du groupe des unités de K^*. Ce quotient est déterminé dans [3]. Pour $K \subset \overline{\mathbb{Q}}$, les ${}^K S$ forment encore un système projectif de limite S : si K contient un corps CM, et que K' est le plus grand d'entre eux, la norme $N_{K/K'} : R_{K/\mathbb{Q}}(\mathbb{G}_m) \to R_{K'/\mathbb{Q}}(\mathbb{G}_m)$ induit un isomorphisme de ${}^K S$ avec ${}^{K'} S$.

Il résulte du lemme 2 que ${}^K S$ est le plus grand quotient de S sur lequel $\mathrm{Gal}(\overline{\mathbb{Q}}/K)$ agisse trivialement. Puisque $R_{E/\mathbb{Q}}(\mathbb{G}_m)$ est abélien, la restriction à S d'un morphisme $\rho : {}_K M \to R_{E/\mathbb{Q}}(\mathbb{G}_m)$ se factorise par ${}^K S$. Nous noterons ρ_1 le morphisme composé

$$\rho_1 : R_{K/\mathbb{Q}}(\mathbb{G}_m) \longrightarrow {}^K S \xrightarrow{\ \rho\ } R_{E/\mathbb{Q}}(\mathbb{G}_m) .$$

<u>Lemma 4</u>. <u>Quels que soient $K \subset \overline{\mathbb{Q}}$ et E, extensions finies de $\mathbb{Q}$, et $\rho : {}_K M \to R_{E/\mathbb{Q}}(\mathbb{G}_m)$ de type $\{(0,1),(1,0)\}$, il existe un caractère de Hecke algébrique χ de K, à valeurs dans E, de partie algébrique ρ_1, tel que, notant ϕ l'ensemble fini des places où χ est ramifié, pour tout nombre premier ℓ et tout idéal premier $\underline{\underline{p}} \notin \phi$ de K, premier à ℓ, $\rho \circ \pi_\ell(F_{\underline{\underline{p}}}) \in R_{E/\mathbb{Q}}(\mathbb{G}_m)(\mathbb{Q}_\ell) = (E \otimes \mathbb{Q}_\ell)^*$ soit dans E^*, et égal à $\chi(\underline{\underline{p}})$</u>.

Je suis ici la terminologie de [2] Sommes trig. §5 : un caractère de Hecke

χ est regardé comme une fonction d'idéaux, sa partie algébrique $\chi_{alg} : K^* \to E^*$ est caractérisée par une identité $\chi((\alpha)) = \chi_{alg}(\alpha)$ pour $\alpha \equiv 1(\underline{m})$, et $F_{\underline{p}}$ est un Frobenius géométrique.

Pour notre M, la validité du lemme est conséquence de la théorie de Shimura et Taniyama, qui implique que la représentation ℓ-adique attachée à une variété abélienne A sur K, de type CM, à multiplication par E, est décrite par un caractère de Hecke algébrique de K à valeurs dans E, de partie algébrique déterminée par la classe d'isomorphie du K-E-bimodule Lie(A).

A posteriori, le lemme vaudra pour tout morphisme $\rho : {}_K M \to R_{E/\mathbb{Q}}(\mathbb{G}_m)$, quel que soit son type de Hodge. Un cas particulier intéressant est celui où $K = E = \mathbb{Q}$, et où ρ correspond au motif de Tate $\mathbb{Z}(1)$, de type de Hodge $(-1,-1)$; l'action de Galois sur $H_B(\mathbb{Z}(1)) \otimes \mathbb{Q}_\ell$ est son action sur $\mathbb{Q}_\ell(1)$. Le caractère de Hecke est $(p) \longrightarrow p^{-1}$, car F_p (géométrique) agit par p^{-1}. Sa partie algébrique est $u \longrightarrow u^{-1} : \mathbb{Q}^* \to \mathbb{Q}^*$. Cet exemple est un test qu'on a pris les bons signes.

Soit $K \subset \overline{\mathbb{Q}}$ CM. Dans les sections D et E, je montrerai que le lemme 4 détermine uniquement l'extension ${}^{K}_{K}M$ de $\mathrm{Gal}(\overline{\mathbb{Q}}/K)$ par ${}^{K}S$, munie de π. Elle se déduit par inflation de $\mathrm{Gal}(\overline{\mathbb{Q}}/K)^{ab}$ à $\mathrm{Gal}(\overline{\mathbb{Q}}/K)$ de l'extension E de $\mathrm{Gal}(\overline{\mathbb{Q}}/K)^{ab}$ par ${}^{K}S$, limite projective des S_m introduits par J.-P. Serre [3].

Lemme 5. Soit $c \in \mathrm{Gal}(\overline{\mathbb{Q}}/\mathbb{Q})$ la conjugaison complexe. Alors, $\pi(c) \in M(\mathbb{A}^f)$ est dans $M(\mathbb{Q})$.

En effet, quel que soit V_1 dans $(CM)_{\mathbb{Q}}$, l'action de c sur $H_B(V_1) \otimes \mathbb{A}^f$ provient d'une action de c sur $H_B(V_1)$. A savoir, si V_1 est le H^i d'une variété algébrique X, l'automorphisme de $H^i(X(\mathbb{C}),\mathbb{Q})$ induit par la conjugaison complexe $X(\mathbb{C}) \to X(\mathbb{C})$.

(A) Preuve du lemme 1.

Disons qu'une $\mathbb{Q}$-structure de Hodge est de type CM si elle est polarisable et que son groupe de Mumford-Tate est abélien. Si H est polarisable, pour que H soit de type CM, il faut et il suffit que le commutant de End(H) dans $\mathrm{End}(H_{\mathbb{Q}})$ soit commutatif. Soit (Hodge CM) la catégorie tannakienne des $\mathbb{Q}$-structures de Hodge de type CM.

Proposition A.1. Le foncteur $H_B : (CM)_{\overline{\mathbb{Q}}} \to$ (Hodge CM) est une équivalence de catégories

Il est pleinement fidèle par la théorie des cycles de Hodge absolu.

Le foncteur fibre "$\mathbb{Q}$-espace vectoriel sous-jacent" de la catégorie (Hodge CM) a pour groupe d'automorphismes le groupe de Serre connexe déjà introduit, dont on donnera ci-dessous trois descriptions équivalentes. La première paraphrase la définition de (Hodge CM), et les autres s'en déduisent. La seconde est celle déjà donnée. Les variétés abéliennes sur $\mathbb{C}$, prises à isogénie près, s'identifiant aux structures de Hodge polarisables de type $\{(0,1),(1,0)\}$, la troisième description de S ramène au lemme élémentaire A.2 la preuve de ce que les variétés abéliennes de type CM sur $\overline{\mathbb{Q}}$ (ou sur $\mathbb{C}$, cela revient au même) fournissent un système fidèle de representations du groupe de Serre connexe. Cette fidélité implique la surjectivité essentielle du foncteur H_B.

(a) Soit I la catégorie filtrante des paires (T,μ) où T est un tore (sur $\mathbb{Q}$) et $\mu : \mathbb{C}_m \longrightarrow T$ un morphisme défini sur $\mathbb{C}$, tel que $w := (\mu\overline{\mu})^{-1}$ soit défini sur $\mathbb{Q}$ et que $[T/w(\mathbb{C}_m)]_{\mathbb{R}}$ soit compact. Pour (T,μ) dans I et V une représentation de T, on définit une bigraduation de Hodge sur $V\otimes\mathbb{C}$ par la règle $\mu(z)*\overline{\mu(z)}*v^{p,q} = z^{-p}\overline{z}^{-q}v^{p,q}$. Elle est dans (Hodge CM), i.e. les composantes isobares $\bigoplus_{p+q=n} V^{pq}$ sont définies sur $\mathbb{Q}$ et définissent des structures de Hodge de type CM. On a

$$S = \text{lim proj } T \qquad \text{(limite projective sur } I\text{)} .$$

Il est loisible de ne considérer que les tores T tels que μ ne se factorise par aucun sous-tore défini sur $\mathbb{Q}$. La catégorie filtrante I se réduit alors à un ensemble ordonné filtrant.

(b) Voici un système cofinal de tores comme en (a), indexé par les sous-corps CM de $\overline{\mathbb{Q}}$. Soit (K,L) CM, avec $K\subset\overline{\mathbb{Q}}$. On prend le tore

$$^{K}S := R_{K/\mathbb{Q}}(\mathbb{C}_m)/\text{Ker}(N_{L/\mathbb{Q}}:R_{L/\mathbb{Q}}(\mathbb{C}_m)\longrightarrow\mathbb{C}_m) \quad ,$$

et on définit μ comme se factorisant par le morphisme ε^{-1} de $\mathbb{C}_m$ dans $(R_{K/\mathbb{Q}}(\mathbb{C}_m))_{\mathbb{C}}$. Le morphisme ε a été défini avant le lemme 1.

On a

$$S = \text{lim proj } {}^{K}S \qquad \text{(limite projective sur } K);$$

les morphismes de transitions sont induits par les morphismes norme $N_{K'/K}$. Le diagramme

$$\begin{array}{ccccccccc} 0 & \longrightarrow & \text{Ker}(N_{L/\mathbb{Q}}:R_{L/\mathbb{Q}}(\mathbb{C}_m)\longrightarrow\mathbb{C}_m) & \longrightarrow & R_{K/\mathbb{Q}}(\mathbb{C}_m) & \longrightarrow & {}^{K}S & \longrightarrow & 0 \\ & & \big\downarrow & & \big\downarrow \text{inj}_1 & & \| & & \\ 0 & \longrightarrow & R_{L/\mathbb{Q}}:R_{L/\mathbb{Q}}(\mathbb{C}_m) & \xrightarrow{(\text{inj},N_{L/\mathbb{Q}})} & R_{K/\mathbb{Q}}(\mathbb{C}_m)\times\mathbb{C}_m & \longrightarrow & {}^{K}S & \longrightarrow & 0 \end{array}$$

fournit une suite exacte

(A.1.1) $0 \longrightarrow L^* \longrightarrow K^* \times \mathbb{Q}^* \longrightarrow {}^K S(\mathbb{Q}) \longrightarrow 0$

Pour K galoisien, $\mathrm{Hom}_{\mathbb{C}}(\mathbb{G}_m, {}^K S)$ est un $\mathrm{Gal}(K/\mathbb{Q})$-module. C'est le quotient du $\mathbb{Z}[\mathrm{Gal}(K/\mathbb{Q})]$-module libre engendré par μ par les relations exprimant que $c\mu+\mu$ ($c \in \mathrm{Gal}(K/\mathbb{Q})$ est le conjugaison complexe) est fixé par $\mathrm{Gal}(K/\mathbb{Q})$.

(c) Le groupe $\mathrm{Hom}_{\mathbb{C}}(S,\mathbb{G}_m)$ des caractères de S s'en déduit par passage à la limite : l'application $\chi \longmapsto \langle \chi, \sigma\mu \rangle$ l'identifie au groupe des fonctions $n(\sigma)$ sur $\mathrm{Gal}(\overline{\mathbb{Q}}/\mathbb{Q})$ à valeurs dans $\mathbb{Z}$, se factorisant par un $\mathrm{Gal}(K/\mathbb{Q})$, avec $K \subset \overline{\mathbb{Q}}$ galoisiens sur $\mathbb{Q}$ et CM, et telles que la fonction $\sigma \longmapsto n(\sigma)+n(c\sigma)$ soit constante.

Soient $K \subset \overline{\mathbb{Q}}$ galoisiens et CM, et χ un caractère, défini sur $\mathbb{C}$, de ${}^K S$. Il est déjà défini sur un corps de nombre $E \subset \overline{\mathbb{Q}}$, qu'on peut prendre CM (et même égal à K). Il fournit $\chi_1 : {}^K S \longrightarrow R_{E/\mathbb{Q}}(\mathbb{G}_m)$, (avec, sur $\mathbb{C}$, $\chi = \varepsilon' \circ \chi_1$). Soit H un E-espace vectoriel de rang 1. Le morphisme χ_1 fournit une représentation de S sur H. Pour que $H \otimes \mathbb{C}$ soit de type de Hodge $\{(0,1),(1,0)\}$, il faut et il suffit que la fonction $n(\sigma)$ qui définit χ ne prenne que les valeurs 0 et 1, et vérifie $n(\sigma)+n(c\sigma)=1$. Comme déjà expliqué, ceci ramène A.1 au

<u>Lemme A.2</u>. <u>Le groupe des fonctions</u> $n(\sigma)$ <u>sur</u> $\mathrm{Gal}(K/\mathbb{Q})$, <u>telles que la fonction</u> $n(\sigma)+n(c\sigma)$ <u>soit constante, est engendré par l'ensemble de celles qui ne prennent que les valeurs</u> 0 <u>et</u> 1 <u>et vérifient</u> $n(\sigma)+n(c\sigma)=1$.

Pour déduire le lemme 1 de la proposition A.1, on observe que le noyau M^o de (1') : $M \longrightarrow \mathrm{Gal}(\overline{\mathbb{Q}}/\mathbb{Q})$ est le schéma en groupe des automorphismes du foncteur fibre H_B de $(CM)_{\overline{\mathbb{Q}}}$ (cf. les explications précédant le lemme 4).

(B) <u>Calcul de l'action de</u> $\mathrm{Gal}(\overline{\mathbb{Q}}/\mathbb{Q})$ <u>sur</u> M^o.

Soit F une clôture algébrique de $\mathbb{Q}$. Chaque plongement complexe σ de F définit un foncteur fibre H_σ de $(CM)_F$, et <u>Aut</u>(H_σ), muni de μ_σ défini par Hodge, vient d'être calculé. C'est un groupe algébrique <u>commutatif</u>. D'après la théorie générale de Saavedra, il est donc indépendant du foncteur fibre (à isomorphisme unique près) : c'est un groupe algébrique sur $\mathbb{Q}$, ne dépendant que de F. Notons le $M^o_{[F]}$. Pour tout foncteur fibre : $(CM)_F \xrightarrow{\omega}$ (vectoriels sur k), on a $M^o_{[F]} \otimes_{\mathbb{Q}} k = \underline{\mathrm{Aut}}_k(\omega)$.

On vérifie que l'action cherchée de $\mathrm{Gal}(\overline{\mathbb{Q}}/\mathbb{Q})$ sur M^o est simplement l'action par transport de structure de $\mathrm{Aut}(F)$ sur $M^o_{[F]}$, pour $F = \overline{\mathbb{Q}}$. Ceci a un sens, puisque $M^o_{[F]}$ ne dépend que de F.

Nous allons utiliser le foncteur fibre H_{DR} (cohomologie de De Rham) : $(CM)_F \longrightarrow$(vectoriels sur F). Ce foncteur fibre est filtré, par la filtration de Hodge.

Proposition B.1. Le foncteur fibre H_{DR} de $(CM)_F$ admet une unique bigraduation, $H_{DR}(V) = \oplus H_{DR}^{pq}(V)$, telle que la filtration de Hodge soit donnée par $\bigoplus_{p' \geq p} H_{DR}^{p'q'}(V)$, et que pour V de poids n on ait $H_{DR}^{pq} = 0$ pour $p+q \neq n$.

Pour V dans $(CM)_F$ de poids n, montrons que la filtration de Hodge de $H_{DR}(V)$ admet un unique scindage stable sous $End(H)$. Il suffit de le vérifier pour $H_{DR}(H) \otimes_{F,\sigma} \mathbb{C}$ pour σ un plongement de F dans $\mathbb{C}$. Existence : les H^{pq} conviennent. Unicité : puisque μ, donnant la p-graduation, est dans le commutant de $End(H)$, donc dans le centre de $End(H) \otimes \mathbb{C}$ puisque H est de type CM, les $End(H) \otimes \mathbb{C}$-modules H^{pq} sont disjoints. Ce scindage fournit la bigraduation cherchée. Sa construction montre que :

Proposition B.2. Après extension des scalaires à $\mathbb{C}$, relativement à n'importe quel plongement complexe de F, la bigraduation ci-dessus devient celle par les H^{pq}

La proposition B.1 fournit

$$\mu_{DR} : G_m \longrightarrow M^o_{[F]} \quad \text{sur } F \text{ canonique,}$$

donc fixé par $Aut(F)$, tel que pour chaque plongement complexe σ de F, le μ_σ correspondant soit $\sigma(\mu_{DR})$.

$$\sigma : Hom_F(G_m, M^o_{[F]}) \longrightarrow Hom_{\mathbb{C}}(\mathbb{G}_m, M^o_{[F]}) \quad .$$

On aurait pu aussi invoquer la théorie générale de Saavedra des foncteurs fibres filtrés pour construire μ_{DR}. Peu importe.

Pour α un automorphisme de F, on a $\alpha(\mu_\sigma) = \mu_{\sigma\alpha^{-1}}$, et pour $\alpha \in Gal(\overline{\mathbb{Q}}/\mathbb{Q})$ (agissant sur $Hom_{\mathbb{Q}}(\mathbb{G}_m, M^o_{[F]})$ car $M^o_{[F]}$ est défini sur $\mathbb{Q}$) : $\alpha(\mu_\sigma) = \alpha\sigma(\mu_{DR}) = \mu_{\alpha\sigma}$. Ceci fournit le calcul du titre. Voici une bonne façon de l'écrire.

Le groupe $Hom_{\mathbb{C}}(M^o_{[F]}, \mathbb{G}_m)$ des caractères de M^o_F s'identifie canoniquement (par $\chi \longmapsto \langle \chi, \mu_\sigma \rangle$) au groupe des fonctions entières sur l'ensemble des plongements complexes de F dans $\mathbb{C}$

(a) factorisables par un $Hom(K, \mathbb{Q})$, pour $K \subset F$ un corps CM, et

(b) telles que $n(\sigma)+n(c\sigma)=C^{te}$, pour c la conjugaison complexe.

Dans cette description, l'action de $Aut(F)$ est bien sûr par transport de structure ; celle, galoisienne, de $Gal(\overline{\mathbb{Q}}/\mathbb{Q})$ aussi.

Ceci achève la preuve des lemmes 1 et 2 ; les troncations mentionnées à la suite du lemme 2 ont maintenant un sens.

(C) Calcul de torseurs.

(a) Explicitons ce qu'est (3'). Le groupe M est limite projective de ses quotients de type fini. Un tel quotient, M' est extension d'un quotient fini de $\mathrm{Gal}(\overline{\mathbb{Q}}/\mathbb{Q})$, $\mathrm{Gal}(E/\mathbb{Q})$, par un quotient de type fini M'^{o} de S

$$M'^{o} \longrightarrow M' \longrightarrow \mathrm{Gal}(E/\mathbb{Q}) \quad .$$

Pour M' de plus en plus grand, on a $\varinjlim E = \overline{\mathbb{Q}}$ et $\varprojlim M'^{o} = S$. C'est en ce sens que M est extension de $\mathrm{Gal}(\overline{\mathbb{Q}}/\mathbb{Q})$ par S. La structure (3') fournit des diagrammes commutatifs

$$\begin{array}{ccccc} M'^{o}(\mathbb{A}^{f}) & \longrightarrow & M'(\mathbb{A}^{f}) & \longrightarrow & \mathrm{Gal}(E/\mathbb{Q}) \\ & & & \nwarrow^{\pi} & \uparrow \\ & & & & \mathrm{Gal}(\overline{\mathbb{Q}}/\mathbb{Q}) \end{array} \quad ,$$

avec π continu pour les topologies de Krull et adéliques. C'est en ce sens que $\pi : \mathrm{Gal}(\overline{\mathbb{Q}}/\mathbb{Q}) \longrightarrow M(\mathbb{A}^{f})$ est continu.

(b) La trivialité ℓ-adique (a) a pour analogue réel l'existence sur $(CM)_{\mathbb{Q}}$ d'une polarisation : pour tout motif M, de poids n, il existe $\Psi : M \otimes M \longrightarrow \mathbb{Q}(-n)$ qui, après extension du corps de base à $\mathbb{C}$, fournisse une polarisation de la structure de Hodge $H_B(M)$.

Traduisons et spécialisons au poids 0. Pour chaque représentation linéaire V de M, de poids 0, il existe une forme bilinéaire symétrique invariante $\phi : V \otimes V \longrightarrow \mathbb{Q}$, telle que

$$\phi(x, h(i)x) > 0 \quad \text{pour} \quad x \neq 0 \quad \text{dans} \quad V \otimes \mathbb{R} \quad .$$

D'après mon article "La conjecture de Weil pour les surfaces K3", Inv. Math. <u>15</u> p. 206-226, 1972, § 2, le lemme 3 en résulte.

Soit $K \subset \overline{\mathbb{Q}}$ galoisien sur $\mathbb{Q}$ et CM. Le lemme 3 assure que la forme tordue de $[{}^{K}M/w(\mathbb{G}_m)]_{\mathbb{R}}$ définie par l'involution $\mathrm{int}(h(i))$ a un point réel dans chaque composante connexe réelle. D'après Hilbert 90, la même assertion vaut pour la forme tordue de $[{}^{K}M]_{\mathbb{R}}$ définie par l'involution $\mathrm{int}\, h(i)$. Noter que $\mathrm{int}(h(i))$ est bien une involution, car $h(-1) = w(-1)$ est central.

Montrons que les lemmes 1 à 3 suffisent à déterminer la classe d'isomorphie des ${}^{K}S$-torseurs ${}^{K}P_{\sigma}$ définis dans le laius qui suit le lemme 3. Soit $\sigma \in \mathrm{Gal}(\overline{\mathbb{Q}}/\mathbb{Q})$, et fixons une clôture algébrique F de $\mathbb{Q}$. La classe à déterminer vit dans $H^{1}(\mathrm{Gal}(F/\mathbb{Q}), S^{o}(F))$. Sur $\mathbb{Q}_{\ell}$, elle devient triviale, par (a) ci-dessus. Sur $\mathbb{R}$, on la calcule : par (b), il existe $u \in P_{\sigma}(\mathbb{C})$ tel que $\mathrm{int}(h(i))(\overline{u}) = u$, i.e.

$$h(i)\overline{u}h(i)^{-1} = h(i)\overline{u}h(-i)\overline{u}^{-1}\overline{u} = h(i)h(-i)^{\sigma}\overline{u} = u$$
$$u\overline{u}^{-1} = h(i)h(-i)^{\sigma} \quad .$$

Il ne reste qu'à invoquer la

Proposition C.1. Si (K,L) est CM, le groupe algébrique ${}^K S$ vérifie le principe de Hasse.

Preuve : La suite exacte de groupes algébriques sur $\mathbb{Q}$

$$0 \longrightarrow R_{L/\mathbb{Q}}(\mathbb{G}_m) \longrightarrow R_{K/\mathbb{Q}}(\mathbb{G}_m) \times \mathbb{G}_m \longrightarrow {}^K S^o \longrightarrow 0$$

fournit une suite exacte

$$0 \longrightarrow H^1(\mathrm{Gal}(F/\mathbb{Q}), {}^K S^o(F)) \longrightarrow \mathrm{Br}(L) \longrightarrow \mathrm{Br}(K) \times \mathrm{Br}(\mathbb{Q}) ,$$

et de même localement, et Br(L) vérifie le principe de Hasse.

(D) Calcul de $M \otimes \mathbb{Q}_\ell$.

Soient K une extension finie de $\mathbb{Q}$, $\overline{K}$ une clôture algébrique de K, et V un espace vectoriel sur une extension finie E_λ de $\mathbb{Q}_\ell$. Une représentation λ-adique $\rho : \mathrm{Gal}(\overline{K}/K) \longrightarrow GL(V)$ est dite de type CM si elle se factorise par $\mathrm{Gal}(\overline{K}/K)^{ab}$ et qu'il existe un conducteur $\underline{\underline{m}}$, et un morphisme, défini sur E_λ, ρ_{alg}, de $R_{K/\mathbb{Q}}(\mathbb{G}_m)$ dans GL(V) tels que la représentation ρ soit non ramifiée en dehors de $\underline{\underline{m}}$ et des places divisant ℓ et que pour tout idéal $a = \Pi\, \underline{\underline{p}}_i^{n_i}$ de K, premier à $\underline{\underline{m}}$, et engendré par un élément $\alpha \equiv 1(\underline{\underline{m}})$, on ait $\Pi\, \rho(F_{\underline{\underline{p}}_i})^{n_i} = \rho_{alg}(\alpha)$. Ce morphisme ρ_{alg} est unique. Il est trivial sur les unités $u \equiv 1\ (\underline{\underline{m}})$. Il se factorise donc par ${}^K S$ (défini avant le lemme 4). Ces représentations sont celles que J.-P. Serre [3] appelle localement algébriques. On dit que ρ_{alg} est la partie algébrique de ρ. Les $F_{\underline{\underline{p}}_i}$ sont ici les Frobenius géométriques.

La représentation ρ est dite potentiellement de type CM si elle devient de type CM sur une extension finie de K. On vérifie qu'une représentation potentiellement de type CM qui se factorise par $\mathrm{Gal}(\overline{K}/K)^{ab}$ est de type CM.

Une représentation linéaire ρ de M sur V est un morphisme du groupe algébrique $M \otimes_{\mathbb{Q}} E_\lambda$, déduit de M par extension des scalaires de $\mathbb{Q}$ à E_λ, dans GL(V). Une telle représentation définit une représentation λ-adique

$$\rho \circ \pi_\ell : \mathrm{Gal}(\overline{\mathbb{Q}}/\mathbb{Q}) \longrightarrow M(\mathbb{Q}_\ell) \subset M(E_\lambda) \longrightarrow GL(V) .$$

Nous allons déduire du lemme 4 que :

Proposition D.1. Le foncteur $\rho \longmapsto \rho \circ \pi_\ell$ est une équivalence de la catégorie des représentations linéaires de $M \otimes_{\mathbb{Q}} E_\lambda$ avec celle des représentations λ-adiques potentiellement de type CM de $\mathrm{Gal}(\overline{\mathbb{Q}}/\mathbb{Q})$.

Proposition D.2. Soient V un espace vectoriel sur E_λ, ρ une représentation linéaire de M sur V et $K \subset \overline{\mathbb{Q}}$ tel que $\rho \circ \pi_\ell$ soit de type CM sur K. Alors,

$\rho|S$ et $(\rho\circ\pi_\ell)_{alg}$ se factorisent par le même morphisme de $^K S$ dans $GL(V)$.

Montrons que, quel que soient $K \subset \bar{\mathbb{Q}}$ et V , espace vectoriel sur une extension finie E_λ de $\mathbb{Q}_\lambda$, une représentation linéaire ρ de M sur V définit une représentation λ-adique potentiellement de type CM de $Gal(\bar{\mathbb{Q}}/K)$. Deux réductions : (a) pour le vérifier pour ρ , il suffit de le vérifier après avoir remplacé K par une extension K' , $_K M$ par $_{K'}M$, E_λ par une extension E'_λ et V par $V \otimes_{E_\lambda} E'_\lambda$. (b) il suffit de le vérifier pour une famille de représentations les engendrant toutes. Combinant ces réductions, on voit qu'il suffit de vérifier l'assertion pour une famille de représentations des $_K M$, fidèle sur S . Que celles considérées dans le lemme 4 (plutôt, celle qui s'en déduisent par extension des scalaires de E à ses complétés E_λ) suffisent à la tâche résulté du lemme A.2, et des explications qui précèdent. Elles vérifient par hypothèse D.1 et D.2 et ceci prouve déjà D.2.

Soient (ℓ-CM) la $\mathbb{Q}_\ell$-catégorie tannakienne des représentations ℓ-adiques potentiellement de type CM de $Gal(\bar{\mathbb{Q}}/\mathbb{Q})$, et A le schéma en groupe des automorphismes du foncteur fibre "espace vectoriel sous-jacent". Il s'envoie dans le groupe profini (vu comme groupe pro-algébrique) $Gal(\bar{\mathbb{Q}}/\mathbb{Q})$, correspondant à la sous-catégorie des représentations qui se factorisent par un quotient fini de $Gal(\bar{\mathbb{Q}}/\mathbb{Q})$.

Pour (V,ρ) dans (ℓ-CM) , la construction "partie algébrique de la restriction de ρ à $Gal(\bar{\mathbb{Q}}/K)$, pour K assez grand" fournit une action de S sur V . Cette construction est compatible au produit tensoriel, et fournit donc, sur $\mathbb{Q}_\ell$, un morphisme de S dans A . Si la partie algébrique de $\rho|Gal(\bar{\mathbb{Q}}/K)$ est triviale, ρ se factorise par un quotient fini de $Gal(\bar{\mathbb{Q}}/\mathbb{Q})$. De là résulte que A est une extension

$$S\otimes\mathbb{Q}_\ell \longrightarrow A \longrightarrow Gal(\bar{\mathbb{Q}}/\mathbb{Q}) \quad .$$

Au foncteur qui à une représentation linéaire ρ de $M\otimes\mathbb{Q}_\ell$ associe la représentation ℓ-adique $\rho\circ\pi_\ell$ de $Gal(\bar{\mathbb{Q}}/\mathbb{Q})$ correspond un morphisme de A dans $M\otimes\mathbb{Q}_\ell$. D'après D.2, le diagramme

$$\begin{array}{ccccc} S\otimes\mathbb{Q}_\ell & \longrightarrow & A & \longrightarrow & Gal(\bar{\mathbb{Q}}/\mathbb{Q}) \\ \| & & \downarrow & & \| \\ S\otimes\mathbb{Q}_\ell & \longrightarrow & M\otimes\mathbb{Q}_\ell & \longrightarrow & Gal(\bar{\mathbb{Q}}/\mathbb{Q}) \end{array}$$

est commutatif. Le morphisme $A \longrightarrow M\otimes\mathbb{Q}_\ell$ est donc un isomorphisme. Ceci prouve D.1 .

Le fait que chaque représentation linéaire de A soit une représentation ℓ-adique de $Gal(\bar{\mathbb{Q}}/\mathbb{Q})$ fournit une section (d'image de Zariski-dense) $\pi_\ell : Gal(\bar{\mathbb{Q}}/\mathbb{Q}) \longrightarrow A(\mathbb{Q}_\ell)$. Le diagramme

$$\begin{array}{ccc} A(\mathbb{Q}_\ell) & \xleftarrow{\quad\pi_\ell\quad} & Gal(\bar{\mathbb{Q}}/\mathbb{Q}) \\ \| & & \| \\ M\otimes\mathbb{Q}_\ell & \xleftarrow{\quad\pi_\ell\quad} & Gal(\bar{\mathbb{Q}}/\mathbb{Q}) \end{array}$$

est commutatif.

(E) Calcul de ${}_K^K M$.

Soient K une extension finie de $\mathbb{Q}$, et K^{ab} la réunion des extensions abéliennes de K contenues dans une clôture algébrique $\overline{K}$. Le groupe $Gal(K^{ab}/K) = Gal(\overline{K}/K)^{ab}$ ne dépend que de K , à isomorphisme unique près. Dans [3], J.-P. Serre a introduit le schéma en groupes de type multiplicatif S_K , dont les caractères définis sur un corps E s'identifient aux caractères de Hecke algébriques de K à valeurs dans E . Plus exactement, il a introduit des groupes S_m , dont S_K est la limite projective. Le groupe S_K est une extension

$$^K S \longrightarrow S_K \longrightarrow Gal(K^{ab}/K)$$

et on dispose d'une section continue $\pi : Gal(K^{ab}/K) \longrightarrow S_1(\mathbb{A}^f)$, de composantes π_ℓ .

Les morphismes $^K S \longrightarrow S_K$, $S_K \longrightarrow Gal(K^{ab}/K)$, et π sont caractérisés comme suit. Soit χ un caractère de Hecke algébrique de K , à valeurs dans E , non ramifié en dehors d'un ensemble fini ϕ de places de K et χ_1 le caractère défini sur E correspondant de S_K .

a) la partie algébrique χ_{alg} de χ est le composé (défini sur E) $R_{K/\mathbb{Q}}(\mathbb{G}_m) \longrightarrow {}^K S \xrightarrow{\chi_1 | {}^K S} \mathbb{G}_m$.

b) Si χ_{alg} est trivial, i.e. si χ est d'ordre fini, χ_1 est le composé de la projection de S_K sur $Gal(K^{ab}/K)$, et du caractère χ' de $Gal(K^{ab}/K)$, non ramifié en dehors de ϕ , tel que $\chi'(F_{\underline{p}}) = \chi(\underline{p})$ pour $\underline{p} \notin \phi$.

c) $\chi_1 \circ \pi_\ell : Gal(K^{ab}/K) \longrightarrow (E \times \mathbb{Q}_\ell)^*$ est non ramifié en dehors de ϕ et des places divisant ℓ , et $\chi_1 \circ \pi_\ell(F_{\underline{p}}) = \chi(\underline{p})$ pour $\underline{p} \notin \phi$, $\underline{p} \nmid \ell$.

Supposons que $K \subset \overline{\mathbb{Q}}$ et faisons $\overline{K} = \overline{\mathbb{Q}}$. Soit ${}_K A$ l'image inverse de $Gal(\overline{\mathbb{Q}}/K)$ dans le groupe A défini en (D) . Ses représentations linéaires s'identifient aux représentations ℓ-adiques potentiellement de type CM de $Gal(\overline{\mathbb{Q}}/K)$. Chaque représentation linéaire de S_K sur $\mathbb{Q}_\ell$ fournit, par composition avec π_ℓ , une représentation de type CM de $Gal(\overline{\mathbb{Q}}/K)$. De là un morphisme de ${}_K A$ dans $S_K \otimes \mathbb{Q}_\ell$ donnant lieu à des diagrammes commutatifs :

$$\begin{array}{ccccc} S\otimes\mathbb{Q}_\ell & \longrightarrow & {}_K A & \longrightarrow & Gal(\overline{\mathbb{Q}}/K) \\ \downarrow & & \downarrow & & \downarrow \\ {}^K S\otimes\mathbb{Q}_\ell & \longrightarrow & S_K\otimes\mathbb{Q}_\ell & \longrightarrow & Gal(K^{ab}/K) \end{array} \quad , \quad \begin{array}{ccc} {}_K A(\mathbb{Q}_\ell) & \xleftarrow{\pi_\ell} & Gal(\overline{\mathbb{Q}}/K) \\ \downarrow & & \downarrow \\ S_K(\mathbb{Q}_\ell) & \xleftarrow{\pi_\ell} & Gal(K^{ab}/K) . \end{array}$$

Le noyau de la projection de S dans $^K S$ est un sous-groupe invariant de ${}_K M$. Notons ${}_K^K M$ le quotient correspondant de ${}_K M$; c'est une extension de $Gal(\overline{\mathbb{Q}}/K)$

par ${}^{K}S$. Remplaçant ${}_{K}A$ par le groupe isomorphe ${}_{K}M \otimes \mathbb{Q}_\ell$, puis par son quotient ${}_{K}^{K}M \otimes \mathbb{Q}_\ell$, on obtient des diagrammes commutatifs

$$\begin{array}{ccccc} {}^{K}S\otimes\mathbb{Q}_\ell & \longrightarrow & {}_{K}^{K}M\otimes\mathbb{Q}_\ell & \longrightarrow & \mathrm{Gal}(\bar{\mathbb{Q}}/K) \\ \| & & \downarrow & & \downarrow \\ {}^{K}S\otimes\mathbb{Q}_\ell & \dashrightarrow & S_K\otimes\mathbb{Q}_\ell & \dashrightarrow & \mathrm{Gal}(K^{ab}/K) \end{array} \qquad \begin{array}{ccc} {}_{K}^{K}M(\mathbb{Q}_\ell) & \xleftarrow{\pi_\ell} & \mathrm{Gal}(\bar{\mathbb{Q}}/K) \\ \downarrow & & \downarrow \\ S_K(\mathbb{Q}_\ell) & \xleftarrow{\pi_\ell} & \mathrm{Gal}(K^{ab}/K) \end{array}$$

Soit $\tilde{S}_K$ le schéma en groupe sur $\mathbb{Q}$ produit fibré sur $\mathrm{Gal}(K^{ab}/K)$ de S_K et $\mathrm{Gal}(\bar{\mathbb{Q}}/K)$. C'est l'inflation à $\mathrm{Gal}(\bar{\mathbb{Q}}/K)$ de l'extension S_K de $\mathrm{Gal}(K^{ab}/K)$ par ${}^{K}S$. On dispose cette fois de $\pi_\ell : \mathrm{Gal}(\bar{\mathbb{Q}}/K) \longrightarrow \tilde{S}_K(\mathbb{Q}_\ell)$, et d'isomorphismes

$$\begin{array}{ccccc} {}^{K}S\otimes\mathbb{Q}_\ell & \longrightarrow & {}_{K}^{K}M\otimes\mathbb{Q}_\ell & \longrightarrow & \mathrm{Gal}(\bar{\mathbb{Q}}/K) \\ \| & & \downarrow f_\ell & & \| \\ {}^{K}S\otimes\mathbb{Q}_\ell & \longrightarrow & \tilde{S}_K\otimes\mathbb{Q}_\ell & \longrightarrow & \mathrm{Gal}(\bar{\mathbb{Q}}/K) \end{array} \qquad \begin{array}{ccc} {}_{K}^{K}M(\mathbb{Q}_\ell) & \xleftarrow{\pi_\ell} & \mathrm{Gal}(\bar{\mathbb{Q}}/K) \\ \downarrow & & \| \\ \tilde{S}_K(\mathbb{Q}_\ell) & \xleftarrow{\pi_\ell} & \mathrm{Gal}(\bar{\mathbb{Q}}/K) \end{array} .$$

Nous allons déduire du lemme 4 la

<u>Proposition E.1</u> . <u>L'isomorphisme</u> $f_\ell : {}_{K}^{K}M \otimes \mathbb{Q}_\ell \longrightarrow \tilde{S}_K \otimes \mathbb{Q}_\ell$ <u>ci-dessus est défini sur</u> $\mathbb{Q}$, i.e. <u>provient par extension des scalaires d'un isomorphisme de schémas en groupes sur</u> $\mathbb{Q}$: $f : {}_{K}^{K}M \xrightarrow{\sim} \tilde{S}_K$.

<u>Lemme E.2</u> . <u>Il existe des quotients de type fini abéliens</u> ${}^{K}M'$ <u>de</u> ${}_{K}^{K}M$ <u>et</u> $\tilde{S}'_K$ <u>de</u> $\tilde{S}_K$, <u>fidèles sur</u> ${}^{K}S$, <u>tels que</u> f_ℓ <u>induise par passage au quotient un isomorphisme</u> $f'_\ell : M'\otimes\mathbb{Q}_\ell \longrightarrow \tilde{S}'_K\otimes\mathbb{Q}_\ell$.

Soit M^1 un quotient de type fini de ${}_{K}^{K}M$, fidèle sur ${}^{K}S$: c'est une extension

$$ {}^{K}S \longrightarrow M^1 \dashrightarrow \mathrm{Gal}(E^1/K) .$$

Soit E^2 une extension de E^1 galoisienne sur K et M^2 le quotient de ${}_{K}^{K}M$ déduit de M^1 par inflation de $\mathrm{Gal}(E^1/K)$ à $\mathrm{Gal}(E^2/K)$. Pour E^2 assez grand, le sous-schéma en groupe des commutateurs de M^2 ne rencontre pas ${}^{K}S$: c'est clair après extension des scalaires à $\mathbb{Q}_\ell$, vu la structure connue de ${}_{K}^{K}M\otimes\mathbb{Q}_\ell$, et donc vrai sur $\mathbb{Q}$. Passant au quotient, on obtient un quotient de type fini abélien M^3 de ${}_{K}^{K}M$, fidèle sur ${}^{K}S$:

$$ {}^{K}S \longrightarrow M^3 \dashrightarrow \mathrm{Gal}(E^3/K) .$$

Soit S_K^1 un quotient de type fini de S_K , fidèle sur ${}^{K}S$:

$$ {}^{K}S \longrightarrow S_K^1 \dashrightarrow \mathrm{Gal}(F^1/K) .$$

Soit enfin E une extension abélienne de K , contenant E^3 et F^1 , et M' et S' déduits de M^3 et S^1_K par inflation à Gal(E/K) . Pour E assez grand, M' et S' répondent aux exigences E.2.

Tout caractère, défini sur $\mathbb{C}$, de ${}^K S$, se prolonge à M' et est défini sur un corps de nombre E . Pour E assez grand, le lemme A.2 et les explications qui précèdent, assurent donc l'existence d'une famille fidèle de caractères $\chi : M' \longrightarrow R_{E/\mathbb{Q}}(\mathbb{G}_m)$, du type considérés dans le lemme 4 . Les représentations ℓ-adiques $\chi \circ \pi_\ell$ sont non ramifiées en dehors d'un ensemble fini ϕ de places de K , et le lemme 4 assure qu'elles sont données par des caractères de Hecke algébriques. En particulier, pour p non dans S , $\chi \circ \pi_\ell(F_p) \in (E \otimes \mathbb{Q}_\ell)^*$ est dans E^* . La famille (χ) étant fidèle, il en résulte que $\pi_\ell(F_p) \in M'(\mathbb{Q}_\ell)$ est dans $M'(\mathbb{Q})$. Pour S' aussi, les $\pi_\ell(F_p) \in S'(\mathbb{Q}_\ell)$ sont dans $S'(\mathbb{Q})$. Les $\pi_\ell(F_p)$ étant Zariski-dense, f'_ℓ envoie donc une partie Zariski-dense de $M'(\mathbb{Q})$ dans $S'(\mathbb{Q})$ - et est défini sur $\mathbb{Q}$. Le morphisme f_ℓ , qui s'en déduit par inflation, l'est aussi, et ceci prouve E.1.

(F) Unicité de M .

Nous nous proposons de vérifier que deux schémas en groupes M' et M'' , munis de structures (1) à (3) , et vérifiant les lemmes 1 à 5 , sont isomorphes.

(F.1) où on utilise les lemmes 1 et 2.

Soit Δ_1 la différence des extensions M' et M'' de $Gal(\overline{\mathbb{Q}}/\mathbb{Q})$ par S . Pour obtenir Δ_1 , on prend l'image inverse dans $M' \times M''$ de l'image diagonale de $Gal(\overline{\mathbb{Q}}/\mathbb{Q})$ dans $Gal(\overline{\mathbb{Q}}/\mathbb{Q}) \times Gal(\overline{\mathbb{Q}}/\mathbb{Q})$, et on divise par l'image diagonale de S dans $S \times S$. Cette image est un sous-groupe invariant, car l'action de $Gal(\overline{\mathbb{Q}}/\mathbb{Q})$ sur S est la même pour M' et M'' . L'extension obtenue de $Gal(\overline{\mathbb{Q}}/\mathbb{Q})$ par S est munie, par différence, d'une section $\pi : Gal(\overline{\mathbb{Q}}/\mathbb{Q}) \longrightarrow \Delta_1(A^f)$.

Pour $K \subset \overline{\mathbb{Q}}$ galoisien sur $\mathbb{Q}$ et CM, on notera ${}^K\Delta_1$ le quotient de Δ_1 extension de $Gal(\overline{\mathbb{Q}}/\mathbb{Q})$ par le quotient ${}^K S$ de S . C'est la différence de ${}_K M'$ et ${}_K M''$.

Que M' et M'' , munis de (1)(2)(3), soient isomorphes équivaut à ce que π provienne d'un scindage $Gal(\overline{\mathbb{Q}}/\mathbb{Q}) \dashrightarrow \Delta_1$ de l'extension Δ_1 ·, faisant de Δ_1 un produit semi-direct. Il revient au même de demander que pour chaque K comme ci-dessus et chaque nombre premier ℓ, l'application continue déduite de π , $\pi_\ell : Gal(\overline{\mathbb{Q}}/\mathbb{Q}) \longrightarrow {}^K\Delta_1(\mathbb{Q}_\ell)$, provienne d'un scindage $Gal(\overline{\mathbb{Q}}/\mathbb{Q}) \longrightarrow {}^K\Delta_1$ de ${}^K\Delta_1$, et que celui-ci soit indépendant de ℓ .

(F.2) où on utilise les lemmes 3 et 4 .

Pour K comme plus haut, on a vu (F.1) que ${}^K_K M'$ et ${}^K_K M''$ sont isomorphes. En d'autres termes, l'image inverse ${}^K_K\Delta_1$ de $Gal(\overline{\mathbb{Q}}/K)$ dans ${}^K\Delta_1$ est un produit semi-direct de $Gal(\overline{\mathbb{Q}}/K)$ par ${}^{K}S$, même un produit, car $Gal(\overline{\mathbb{Q}}/K)$ agit trivialement sur ${}^{K}S$ et les $\pi_\ell : Gal(\overline{\mathbb{Q}}/K) \longrightarrow {}^{K}_{K}\Delta_1(\mathbb{Q}_\ell)$ proviennent tous d'un même scindage ${}^{K}_{K}\pi_{\mathbb{Q}} : Gal(\overline{\mathbb{Q}}/K) \longrightarrow {}^{K}_{K}\Delta_1$. L'image de ${}^{K}_{K}\pi_{\mathbb{Q}}$ est un sous-groupe invariant de ${}^{K}_{K}\Delta_1$. Elle est même invariante dans ${}^{K}\Delta$: il suffit de le vérifier après extension des scalaires à $\mathbb{Q}_\ell$, et, là, d'observer qu'elle est normalisée par $\pi_\ell(Gal(\overline{\mathbb{Q}}/\mathbb{Q}))$, puisque π_ℓ prolonge ${}^{K}_{K}\pi_{\mathbb{Q}}$. Passons au quotient. On obtient un groupe algébrique ${}^{K}\Delta_2$, extension de $Gal(K/\mathbb{Q})$ par ${}^{K}S$:

$$ {}^{K}S \longrightarrow {}^{K}\Delta_2 \dashrightarrow Gal(K/\mathbb{Q}) \quad , $$

muni de sections $\pi_\ell : Gal(K/\mathbb{Q}) \dashrightarrow \Delta_2(\mathbb{Q}_\ell)$, dont $({}^{K}\Delta_1, \pi)$ se déduit par inflation de $Gal(K/\mathbb{Q})$ à $Gal(\overline{\mathbb{Q}}/\mathbb{Q})$. Pour que M' et M" , munis de (1)(2)(3) , soient isomorphes, il faut et il suffit que ces sections π_ℓ soient rationnelles et indépendantes de ℓ .

Le lemme 3 implique (C) que pour chaque $\sigma \in Gal(K/\mathbb{Q})$, le ${}^{K}S$-torseur image inverse de σ a un point rationnel : ${}^{K}\Delta_2(\mathbb{Q})$ est extension de $Gal(K/\mathbb{Q})$ par ${}^{K}S(\mathbb{Q})$.

Pour $K' \supset K$, on dispose de diagrammes commutatifs

(F.2.1)

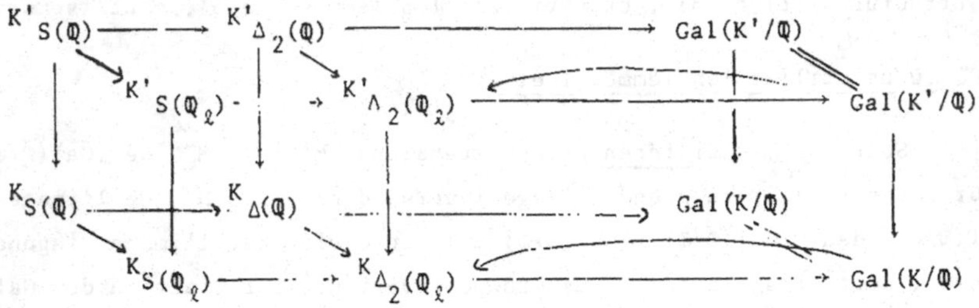

(F.3) <u>où on utilise le lemme 5</u> .

Soient K comme plus haut, et $c \in Gal(K/\mathbb{Q})$ la conjugaison complexe. Puisque K est CM , c est central dans $Gal(K/\mathbb{Q})$. Le lemme 5 assure que $\pi_\ell(c) \in {}^{K}\Delta_2(\mathbb{Q}_\ell)$ est dans ${}^{K}\Delta_2(\mathbb{Q})$ et indépendant de ℓ . On écrira $\pi(c)$ pour $\pi_\ell(c)$.

<u>Lemme F.4 . Le centralisateur ${}^{K}\Delta_3$ de $\pi(c)$ dans ${}^{K}\Delta_2(\mathbb{Q})$ s'envoie</u> sur $Gal(K/\mathbb{Q})$.

Notons $\mathbb{Z}/2$ le groupe $\{1,c\}$. Il agit sur ${}^{K}S(\mathbb{Q})$. Quel que soit σ dans $Gal(K/\mathbb{Q})$, l'image inverse P_σ de σ dans ${}^{K}\Delta_2(\mathbb{Q})$ est un torseur sous ${}^{K}S(\mathbb{Q})$, agissant par translations à gauche. Elle est stable par $\pi(c)$-conjugaison, et l'action de $\pi(c)$ en fait un ${}^{K}S(\mathbb{Q})$-torseur $\mathbb{Z}/2$-équivariant. Il nous faut montrer que l'action par automorphismes intérieurs de $\pi(c)$ sur P_σ a un point fixe, i.e. que la classe de ce torseur dans $H^1(\mathbb{Z}/2, {}^{K}S(\mathbb{Q}))$ est triviale. Quel que soit ℓ , l'image inverse de σ dans ${}^{K}\Delta_2(\mathbb{Q}_\ell)$ est le ${}^{K}S(\mathbb{Q}_\ell)$-torseur $\mathbb{Z}/2$-équivariant $P_{\sigma,\ell}$ déduit de P_σ en poussant par ${}^{K}S(\mathbb{Q}) \longrightarrow {}^{K}S(\mathbb{Q}_\ell)$. Sa classe est triviale, car $\pi_\ell(\sigma)$ commute à $\pi(c)$. Le lemme F.4 résulte donc du

<u>Lemme F.5</u>. $H^1(\mathbb{Z}/2, {}^K S(\mathbb{Q}))$ <u>s'injecte dans le produit des</u> $H^1(\mathbb{Z}/2, {}^K S(\mathbb{Q}_\ell))$.

Ecrivons $H^*(G)$ pour la cohomologie de $\mathbb{Z}/2$ à valeurs dans un groupe abélien G muni d'une action de $\mathbb{Z}/2$. Soit L le corps totalement réel dont K est extension quadratique totalement imaginaire. La suite exacte (A.1.1) est $\mathbb{Z}/2$-équivariante. Elle fournit une suite exacte longue

$$0 \longrightarrow L^* \longrightarrow L^* \times \mathbb{Q}^* \longrightarrow H^0({}^K S(\mathbb{Q})) \longrightarrow$$

$$\longrightarrow H^1(L^*) \longrightarrow H^1(K^*) \times H^1(\mathbb{Q}^*) \longrightarrow H^1({}^K S(\mathbb{Q})) \longrightarrow H^2(L^*) \quad .$$

On a $H^1(L^*) = \mathrm{Hom}(\mathbb{Z}/2, L^*) = \mu_2(L)$, $H^1(\mathbb{Q}^*) = \mu_2(\mathbb{Q})$, $H^1(K^*) = 0$ par Hilbert 90 et $H^2(L^*) = L^*/L^{*2}$. Ceci ramène la suite exacte à

(F.5.1) $$0 \longrightarrow \mathbb{Q}^* \longrightarrow H^0({}^K S(\mathbb{Q})) \longrightarrow \mu_2(L) \xrightarrow{N_{L/\mathbb{Q}}} \mu_2(\mathbb{Q}) \to H^1({}^K S(\mathbb{Q})) \to L^*/L^{*2} .$$

Pour ${}^K S(\mathbb{Q}_\ell)$, on trouve de même, notant L_ℓ le complété ℓ-adique $L \otimes \mathbb{Q}_\ell$ de L , une suite exacte

(F.5.2) $$0 \to \mathbb{Q}_\ell^* \to H^0({}^K S(\mathbb{Q}_\ell)) \to \mu_2(L_\ell) \to \mu_2(\mathbb{Q}_\ell) \longrightarrow H^1({}^K S(\mathbb{Q}_\ell)) \to L_\ell^*/L_\ell^{*2} ,$$

et des diagrammes commutatifs

$$\begin{array}{ccccccc} 0 & \to & \mathrm{coker}(\mu_2(L) \to \mu_2(\mathbb{Q})) & \longrightarrow & H^1({}^K S(\mathbb{Q})) & \longrightarrow & L^*/L^{*2} \\ & & \downarrow & & \downarrow & & \downarrow \\ 0 & \to & \mathrm{coker}(\mu_2(L_\ell) \to \mu_2(\mathbb{Q}_\ell)) & \longrightarrow & H^1({}^K S(\mathbb{Q}_\ell)) & \longrightarrow & L_\ell^*/L_\ell^{*2} \end{array} .$$

Le quotient L^*/L^{*2} s'injecte dans le produit des L_ℓ^*/L_ℓ^{*2} , car un élément de L^* qui n'a pas de racine carrée n'en a pas localement, en la moitié des places de L . Le conoyau $\mathrm{coker}(\mu_2(L) \to \mu_2(\mathbb{Q}))$ est nul pour L de degré impair et, réduit à ± 1 pour L de degré pair. Si L est de degré pair, il existe ℓ tels que les complétés L_λ de L $(\lambda|\ell)$ soient de degrés pairs sur $\mathbb{Q}_\ell$, et pour un tel ℓ le morphisme

$$\mathrm{coker}(\mu_2(L) \longrightarrow \mu_2(\mathbb{Q})) \longrightarrow \mathrm{coker}(\mu_2(L_\ell) \to \mu_2(\mathbb{Q}))$$

est un isomorphisme. En particulier, $\mathrm{coker}(\mu_2(L) \to \mu_2(\mathbb{Q}))$ s'injecte dans le produit des $\mathrm{coker}(\mu_2(L_\ell) \to \mu_2(\mathbb{Q}_\ell))$, et $H^1({}^K S(\mathbb{Q}))$ s'injecte donc dans le produit des $H^1({}^K S(\mathbb{Q}_\ell))$.

Notons par c-1 en exposant un lieu fixe par c . Si ${}^K \Delta_{3,\ell}$ est le centralisateur de $\pi_\ell(c)$ dans ${}^K \Delta_2(\mathbb{Q}_\ell)$, on a un diagramme

(F.5.3) $$\begin{array}{ccccc} {}^K S(\mathbb{Q})^{c-1} & \longrightarrow & {}^K \Delta_3 & \longrightarrow & \mathrm{Gal}(K/\mathbb{Q}) \\ \downarrow & & \downarrow & & \| \\ {}^K S(\mathbb{Q}_\ell)^{c-1} & \longrightarrow & {}^K \Delta_{3,\ell} & \longrightarrow & \mathrm{Gal}(K/\mathbb{Q}) \end{array} .$$

Soient ${}^K\Delta_4 = {}^K\Delta_3/\{1,\pi(c)\}$ et ${}^K\Delta_{4,\ell} = {}^K\Delta_{3,\ell}/\{1,\pi_\ell(c)\}$. Le diagramme (F.5.3) se déduit par inflation de $\mathrm{Gal}(L/K) = \mathrm{Gal}(K/\mathbb{Q})/\{1,c\}$ à $\mathrm{Gal}(K/\mathbb{Q})$ de

(F.5.4)
$$\begin{array}{ccccc} {}^K S(\mathbb{Q})^{c-1} & \longrightarrow & {}^K\Delta_4 & \longrightarrow & \mathrm{Gal}(L/\mathbb{Q}) \\ \downarrow & & \downarrow & & \| \\ {}^K S(\mathbb{Q}_\ell)^{c-1} & \longrightarrow & {}^K\Delta_{4,\ell} & \rightleftarrows & \mathrm{Gal}(L/\mathbb{Q}) \end{array} .$$

Il s'agit encore de montrer que π_ℓ tombe dans ${}^K\Delta_4$ et est indépendant de L . Pour $K' \supset K$, on a encore des diagrammes commutatifs de type (F.2.1) .

(F.6) <u>où on conclut</u>.

Fixons un corps quadratique imaginaire K_o , par exemple $\mathbb{Q}(i)$. Il suffit de vérifier que π_ℓ tombe dans ${}^K\Delta_4$, et est indépendant de ℓ , pour $K \supset K_o$: ces corps K sont arbitrairement grands.

Pour $K = K_o$, on a $L = \mathbb{Q}$ et le diagramme (F.5.4) est trivial. Pour $K \supset K_o$ quelconque, on a donc $\pi_\ell(\mathrm{Gal}(L/\mathbb{Q})) \subset \mathrm{Ker}({}^K\Delta_{4,\ell} \longrightarrow {}^{K_o}\Delta_{4,\ell})$. Posons ${}^K\Delta_5 = \mathrm{Ker}({}^K\Delta_4 \longrightarrow {}^{K_o}\Delta_4)$ (resp. ${}^K\Delta_{5,\ell} = \mathrm{Ker}({}^K\Delta_{4,\ell} \longrightarrow {}^{K_o}\Delta_{4,\ell})$. C'est une extension de $\mathrm{Gal}(L/\mathbb{Q})$ par $\mathrm{Ker}({}^K S(\mathbb{Q})^{c-1} \longrightarrow {}^{K_o}S(\mathbb{Q})^{c-1})$ (resp. $\mathrm{Ker}({}^K S(\mathbb{Q}_\ell)^{c-1} \longrightarrow {}^{K_o}(S(\mathbb{Q}_\ell)^{c-1}))$.

La suite (F.5.1) fournit un diagramme commutatif

$$\begin{array}{ccccccccc} 0 & \longrightarrow & \mathbb{Q}^* & \longrightarrow & {}^K S(\mathbb{Q})^{c-1} & \longrightarrow & \mathrm{Ker}(N_{L/\mathbb{Q}} : \mu_2(L) \longrightarrow \mu_2(\mathbb{Q})) & \longrightarrow & 0 \\ & & \| & & \downarrow & & \downarrow & & \\ 0 & \longrightarrow & \mathbb{Q}^* & \longrightarrow & {}^{K_o}S(\mathbb{Q})^{c-1} & \longrightarrow & \mathrm{Ker}(N_{\mathbb{Q}/\mathbb{Q}} : \mu_2(L) \longrightarrow \mu_2(\mathbb{Q})) & = & 0 \end{array} ,$$

d'où un isomorphisme

$$\mathrm{Ker}({}^K S(\mathbb{Q})^{c-1} \longrightarrow {}^{K_o}S(\mathbb{Q})^{c-1}) \simeq \mathrm{Ker}(N_{L/\mathbb{Q}} : \mu_2(L) \longrightarrow \mu_2(\mathbb{Q})) .$$

Localement, on trouve de même

$$\mathrm{Ker}({}^K S(\mathbb{Q}_\ell)^{c-1} \longrightarrow {}^{K_o}S(\mathbb{Q}_\ell)^{c-1}) = \mathrm{Ker}(N_{L/\mathbb{Q}} : \mu_2(L_\ell) \longrightarrow \mu_2(\mathbb{Q}_\ell))$$

et (F.5.4) se ramène au sous diagramme

(F.6.1)
$$\begin{array}{ccccccc} \mathrm{Ker}(\mu_2(L) & \longrightarrow & \mu_2(\mathbb{Q})) & \longrightarrow & {}^K\Delta_5 & \longrightarrow & \mathrm{Gal}(L/\mathbb{Q}) \\ & \big\downarrow & & & \big\downarrow & & \| \\ \mathrm{Ker}(\mu_2(L_\ell) & \longrightarrow & \mu_2(\mathbb{Q}_\ell)) & \longrightarrow & {}^K\Delta_{5,\ell} & \overset{\pi_\ell}{\rightleftarrows} & \mathrm{Gal}(L/\mathbb{Q}) \end{array} .$$

Pour L' une extension totalement réelle de L , et $K' = L' \otimes K_o \supset K = L \otimes K_o$, on dispose d'un diagramme commutatif

$$\begin{array}{ccccc}
\mathrm{Ker}(\mu_2(L'_\ell) & \longrightarrow \mu_2(\mathbb{Q}_\ell)) \longrightarrow & {}^{K'}\Delta_{5,\ell} & \underset{\longrightarrow}{\overset{\pi_\ell}{\longleftarrow}} & \mathrm{Gal}(L'/\mathbb{Q}) \\
\downarrow & & \downarrow & & \downarrow \\
\mathrm{Ker}(\mu_2(L_\ell) & \longrightarrow \mu_2(\mathbb{Q}_\ell)) \longrightarrow & {}^{K}\Delta_{5,\ell} & \underset{\longrightarrow}{\overset{\pi_\ell}{\longleftarrow}} & \mathrm{Gal}(L/\mathbb{Q})
\end{array} .$$

Choisissons L' tel que pour chaque place λ' de L' divisant ℓ, se projetant sur une place λ de L, le degré $[L'_{\lambda'} : L_\lambda]$ soit pair. Le morphisme $N_{L'/L}$ de $\mu_2(L'_\ell)$ dans $\mu_2(L_\ell)$ est alors 0, et a fortiori le morphisme de $\mathrm{Ker}(\mu_2(L'_\ell) \longrightarrow \mu_2(\mathbb{Q}_\ell))$ dans $\mathrm{Ker}(\mu_2(L_\ell) \longrightarrow \mu_2(\mathbb{Q}_\ell))$. Pour $\sigma \in \mathrm{Gal}(L/\mathbb{Q})$, image de $\sigma' \in \mathrm{Gal}(L'/\mathbb{Q})$, $\pi_\ell(\sigma)$ est donc l'image de n'importe quel élément de ${}^{K'}\Delta_{5,\ell}$ au-dessus de σ'. En particulier, $\pi_\ell(\sigma)$ est l'image de n'importe quel élément de ${}^{K'}\Delta_5$ au-dessus de σ' : il est dans ${}^{K}\Delta_5$ et, prenant L' convenant pour deux nombres premiers à la fois, on voit qu'il est indépendant de ℓ. Ceci termine la démonstration.

B I B L I O G R A P H I E

[1] P. Deligne. Valeurs de fonctions L et périodes d'intégrales. Proc. Symp. in Pure Math. 33 part 2, p. 313-346. AMS 1979.

[2] SGA$4\frac{1}{2}$. Lecture Notes in Mathematics 569. Springer Verlag.

[3] J.-P. Serre. Abelian ℓ-adic representations and elliptic curves. Benjamin 1968.

[4] G. Shimura. Algebraic number fields and symplectic discrete groups. Ann. of Math. 86 3 (1967) p. 503-592.

V. CONJUGATES OF SHIMURA VARIETIES

J. S. Milne and K.-y. Shih.

Introduction

Introduction: In the first three sections we review the definition of a Shimura variety of abelian type, describe how certain Shimura varieties are moduli varieties for abelian varieties with Hodge cycles and level structure, and prove a result concerning reductive groups that will frequently enable us to replace one such group by a second whose derived group is simply connected.

To be able to discuss the results in the remaining sections both concisely and precisely, we shall assume throughout the rest of the introduction that a pair (G,X) defining a Shimura variety Sh(G,X) satisfies the following additional conditions (Deligne [2, 2.1.1.4, 2.1.1.5]):

(0.1) for any $h \in X$, the weight $w_h : \mathbb{G}_m \to G_{\mathbb{R}}$ is defined over $\mathbb{Q}$;

(0.2) ad $h(i)$ is a Cartan involution on $(G/w(\mathbb{G}_m))_{\mathbb{R}}$.

These conditions imply that for any special $h \in X$, the associated cocharacter $\mu = \mu_h$ factors through the Serre group: $\mu = \rho_\mu \circ \mu_{can}$, $\rho_\mu : S \to G$. Thus to any such h and any representation of G there is associated a representation of S, and hence an object in the category of motives generated by abelian varieties of CM-type over $\mathbb{C}$.

Consider the Taniyama group

$$1 \to S \to \underset{\sim}{T} \overset{\pi}{\to} \mathrm{Gal}(\bar{\mathbb{Q}}/\mathbb{Q}) \to 1$$

$$\underset{\sim}{T}(\mathbb{A}^f) \underset{\pi}{\overset{sp}{\rightleftarrows}} \mathrm{Gal}(\bar{\mathbb{Q}}/\mathbb{Q}), \quad \pi \circ sp = 1 .$$

For any $\tau \in \mathrm{Gal}(\bar{\mathbb{Q}}/\mathbb{Q})$, ${}^{\tau}S \overset{df}{=} \pi^{-1}(\tau)$ is an S-torsor with a distinguished $\mathbb{A}^f$-point $sp(\tau)$. If $h \in X$ is special, we can use ρ_μ, $\mu = \mu_h$, to transform the adjoint action of G on itself into an action of S on G. We can then use ${}^{\tau}S$ to twist G, and so define ${}^{\tau,\mu}G = {}^{\tau}S \times {}^{S}G$. Thus ${}^{\tau,\mu}G$ is a $\mathbb{Q}$-rational algebraic group such that ${}^{\tau,\mu}G(\bar{\mathbb{Q}}) = \{s.g \mid s \in {}^{\tau}S(\bar{\mathbb{Q}}),\ g \in G(\bar{\mathbb{Q}})\}/\sim$ where $ss_1.g \sim s.\rho_\mu(s_1)g\rho_\mu(s_1)^{-1}$, all $s_1 \in S(\bar{\mathbb{Q}})$. Let $T \subset G$ be a $\mathbb{Q}$-rational torus through which h factors. Then ${}^{\tau,\mu}T \overset{df}{=} {}^{\tau}S \times {}^{S}T = T$, and so T is also a subgroup of ${}^{\tau,\mu}G$. Define ${}^{\tau}h$ to be the homomorphism $\mathbb{S} \to {}^{\tau,\mu}G$ with associated cocharacter $\tau\mu : \mathbb{G}_m \to T \subset {}^{\tau,\mu}G$, and let ${}^{\tau,\mu}X$ be the ${}^{\tau,\mu}G(\mathbb{R})$-

conjugacy class containing ${}^{\tau}h$. The point $sp(\tau)$ provides us with a canonical isomorphism $g \mapsto {}^{\tau,\mu}g \stackrel{df}{=} sp(\tau).g : G(\mathbb{A}^f) \to {}^{\tau,\mu}G(\mathbb{A}^f)$. The pair $({}^{\tau,\mu}G, {}^{\tau,\mu}X)$ defines a Shimura variety, and the first part of the Langlands's conjecture states the following.

<u>Conjecture</u> C . (a) For any special $h \in X$, with $\mu_h = \mu$, there is an isomorphism $\phi_{\tau,\mu} : \tau\, Sh(G,X) \to Sh({}^{\tau,\mu}G, {}^{\tau,\mu}X)$ such that

$$\phi_{\tau,\mu}(\tau[h,1]) = [{}^{\tau}h,1]$$

$$\phi_{\tau,\mu} \circ \tau\, \mathcal{J}(g) = \mathcal{J}({}^{\tau,\mu}g) \circ \phi_{\tau,\mu},\ g \in G(\mathbb{A}^f)\ ,\ \mathcal{J}(g) =$$

Hecke operator.

In order to compare the isomorphisms ϕ corresponding to two different special points, it is necessary to construct some isomorphisms. For this the following two lemmas are useful.

<u>Lemma</u> 0.3. Let G be a reductive group over $\mathbb{Q}$ such that G^{der} is simply connected. Two elements of $H^1(\mathbb{Q},G)$ are equal if their images in $H^1(\mathbb{Q},G/G^{der})$ and $H^1(\mathbb{R},G)$ are equal.

<u>Lemma</u> 0.4. Let (G_1,X_1) and (G_2,X_2) define Shimura varieties, and suppose there are given:

$f_1 : G_1 \xrightarrow{\approx} G_2$ mapping X_1 into X_2 ;

$f_2 : G_1(\mathbb{A}^f) \xrightarrow{\approx} G_2(\mathbb{A}^f)$;

$\beta \in G_1(\mathbb{A}^f)$ such that $f_1 \circ \underset{\sim}{ad}\, \beta^{-1} = f_2$.

Then $\phi \overset{df}{=} Sh(f_1) \circ \mathcal{J}(\beta) : Sh(G_1, X_1) \xrightarrow{\approx} Sh(G_2, X_2)$ has the following properties:

$\phi[h, \beta^{-1}] = [f_1 \circ h, 1]$, all $h \in X$;

$\phi \circ \mathcal{J}(g) = \mathcal{J}(f_2(g)) \circ \phi$, all $g \in G_1(\mathbb{A}^f)$.

Moreover, if f_1 is replaced with $f_1 \circ \underset{\sim}{ad}\, q$, $q \in G_1(\mathbb{Q})$, and β with βq, then ϕ is unchanged.

Let h and h' be special points of X with cocharacters μ and μ' . A direct calculation shows that $\rho_{\mu *}({}^{\tau}S)$ and $\rho_{\mu' *}({}^{\tau}S)$ have the same image in $H^1(\mathbb{R}, G)$, and they become equal in $H^1(\mathbb{Q}, G/G^{der})$ because ρ_μ and $\rho_{\mu'}$ define the same map to G/G^{der} . There is therefore a $\mathbb{Q}$-rational isomorphism $f : \rho_{\mu *}({}^{\tau}S) \to \rho_{\mu' *}({}^{\tau}S)$ which, because ${}^{\tau,\mu}G \overset{df}{=} {}^{\tau}S \times {}^{S}G = {}^{\tau}S \times {}^{S}G \times {}^{G}G = \rho_{\mu *}({}^{\tau}S) \times {}^{G}G$, can be transferred into an isomorphism $f_1 : {}^{\tau,\mu}G \to {}^{\tau,\mu'}G$ which is uniquely determined up to composition with $\operatorname{ad} q$, $q \in {}^{\tau,\mu}G(\mathbb{Q})$; it maps ${}^{\tau,\mu}X$ into ${}^{\tau,\mu'}X$. Let $f_2 : {}^{\tau,\mu}G(\mathbb{A}^f) \to {}^{\tau,\mu'}G(\mathbb{A}^f)$ be $sp(\tau).g \mapsto sp(\tau).g$. Then there is a $\beta \in {}^{\tau,\mu}G(\mathbb{A}^f)$ satisfying $f_1 \circ \underset{\sim}{ad}\, \beta^{-1} = f_2$ whose definition depends on the choice of f_1 : if f_1 is changed to $f_1 \circ \underset{\sim}{ad}\, q$ then β is changed to βq . There is therefore a well-defined map $\phi(\tau;\mu',\mu) : Sh({}^{\tau,\mu}G, {}^{\tau,\mu}X) \to Sh({}^{\tau,\mu'}G, {}^{\tau,\mu'}X)$ such that $\phi(\tau;\mu',\mu)\circ \mathcal{J}({}^{\tau,\mu}g) = \mathcal{J}({}^{\tau,\mu'}g)\circ \phi(\tau;\mu',\mu)$.

Conjecture C. (b) For special $h,h' \in X$, the maps $\phi_{\tau,\mu}$ and $\phi_{\tau,\mu'}$ satisfy $\phi(\tau;\mu',\mu)\circ\phi_{\tau,\mu} = \phi_{\tau,\mu'}$.

If τ fixes the reflex field $E(G,X)$ of $Sh(G,X)$, then Shimura's conjecture asserting the existence of a canonical model for $Sh(G,X)$ over $E(G,X)$ shows that $\tau\, Sh(G,X) \approx Sh(G,X)$ canonically. This suggests that, for τ fixing $E(G,X)$, there should exist a canonical isomorphism $\phi(\tau;\mu) : Sh(G,X) \to Sh({}^{\tau,\mu}G, {}^{\tau,\mu}X)$. Again (0.3) and the result in §3 enable one to show that, in this case, $\rho_{\mu*}({}^{\tau}S) \in H^1(\mathbb{Q},G)$ is trivial. This allows us to define an isomorphism $f_1 : (G,X) \xrightarrow{\approx} ({}^{\tau,\mu}G, {}^{\tau,\mu}X)$ such that the conditions of (0.4) are satisfied for f_1, $f_2 = (g \mapsto sp(\tau).g)$, and a certain $\beta \in G(\mathbb{A}^f)$. Thus the canonical isomorphism $\phi(\tau;\mu)$ exists.

Theorem 0.5. Let $\tau \in Aut(\mathbb{C})$ fix $E(G,X)$.

(a) Let $h \in X$ be special and let $\mu = \mu_h$. Choose elements $a(\tau) \in {}^{\tau}S(\overline{\mathbb{Q}})$ and $c(\tau) \in \rho_{\mu*}({}^{\tau}S)(\mathbb{Q})$, and let $v \in G(\overline{\mathbb{Q}})$ and $\alpha \in G(\mathbb{A}^f)$ be such that $\rho_\mu(a(\tau)) = c(\tau)v$ and $\rho_\mu(sp(\tau)) = c(\tau)\alpha$. Then the element $[\underset{\sim}{ad}(v)\circ{}^{\tau}h, \alpha]$ of $Sh(G,X)$ is independent of the choice of $a(\tau)$ and $c(\tau)$.

(b) Assume that $Sh(G,X)$ has a canonical model; then conjecture C is true for τ and $Sh(G,X)$ if and only if

$$\tau[h,1] = [\mathrm{ad}\,v\circ{}^{\tau}h,\ \alpha]$$

for all special $h \in X$.

(c) If conjecture C is true for $Sh(G,X)$ and all τ fixing $E(G,X)$ then $Sh(G,X)$ has a canonical model $(M(G,X), M(G,X)_{\mathbb{C}} \xrightarrow[\approx]{f} Sh(G,X))$; moreover, $f\circ(\tau f)^{-1} = \phi(\tau,\mu)^{-1}\circ\phi_{\tau,\mu}$ for every μ corresponding to a special h .

Let A be an abelian variety over $\mathbb{C}$ with complex multiplication by a CM-field F (so that $V \overset{df}{=} H_1(A, \mathbb{Q})$ is of dimension 1 over F) . Write T for $\mathrm{Res}_{F/\mathbb{Q}}\, \mathbb{G}_m$, and let $h : \mathbb{S} \to T_{\mathbb{R}}$ be the homomorphism defined by the Hodge structure on V . The main theorem of complex multiplication describes the action of $\mathrm{Gal}(\overline{\mathbb{Q}}/E(G,X))$ on $Sh(T, \{h\})$ arising from its identification with a moduli variety. From conjecture C for $Sh(CSp(V), S^{\pm})$ one can deduce a description of the action of the whole of $\mathrm{Gal}(\overline{\mathbb{Q}}/\mathbb{Q})$ on $\bigcup_{\tau \in \mathrm{Hom}(F,\overline{\mathbb{Q}})} Sh(T,\{{}^{\tau}h\}) \subset Sh(CSp(V),S^{\pm})$. This suggests a conjecture (conjecture CM) stated purely in terms of abelian varieties of CM-type.

<u>Proposition</u> 0.6. Conjecture CM is true if and only if conjecture C is true for all Shimura varieties of the form $Sh(CSp(V),S^{\pm})$.

It is possible to restate conjecture C for connected Shimura varieties. For this it is first necessary to show that, for a connected Shimura variety $Sh^{\circ}(G,G',X^{+})$, special $h, h' \in X^{+}$, and $\tau \in \mathrm{Aut}(\mathbb{C})$, there are maps

$$g \mapsto {}^{\tau,\mu}g\colon G(\mathbb{Q})^{+} \ (\mathrm{rel}\ G') \to {}^{\tau,\mu}G(\mathbb{Q})^{+} \ (\mathrm{rel}\ {}^{\tau,\mu}G')$$

$$\phi^{\circ}(\tau;\mu',\mu) : Sh^{\circ}({}^{\tau,\mu}G, {}^{\tau,\mu}G', X^{+}) \to Sh^{\circ}({}^{\tau,\mu'}G, {}^{\tau,\mu'}G', X^{+})$$

compatible with those defined for nonconnected Shimura varieties.

Conjecture C°

(a) For any special $h \in X^{+}$, with $\mu = \mu_h$, there is an isomorphism

$$\phi^{\circ}_{\tau,\mu} : \tau\, Sh^{\circ}(G,G',X^{+}) \to Sh^{\circ}({}^{\tau,\mu}G, {}^{\tau,\mu}G', {}^{\tau}X^{+})$$

such that $\phi^{\circ}_{\tau,\mu}(\tau[h]) = [{}^{\tau}h]$

$$\phi^{\circ}_{\tau,\mu} \circ \tau(\gamma .) = {}^{\tau}\gamma . \circ \phi^{\circ}_{\tau,\mu} , \quad \gamma \in G(\mathbb{Q})^{+\wedge} \ (\mathrm{rel}\ G') .$$

(b) For h' a second special element and $\mu' = \mu_{h'}$,

$$\phi^{\circ}(\tau;\mu',\mu) \circ \phi_{\tau,\mu} = \phi_{\tau,\mu'} .$$

Proposition 0.7. Conjecture C is true for $Sh(G,X)$ if and only if conjecture C° is true for $Sh^{\circ}(G^{ad}, G^{der}, X^{+})$

Using 0.7) we prove the following.

Theorem 0.8. If conjecture C is true for all Shimura varieties of the form $Sh(CSp(V), S^{\pm})$ then it is true for all Shimura varieties of abelian type.

All of the above continues to make sense if the Taniyama group is replaced by the motivic Galois group (II.6) except that the maps $\phi(\tau;\mu',\mu)$ and $\phi(\tau;\mu)$ are (possibly) different and the conjectures have a (possibly) different meaning. We shall use a tilde to distinguish the objects associated with the motive Galois group from those associated with the Taniyama group. A new fact is that, almost by construction of the motivic Galois group, conjecture $\widetilde{CM}$ is true. Thus $(\widetilde{0.6})$ and $(\widetilde{0.8})$ show that conjecture $\tilde{C}$ is true for all Shimura varieties of CM-type. This has the following consequence.

Theorem 0.9. Let $Sh(G,X)$ be a Shimura variety of abelian type and let $M(G,X)$ be its canonical model. For any μ associated with a special h, there is an isomorphism $g \mapsto g' : G(\mathbb{A}^f) \to {}^{\tau,\mu}G(\mathbb{A}^f)$ such that, if $g' \in {}^{\tau,\mu}G(\mathbb{A}^f)$ is made to act on $\tau M(G,X)$ as $\tau(\mathcal{J}(g))$, then $\tau M(G,X)$ together with this action is a canonical model for $Sh({}^{\tau,\mu}G, {}^{\tau,\mu}X)$.

(0.9) is the original form Langlands's conjecture on Shimura varieties. (${}^{\tau,\mu}G$ is the same for the motivic Galois group and the Taniyama group.) Such a result was first proved for Shimura curves by Doi and Naganuma [1] and for Shimura varieties of primitive type A and C by Shih [2]. A theorem of Kazhdan [1] can be interpreted as saying that the conjugate $\tau\, Sh(G,X)$ of a compact Shimura variety is again a Shimura variety but unfortunately his method gives little information on the pair (G',X') to which the conjugate corresponds.

We would like to thank P. Deligne and R. Langlands for making available to us pre-prints of their work and D. Shelstad for a letter on which we have based Proposition 4.2 and preceding discussion. One of us was fortunate to be able to spend seven months during 1978-79 at I.H.E.S. and have numerous discussions with P. Deligne, which have profoundly influenced this paper.

Notations and conventions.

For Shimura varieties and algebraic groups we generally follow the notations of Deligne [2]. Thus a reductive algebraic group G is always connected, with derived group G^{der}, adjoint group G^{ad}, and centre $Z = Z(G)$. (We assume also that G^{ad} has no factors of type E_8). A central extension is an epimorphism $G \to G'$ whose kernel is contained in $Z(G)$, and a covering is a central extension such that G is connected and the kernel is finite. If G is reductive, then $\rho : \tilde{G} \to G^{der}$ is the universal covering of G^{der}.

A superscript + refers to a topological connected component; for example $G(\mathbb{R})^+$ is the identity connected component of $G(\mathbb{R})$ relative to the real topology, and $G(\mathbb{Q})^+ = G(\mathbb{Q}) \cap G(\mathbb{R})^+$. For G reductive, $G(\mathbb{R})_+$ is the inverse image of $G^{ad}(\mathbb{R})^+$ in $G(\mathbb{R})$ and $G(\mathbb{Q})_+ = G(\mathbb{Q}) \cap G(\mathbb{R})_+$. In contrast to Deligne [2], we use the superscript ^ to denote both completions and closures since we wish to reserve the superscript - for certain negative components.

We write $Sh(G,X)$ for the Shimura variety defined by a pair (G,X) and $Sh^{\circ}(G, G', X^{+})$ for the connected Shimura variety defined by a triple (G, G', X^{+}). The canonical model of $Sh(G,X)$ is denoted by $M(G,X)$.

Vector spaces are finite-dimensional, number fields are of finite degree over $\mathbb{Q}$ (and usually contained in $\mathbb{C}$), and $\overline{\mathbb{Q}}$ is the algebraic closure of $\mathbb{Q}$ in $\mathbb{C}$. If V is a vector space over $\mathbb{Q}$ and R is the $\mathbb{Q}$-algebra, we often write $V(R)$ for $V \otimes R$.

If $x \in X$ and $g \in G(\mathbb{A}^{f})$ then $[x, g]$ denotes the element of $Sh(G,X) = G(\mathbb{Q})\backslash X \times G(\mathbb{A}^{f})/Z(\mathbb{Q})^{\wedge}$ containing (x,g). The Hecke operator $[x,g] \mapsto [x,gg']$ is denoted by $\mathcal{J}(g')$. The symbol $A \overset{df}{=} B$ means A is defined to be B or that A equals B by definition.

For Galois cohomology and torsors (= principal homogeneous spaces) we follow the notations of Serre [1].

For the Taniyama group, we use the same notations as in III; we refer the reader particularly to III. 2.9.

If A is an abelian variety, then

$$V^{f}(A) \overset{df}{=} (\varprojlim \ker(n: A \to A)) \otimes \mathbb{Q}$$

depends functorially on the isogeny class of A. Throughout the article, an abelian variety will be regarded as an object in the category of abelian varieties up to isogeny.

1. Shimura varieties of abelian type.

A Shimura variety $Sh(G,X)$ is defined by a pair (G,X), comprising a reductive group G over $\mathbb{Q}$ and a $G(\mathbb{R})$-conjugacy class X of homomorphisms $\mathbb{S} \to G_{\mathbb{R}}$, that satisfies the following axioms:

(1.1a) the Hodge structure defined on $Lie(G_{\mathbb{R}})$ by any $h \in X$ is of type $\{(-1, 1), (0,0), (1, -1)\}$;

(1.1b) for any $h \in X$, ad $h(i)$ is a Cartan involution on $G^{ad}_{\mathbb{R}}$;

(1.1c) the group G^{ad} has no factor defined over $\mathbb{Q}$ whose real points form a compact group. Then $Sh(G,X)$ has complex points $G(\mathbb{Q}) \backslash X \times G(\mathbb{A}^f)/Z(\mathbb{Q})^\wedge$, where Z is the centre of G and $Z(\mathbb{Q})^\wedge$ the closure of $Z(\mathbb{Q})$ in $Z(\mathbb{A}^f)$.

A connected Shimura variety $Sh^\circ(G,G',X^+)$ is defined by a triple (G,G',X^+) comprising an adjoint group G over $\mathbb{Q}$, a covering G' of G, and a $G(\mathbb{R})^+$-conjugacy class of homomorphisms $\mathbb{S} \to G_{\mathbb{R}}$ such that G and the $G(\mathbb{R})$-conjugacy class of X containing X^+ satisfy (1.1). The topology $\tau(G')$ on $G(\mathbb{Q})$ is that for which the images of the congruence subgroups of $G'(\mathbb{Q})$ form a fundamental system of neighbourhoods of the identity and $Sh^\circ(G,G',X^+)$ has complex points $\varprojlim \Gamma\backslash X^+$ where Γ runs over the arithmetic subgroups of $G(\mathbb{Q})^+$ that are open relative to the topology $\tau(G')$ (Deligne [2, 2.1.8]).

The relation between the two notions of Shimura variety is as follows: let (G,X) be as in the first paragraph and let X^+ be some connected component of X; then X^+ can be regarded as a $G^{ad}(\mathbb{R})^+$-conjugacy class of maps $\mathbb{S} \to G^{ad}_{\mathbb{R}}$ and $Sh^o(G^{ad}, G^{der}, X^+)$ can be identified with the connected component of $Sh(G,X)$ that contains the image of $X^+ \times \{1\}$.

We recall that the reflex field $E(G,X)$ of (G,X) is the subfield of $\mathbb{C}$ that is the field of definition of the $G(\mathbb{C})$-conjugacy class of μ_h, any $h \in X$, (μ_h = restriction of $h_{\mathbb{C}}$ to $\mathbb{G}_m \times 1 \subset \mathbb{S}_{\mathbb{C}}$) and that $E(G,X^+)$ is defined to equal $E(G,X)$ if X^+ is a connected component of X (Deligne [2, 2.2.1]).

The following easy lemma will be needed in comparing the Shimura varieties defined by (G,X) and (G^{ad}, G^{der}, X^+).

<u>Lemma 1.2</u>. Let $G_1 \to G$ be a central extension of reductive groups over $\mathbb{C}$; let M be a $G(\mathbb{C})$-conjugacy class of homomorphisms $\mathbb{G}_m \to G$ and let M_1 be a $G_1(\mathbb{C})$-conjugacy class lifting M. Then $M_1 \to M$ is bijective.

<u>Proof</u>. The map is clearly surjective and so it suffices to show that, for $\mu_1 \in M_1$ lifting $\mu \in M$, the centralizer of μ_1 is the inverse image of the centralizer of μ. Since the centralizer of μ_1 contains the center of G_1, we only have to show the map on centralizers is surjective. We can construct a diagram

$$C \times G_2 \to G_1 \to G$$

in which the first map, and the composite $G_2 \to G$ are coverings. After replacing μ_1 and μ by multiples, we can assume μ_1 lifts to a homomorphism $(\mu',\mu''): \mathbb{G}_m \to C \times G_2$. Then the centralizer of (μ',μ'') maps into the centralizer of μ_1, and onto the centralizer of μ.

Let (G,X) be as in (1.1) with G adjoint and $\mathbb{Q}$-simple; if every $\mathbb{R}$-simple factor of $G_{\mathbb{R}}$ is of one of the types A, B, C, $D^{\mathbb{R}}$, $D^{\mathbb{H}}$, or E (in the sense of Deligne [2, 2.3.8]) then G will be said to be of that type. When G' is a covering of G, we say that (G,G') (or (G,G',X)) is of primitive abelian type if G is of type A, B, C, or $D^{\mathbb{R}}$ and G' is the universal covering of G, or if G is of type $D^{\mathbb{H}}$ and G' is the double covering described in Deligne [2, 2.3.8] (see Milne-Shih [1, Appendix]).

If (G,X) satisfies (1.1) and G is adjoint and $\mathbb{Q}$-simple, then there is a totally real number field F_o and an absolutely simple group G^s over F_o such that $G = \mathrm{Res}_{F_o/\mathbb{Q}} G^s$. For any embedding $v : F_o \hookrightarrow \mathbb{R}$, let $G_v = G^s \otimes_{F_o,v} \mathbb{R}$, and write I_c and I_{nc} for the sets of embeddings for which $G_v(\mathbb{R})$ is compact and noncompact. Let F be a quadratic totally imaginary extension of F_o and let $\Sigma = (\sigma_v)_{v \in I_c}$ be a set of embeddings $\sigma_v : F \hookrightarrow \mathbb{C}$ such that $\sigma_v | F_o = v$; we define h_Σ to be the Hodge structure on F (regarded as a vector space over $\mathbb{Q}$) such that $(F\otimes_{\mathbb{Q}} \mathbb{C})^{-1,0}$, $(F\otimes_{\mathbb{Q}} \mathbb{C})^{0,-1}$ and $(F\otimes_{\mathbb{Q}} \mathbb{C})^{0,0}$ are the direct summands of $F\otimes_{\mathbb{Q}} \mathbb{C} = \mathbb{C}^{\mathrm{Hom}(F,\mathbb{C})}$ corresponding to Σ, $\iota\Sigma$, and

$\{\sigma: F \hookrightarrow \mathbb{C} \mid \sigma|F_o \in I_{nc}\}$.

<u>Proposition 1.3</u>. Let G be a $\mathbb{Q}$-simple adjoint group and assume that (G,G',X) is of primitive abelian type. For any pair (F,Σ) as above there exists a diagram

$$(G,X) \longleftarrow (G_1,X_1) \hookrightarrow (CSp(V),S^{\pm})$$

such that $G_1^{ad} = G$, $G_1^{der} = G'$, and $E(G_1,X_1) = E(G,X)\ E(F^{\times},h_{\Sigma})$.

Proof. This is Deligne [2, 2.3.10].

Let (G,X) satisfy (1.1) with G adjoint, and let G' be a covering of G. We say that (G,G') or (G,G',X) is of <u>abelian type</u> if there exist pairs $(G_i,G_i')_i$ of primitive abelian type such that $G = \Pi G_i$ and G' is a quotient of the covering $\Pi G_i'$ of ΠG_i . If (G,X) satisfies (1.1), we say that G or (G,X) is of <u>abelian type</u> if (G^{ad},G^{der}) is of this type. Finally, we say that a Shimura variety $Sh^{o}(G,G',X^{+})$ or $Sh(G,X)$ is of <u>abelian type</u> if (G,G') or G is.

2. Shimura varieties as moduli varieties.

We shall want to make use of the notion of an absolute Hodge cycle on a variety (Deligne [3, 0.7]) and the important result (see I.2.11) that any Hodge cycle on an abelian variety is an absolute Hodge cycle. Let A be an abelian variety over an algebraically closed field $k \subset \mathbb{C}$; we shall always identify a Hodge cycle on A with its Betti realization. By this we mean the following. Let $V = H_1(A_{\mathbb{C}}, \mathbb{Q})$ (usual Betti homology) and note that V has a natural Hodge structure and that its dual $\check{V} = H^1(A, \mathbb{Q})$. If $H^1_{dR}(A)$ denotes the de Rham cohomology of A over k then there is a canonical isomorphism $H^1_{dR}(A) \otimes_k \mathbb{C} \xrightarrow{\approx} \check{V}(\mathbb{C})$. There is also a canonical isomorphism $V^f(A) \xrightarrow{\approx} V(\mathbb{A}^f)$. A Hodge cycle s on A is to be an element of some space $V^{\otimes m} \otimes \check{V}^{\otimes n}(p)$ such that:

(2.1a) s is of type $(0,0)$ for the Hodge structure defined by that on V;

(2.1b) there is an $s_{dR} \in (H^1_{dR}(A)^{\vee})^{\otimes m} \otimes H^1_{dR}(A)^{\otimes n}$ that corresponds to s under the isomorphism induced by $H^1_{dR}(A) \otimes_k \mathbb{C} \approx V(\mathbb{C})$ and $\mathbb{C} \approx 2\pi i\mathbb{C}$;

(2.1c) there is an $s_{et} \in V^f(A)^{\otimes m} \otimes (V^f(A)^{\vee})^{\otimes n} \otimes (\varprojlim \mu_n(k))^{\otimes p}$ that corresponds to s under the isomorphism induced by $V(\mathbb{A}^f) \approx V^f(A)$ and $2\pi i\hat{\mathbb{Z}} \xrightarrow{\exp} \varprojlim \mu_n(\mathbb{C})$.

Let τ be an automorphism of $\mathbb{C}$; then τA is an abelian variety over $\tau k \subset \mathbb{C}$ and the above-mentioned result of Deligne shows that τs is a well-defined Hodge cycle on τA: it has $(\tau s)_{dR} = s_{dR} \otimes 1 \in H_{dR}(\tau A) = H_{dR}(A) \otimes_{k,\tau} k$ and $(\tau s)_{et} = \tau s_{et}$.

Certain Shimura varieties can be described as parameter spaces for families of abelian varieties. Let (G,X) satisfy (1.1), and assume there is an embedding $(G,X) \hookrightarrow (CSp(V), S^{\pm})$ where V is a vector space over $\mathbb{Q}$, $CSp(V)$ is the group of symplectic similitudes corresponding to some non-degenerate skew-symmetric form ψ on V, and $S^{\pm}$ is the Siegel double space (in the sense of Deligne [2, 1.3.1]). There will be some family of tensors $(s_\alpha)_{\alpha \in J}$ in spaces of the form $V^{\otimes m} \otimes \check{V}^{\otimes n}(p)$ such that $G = \mathrm{Aut}(V, (s_\alpha)) \subset GL(V) \times \mathbb{G}_m$ (see I, Prop. 3.1). We shall always take ψ to be one of the s_α; then the projection $G \to \mathbb{G}_m$ is defined by the action of G on ψ.

Consider triples $(A, (t_\alpha)_{\alpha \in J}, k)$ with A an abelian variety over $\mathbb{C}$, (t_α) a family of Hodge cycles on A, and k is an isomorphism $k\colon V^f(A) \xrightarrow{\approx} (V(\mathbb{A}^f))$ under which t_α corresponds to s_α for each $\alpha \in J$. We define $\mathcal{A}(G,X,V)$ to be the set of isomorphism classes of triples of this form that satisfy the following conditions:

(2.2a) there exists an isomorphism $H_1(A, \mathbb{Q}) \xrightarrow{\approx} V$ under which s_α corresponds to t_α for each $\alpha \in J$;

(2.2b) the map $\mathbb{S} \xrightarrow{h_A} GL(H_1(A, \mathbb{R}))$ defined by the Hodge structure on $H_1(A, \mathbb{R})$, when composed with the map $GL(H_1(A, \mathbb{R})) \to GL(V(\mathbb{R}))$ induced by an isomorphism as in (a), lies in X.

We let $g \in G(\mathbb{A}^f)$ act on a class $[A, (s_\alpha), k] \in \mathcal{A}(G,X,V)$ as follows: $[A, (t_\alpha), k]g = [A, (t_\alpha), g^{-1}k]$.

Proposition 2.3. There is a bijection $Sh(G,X) \xrightarrow{\approx} A(G,X,V)$ commuting with the actions of $G(\mathbb{A}^f)$.

Proof: Corresponding to $[h,g] \in Sh(G,X) = G(\mathbb{Q})\backslash X \times G(\mathbb{A}^f)$, we choose A to be the abelian variety associated with the Hodge structure (V,h) . Thus $H_1(A, \mathbb{Q}) = V$ and the s_α can be regarded as Hodge cycles on A . As $V^f(A) = V(\mathbb{A}^f)$ we can define k to be $V^f(A) = V(\mathbb{A}^f) \xrightarrow{g^{-1}} V(\mathbb{A}^f)$. It is easily checked that the class $[A, (t_\alpha), k] \in \mathcal{A}(G,X,V)$ depends only on the class $[h,g] \in Sh(G,X)$. Conversely, let $(A, (t_\alpha), k)$ represent a class in $\mathcal{A}(G,X,V)$. We choose an isomorphism $f:H_1(A, \mathbb{Q}) \to V$ as in (2.2a) and define h to be $f\, h_A f^{-1}$ (cf. 2.2b) and g to be $V(\mathbb{A}^f) \xrightarrow{k^{-1}} V^f(A) \xrightarrow{f\otimes 1} V(\mathbb{A}^f)$. If f is replaced by qf, then (h,g) is replaced by $(\underset{\sim}{ad}(q) \circ h, qg)$, and $q \in G(\mathbb{Q})$.

Remark 2.4. The above proposition can be strengthened to show that $Sh(G,X)$ is the solution of a moduli problem over $\mathbb{C}$. Since the moduli problem is defined over $E(G,X)$, $Sh(G,X)$ therefore had model over $E(G,X)$ which, because of the main theorem of complex multiplication, is canonical. This is the proof of Deligne [2, 2.3.1] hinted at in the last paragraph of the introduction to that paper. Let K and K_1 be compact open subgroups of $G(\mathbb{A}^f)$ and $CSp(V)(\mathbb{A}^f)$ with K small and K_1 such that $Sh(G,X)_K \to Sh(CSp(V),S^{\pm})_{K_1}$ is injective (see Deligne [1,1.15]). The pullback of the universal family of abelian varieties on $Sh(CSp(V),S^{\pm})_{K_1}$, constructed by Mumford, is universal for families of abelian varieties carrying Hodge cycles (t_α) and a level structure (mod K).

3. A result on reductive groups; applications.

The following proposition will usually be applied to replace a given reduction group by one whose derived group is simply connected.

Proposition 3.1. (cf. Langlands [3, p 228-29]). Let G be a reductive group over a field k of characteristic zero and let L be a finite Galois extension of k that is sufficiently large to split some maximal torus in G. Let $G' \to G^{der}$ be a covering of the derived group of G. Then there exists a central extension defined over k

$$1 \longrightarrow N \longrightarrow G_1 \longrightarrow G \longrightarrow 1$$

such that G_1 is a reductive group, N is a torus whose group of characters $X^*(N)$ is a free module over the group ring $\mathbb{Z}[\mathrm{Gal}(L/k)]$, and $(G_1^{der} \longrightarrow G^{der}) = (G' \longrightarrow G^{der})$.

Proof: The construction of G_1 will use the following result about modules.

Lemma 3.2. Let $\mathfrak{g}$ be a finite group and M a finitely generated $\mathfrak{g}$-module. Then there exists an exact sequence of $\mathfrak{g}$-modules $0 \longrightarrow P_1 \longrightarrow P_0 \longrightarrow M \longrightarrow 0$ in which P_0 is free and finitely generated as a $\mathbb{Z}$-module and P_1 is a free $\mathbb{Z}[\mathfrak{g}]$-module.

Proof: Write M_0 for M regarded as an abelian group, and choose an exact sequence

$$0 \longrightarrow F_1 \longrightarrow F_0 \longrightarrow M_0 \longrightarrow 0$$

of abelian groups with F_0 (and hence F_1) finitely generated and free. On tensoring this sequence with $\mathbb{Z}[\mathfrak{g}]$ we obtain an exact sequence of $\mathfrak{g}$-modules

$$0 \longrightarrow \mathbb{Z}[\mathfrak{g}] \otimes F_1 \longrightarrow \mathbb{Z}[\mathfrak{g}] \otimes F_0 \longrightarrow \mathbb{Z}[\mathfrak{g}] \otimes M_0 \longrightarrow 0$$

whose pull-back relative to the injection

$$(m \longmapsto \Sigma\, g \otimes g^{-1}m): \; M \hookrightarrow \mathbb{Z}[\mathfrak{g}] \otimes M_0$$

has the required properties.

We now prove (3.1). Let T be a maximal torus in G that splits over L and let T' be the inverse image of T under $G' \to G^{der} \subset G$; it is a maximal torus in G'. An application of (3.2) to the $\mathfrak{g} = \mathrm{Gal}(L/k)$-module $M = X_*(T)/X_*(T')$ provides us with the bottom row of the following diagram, and we define Q to be the fibred product of P_0 and $X_*(T)$ over M :

$$\begin{array}{ccccccccc}
 & & & & 0 & & 0 & & \\
 & & & & \downarrow & & \downarrow & & \\
 & & & & X_*(T') & = & X_*(T') & & \\
 & & & & \downarrow & & \downarrow & & \\
0 & \longrightarrow & P_1 & \longrightarrow & Q & \longrightarrow & X_*(T) & \longrightarrow & 0 \\
 & & \| & & \downarrow & & \downarrow & & \\
0 & \longrightarrow & P_1 & \longrightarrow & P_0 & \longrightarrow & M & \longrightarrow & 0 \\
 & & & & \downarrow & & \downarrow & & \\
 & & & & 0 & & 0 & &
\end{array}$$

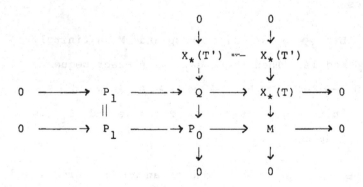

Since the terms of the middle row of the diagram are torsion-free, the $\mathbb{Z}$-linear dual of the sequence is also exact, and hence corresponds, via the functor X^*, to an exact sequence

$$1 \to N \to T_1 \to T \to 1$$

of tori. The map $X_*(T') \to Q = X_*(T_1)$ corresponds to a map $T' \to T_1$ lifting $T' \to T$. Since the kernel of $T' \to T_1$ is finite, the torsion-freeness of $P_0 =$ $\operatorname{coker}(X_*(T') \to X_*(T_1))$ thus implies that $T' \to T_1$ is injective. On forming the pull-back of the above sequence of tori relative to $Z \hookrightarrow T$, where $Z = Z(G)$, we obtain an exact sequence

$$1 \to N \to Z_1 \to Z \to 1 .$$

As T' contains $Z' = Z(G')$, $T' \hookrightarrow T_1$ induces an inclusion $Z' \hookrightarrow Z_1$. The group G can be written as a fibred sum, $G = G *_{\tilde{Z}} Z$, where $\tilde{G}$ is the universal covering group of G^{der} and $\tilde{Z} = Z(\tilde{G})$ (Deligne [2,2.0.1]). We can identify G' with a quotient of $\tilde{G}$. Define $G_1 = \tilde{G} *_{\tilde{Z}} Z_1$. It is easy to check that $Z_1 \to Z$ induces a surjection $G_1 \to G$ with kernel $N \subset Z_1 = Z(G_1)$ and that $\tilde{G} \to G_1$ induces an isomorphism $G' \xrightarrow{\approx} G_1^{der}$. Finally, we note that $X_*(N)$ is a free $\mathbb{Z}[\mathcal{G}]$-module and $X^*(N)$ is the $\mathbb{Z}$-linear dual of $X_*(N)$.

<u>Remark 3.3</u> (a) The torus N in (3.1) is a product of copies of $\operatorname{Res}_{L/k}\mathbb{G}_m$. Thus $H^1(k',N_{k'}) = 0$ for any field $k' \supset k$, and

the sequence $1 \to N(k') \to G_1(k') \to G(k') \to 1$ is exact.

(b) Let $\tilde{T}$ be the inverse image of T (or T') in $\tilde{G}$. Then the maps $\tilde{T} \to T' \hookrightarrow T_1$ and $Z_1 \hookrightarrow T_1$ induce an isomorphism $\tilde{T} *_{\tilde{Z}} Z_1 \xrightarrow{\approx} T_1$. Thus T_1 can be identified with a subgroup of G_1, and the diagram

$$\begin{array}{ccccccccc} 1 & \longrightarrow & N & \longrightarrow & T_1 & \longrightarrow & T & \longrightarrow & 1 \\ & & \| & & \downarrow & & \downarrow & & \\ 1 & \longrightarrow & N & \longrightarrow & G_1 & \longrightarrow & G & \longrightarrow & 1 \end{array}$$

commutes. Obviously T_1 is a maximal torus in G_1.

Application 3.4. Let (G,X) satisfy (1.1), let $h \in X$ be special, and let T be a maximal torus such that h factors through $T_{\mathbb{R}}$. Let $G' \to G^{der}$ be some covering. Take k to be $\mathbb{Q}$ and L to split T, and construct $T_1 \subset G_1 \to G$ as above. Choose some $\mu_1 \in X_*(T_1)$ mapping to $\mu_h \in X_*(T)$. Then μ_1 obviously commutes with $\iota\mu_1$ and so defines a homomorphism h_1: $\mathbb{S} \to T_{\mathbb{R}} \subset G_{\mathbb{R}}$. We let X_1 be the $G(\mathbb{R})$- conjugacy class of maps containing h_1. The pair (G_1,X_1) satisfies (1.1) because, modulo centres, (G_1,X_1) and (G,X) are equal.

It is possible to choose μ_1 so that $E(G_1,X_1) = E(G,X)$. To prove this we first show that the image $\bar{\mu}_h$ of μ_h in M is fixed by $\mathrm{Aut}(\mathbb{C}/E(G,X))$, where $M = X_*(T)/X_*(T')$ is as in the proof of (3.1).

We have to show $\tau\mu_h - \mu_h$ lifts to an element of $X_*(T')$ for any $\tau \in \mathrm{Aut}(\mathbb{C}/E(G,X))$. Since $\tau\mu_h - \mu_h \in X_*(T^{der})$, where $T^{der} = T \cap G^{der}$, and $X_*(T^{der}) \to X_*(T^{ad})$ is injective, where T^{ad} is the image of T in G^{ad}, it suffices to show that the image of $\tau\mu_h - \mu_h$ in $X_*(T^{ad})$ lifts to $X_*(T')$ or, equivalently, to $X_*(G')$. Let $N = \{\mu_h^{ad} \mid h \in X\}$, where μ_h^{ad} is the composite $\mathbb{G}_m \xrightarrow{\mu_h} G \twoheadrightarrow G^{ad}$. Then N is a $G(\mathbb{C})$-conjugacy class of homomorphisms defined over $E = E(G,X)$. For any $\mu \in N$, the identity component of the pull-back of $G' \to G$ by μ is a covering $\pi: \mathbb{G}_m' \to \mathbb{G}_m$ that is independent of μ; it is therefore defined over E, and N lifts to a conjugacy class of N' of maps $\mathbb{G}_m' \to G'$ defined over E. Any two elements of N' restrict to the same element on $\mathrm{Ker}(\pi)$. Thus if $\mu' \in N'$ lifts $\mu \in N$, then $\tau\mu' - \mu'$ factors through $\mathbb{G}_m$ by a map that lifts $\tau\mu - \mu$.

We now use the fact that $X_*(N)$ is a free $\mathrm{Gal}(LE/E)$-module to deduce the existence of a $\mu_1 \in X_*(T_1)$ mapping to $\mu \in X_*(T)$ and whose image $\bar{\mu}_1$ in P_0 is fixed by $\mathrm{Aut}(\mathbb{C}/E)$. The map $G_1 \to G$ induces an isomorphism $W(G_1,T_1) \xrightarrow{\approx} W(G,T)$ of Weyl groups. Let $\tau \in \mathrm{Aut}(\mathbb{C}/E)$ and suppose $\tau\mu = \omega \circ \mu$ with $\omega \in W(G,T)$. If $\omega_1 \in W(G_1,T_1)$ maps to ω, then $\omega_1 \circ \mu_1$ maps to $\tau\mu$ in $X_*(T)$ and $\bar{\mu}_1 = \tau\bar{\mu}_1$ in P_0; thus $\omega_1 \circ \mu_1 = \tau\mu_1$. It follows that τ fixes $E(G_1,X_1)$, and so $E(G,X) \supset E(G_1,X_1)$. The reverse inclusion is automatic.

We can apply this to a triple (G,G',X^+) defining a connected Shimura variety. Thus there exists a pair (G_1,X_1) satisfying (1.1) and such that $(G_1^{ad},G_1^{der},X_1^+) \approx (G,G',X^+)$, $E(G_1,X_1) = E(G,X^+)$, and $X^*(Z(G_1))$ is a free $Gal(L/\mathbb{Q})$-module for some finite Galois extension L of $\mathbb{Q}$ (cf. Deligne [2, 2.7.16]). The last condition implies $G_1(k) \to G_1^{ad}(k) = G(k)$ is surjective for any field $k \supset \mathbb{Q}$.

<u>Application 3.5</u>. Let G be a reductive group over a field k of characteristic zero, and let $\rho\colon \tilde{G} \to G^{der} \subset G$ be the universal covering of G^{der}. When k is a local or global field and k' is a finite extension of k, there is a canonical norm map $N_{k'/k}\colon G(k')/\rho\tilde{G}(k') \to G(k)/\rho\tilde{G}(k)$ (Deligne [2,2.4.8]). We shall use (3.1) to give a more elementary construction of this map.

If G is commutative, $N_{k'/k}$ is just the usual norm map $G(k') \to G(k)$.

Next assume G^{der} is simply connected and let $T = G/G^{der}$. If in the diagram

$$\begin{array}{ccccccc} 1 & \longrightarrow & G(k)/\tilde{G}(k') & \longrightarrow & T(k') & \longrightarrow & H^1(k',\tilde{G}) \\ & & & & \big\downarrow N_{k'/k} & & \\ 1 & \longrightarrow & G(k)/\tilde{G}(k) & \longrightarrow & T(k) & \longrightarrow & H^1(k,\tilde{G}) \end{array}$$

the map $G(k')/\tilde{G}(k') \to H^1(k,\tilde{G})$ is a zero, we can define $N_{L'/k}$ for G to be the restriction of $N_{k'/k}$ for T.

When k is local and nonarchimedean then $H^1(k,\tilde{G}) = 0$, and so the map is zero. When k is local and archimedean we can suppose $k = \mathbb{R}$ and $k' = \mathbb{C}$; then $N_{\mathbb{C}/\mathbb{R}} : T(\mathbb{C}) \to T(\mathbb{R})$ maps into $T(\mathbb{R})^+$, and any element of $T(\mathbb{R})^+$ lifts to an element of $G(\mathbb{R})$ (even to an element of $Z(G)(\mathbb{R})$). When k is global, we can apply the Hasse principle.

In the general case we choose an exact sequence

$$1 \to N \to G_1 \to G \to 1$$

as in (3.1) with G_1^{der} simply connected. From the diagram

$$\begin{array}{ccccccc} N(k') & \longrightarrow & G_1(k')/\tilde{G}(k') & \longrightarrow & G(k')/\rho\tilde{G}(k') & \to & 1 \\ \downarrow N_{k'/k} & & \downarrow N_{k'/k} & & & & \\ N(k) & \longrightarrow & G_1(k)/\tilde{G}(k) & \longrightarrow & G(k)/\rho\tilde{G}(k) & \to & 1 \end{array}$$

we can deduce a norm map for G.

Let k be a number field. If we take the restricted product of the norm maps for the completions of k, and form the quotient by the norm map for k, we obtain the map

$$N_{k'/k}: \pi(G_{k'}) \longrightarrow \pi(G_k)$$

of Deligne [2,2.4.0.1], where $\pi(G_k) = G(\mathbb{A}_k)/(G(k).\rho\tilde{G}(\mathbb{A}_k))$.

<u>Application 3.6.</u> Let G and G' be reductive groups over $\mathbb{Q}$ with adjoint groups having no factors over $\mathbb{Q}$ whose real points are compact. Assume G' is an inner twist of G, so that for some Galois extension L of $\mathbb{Q}$ there is an isomorphism $f : G_L \xrightarrow{\approx} G'_L$ such that, for all $\sigma \in Gal(L/\mathbb{Q})$,

$(\sigma f)^{-1} \circ f = \mathrm{ad}\,\alpha_\sigma$ with $\alpha_\sigma \in G^{ad}(L)$. We shall show that f induces a canonical isomorphism $\pi_0\pi(f): \pi_0\pi(G) \to \pi_0\pi(G')$ with $\pi(-)$ defined as in Deligne [2,2.0.15] (not Deligne [1,2.3]).

If f is defined over $\mathbb{Q}$, for example if G is commutative, then $\pi_0\pi(f)$ exists because $\pi_0\pi$ is a functor.

Next assume that G^{der} is simply connected, and let $\bar{f}$ be the isomorphism from $T = G/G^{der}$ to $T' = G'/G'^{der}$ induced by f . A theorem of Deligne [1,2.4] shows that the vertical arrows in the following diagram are isomorphisms

$$\begin{array}{ccc} \pi_0\pi(G) & \xrightarrow{\pi_0\pi(f)} & \pi_0\pi(G') \\ \downarrow \approx & & \downarrow \approx \\ \pi_0\pi(T) & \xrightarrow{\pi_0\pi(\bar{f})} & \pi_0\pi(T') \end{array} .$$

We define $\pi_0\pi(f)$ to make the diagram commute.

In the general case we choose an exact sequence

$$1 \to N \to G_1 \to G \to 1$$

as in (3.1) with G_1^{der} simply connected. Note that $G_1^{ad} = G^{ad}$ so that we can use the same cocycle to define an inner twist $f_1\colon G_{1L} \to G'_{1L}$. The first case considered above allows us to assume f_1 lifts f . Remark (3.3a) shows that $\pi_0\pi(G_1) \to \pi_0\pi(G)$ is surjective, and we define $\pi_0\pi(f)$ to make the following diagram commute:

$$\begin{array}{ccccccc}
\pi_0\pi(N) & \longrightarrow & \pi_0\pi(G_1) & \longrightarrow & \pi_0\pi(G) & \longrightarrow & 1 \\
\downarrow{\scriptstyle \pi_0\pi(f|N)} & & \downarrow{\scriptstyle \pi_0\pi(f)} & & \vdots{\scriptstyle \pi_0\pi(f)} & & \\
\pi_0\pi(N') & \longrightarrow & \pi_0\pi(G'_1) & \longrightarrow & \pi_0\pi(G') & \longrightarrow & L
\end{array}$$

Note that, if $f : G_L \to G'_L$ and $f' : G'_L \to G''_L$ define G' and G'' as inner twists of G and G', then $\pi_0\pi(f') \circ \pi_0\pi(f) = \pi_0\pi(f' \circ f)$. Also that if f is of the form $\underset{\sim}{ad}\,q : G_L \to G_L$ with $q \in G^{ad}(L)$, then $\pi_0\pi(f) = id$. In the case that G^{der} is simply connected this is obvious because $\underset{\sim}{ad}\,q$ induces id on T, and the general case follows. On combining these two remarks we find that $\pi_0\pi(f)$ is independent of f, because f can only be replaced by $f \circ \underset{\sim}{ad}\,q$ with $q \in G^{ad}(L)$, and $\pi_0\pi(f) \circ \pi_0\pi(\underset{\sim}{ad}\, q) = \pi_0\pi(f)$.

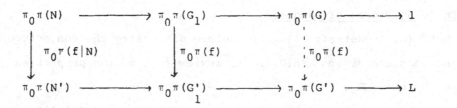

§4. The conjectures of Langlands.

Let (G,X) satisfy (1.1). Before discussing the conjectures of Langlands concerning $Sh(G,X)$ we review some of the properties of (G,X) over $\mathbb{R}$.

Let $h \in X$ be special (in the sense of Deligne [2,2.2.4]), and let T be a $\mathbb{Q}$-rational maximal torus such that h factors through $T_{\mathbb{R}}$. Let $\mu = \mu_h$ be the cocharacter corresponding to h. According to (1.1b) $\underline{ad}\, h(i)$ is a Cartan involution on $G^{ad}_{\mathbb{R}}$, and hence on $G^{der}_{\mathbb{R}}$. Thus $\underline{g}^{der} = \underline{k} \oplus \underline{p}$ where $\underline{g}^{der} = \mathrm{Lie}(G^{der}_{\mathbb{R}}) = \mathrm{Lie}(G_{\mathbb{R}})^{der}$ and $\mathrm{Ad}\, h(i)$ acts as 1 on $\underline{k}$ and -1 on $\underline{p}$. According to (1.1a) there is a decomposition

$$\underline{g}_{\mathbb{C}} = \underline{c}_{\mathbb{C}} \oplus \underline{k}_{\mathbb{C}} \oplus \underline{p}^{+} \oplus \underline{p}^{-}$$

where $\underline{g} = \mathrm{Lie}(G_{\mathbb{R}})$, $\underline{c} = \mathrm{Lie}(Z(G)_{\mathbb{R}})$, $\underline{p}_{\mathbb{C}} = \underline{p}^{+} \oplus \underline{p}^{-}$, and $\mathrm{Ad}\,\mu(z)$ acts as z on $\underline{p}^{+}$ and $\bar{z}$ on $\underline{p}^{-}$. (Thus $\underline{g}^{0,0} = \underline{c}_{\mathbb{C}} + \underline{k}_{\mathbb{C}}$, $\underline{g}^{-1,1} = \underline{p}^{+}$ and $\underline{g}^{1,-1} = \underline{p}^{-}$.) As $T_{\mathbb{C}}$ is a maximal torus in $G_{\mathbb{C}}$, we also have a decomposition

$$\underline{g}_{\mathbb{C}} = \underline{t}_{\mathbb{C}} + \sum_{\alpha \in R} \underline{g}_{\alpha}$$

where $\underline{t} = \mathrm{Lie}(T_{\mathbb{R}})$ and $R \subset \underline{t}^{\vee}_{\mathbb{C}}$ is the set of roots of (G,T). A root α is said to be <u>compact</u> or <u>noncompact</u> according as $\underline{g}_{\alpha} \subset \underline{k}_{\mathbb{C}}$ or $\underline{g}_{\alpha} \subset \underline{p}_{\mathbb{C}}$.

<u>Remark 4.1</u>. If $Y \in \underline{g}_{\alpha}$ then $\mathrm{Ad}(\mu(-1))Y = \alpha(\mu(-1))Y = (-1)^{\langle\alpha,\mu\rangle}Y$. Since $\mathrm{Ad}\mu(-1)$ acts on $\underline{k}_{\mathbb{C}}$ as $+1$ and on $\underline{p}_{\mathbb{C}}$ as -1, this shows that α is compact or noncompact according as $\langle\alpha,\mu\rangle$ is

even or odd.

Note that $T^{der} \overset{df}{=} T \cap G^{der}$ is anisotropic because $\underline{t}^{der} \subset \underline{k}$. Let N be the normalizer of T in G and let $W = N(\mathbb{C})/T(\mathbb{C})$ be the Weyl group. As ι acts as -1 on $R \subset \underline{t}^{der}_{\mathbb{C}}$, it commutes with the action of any reflection s_α. Hence ι acts trivially on W and there is an exact cohomology sequence

$$1 \to T(\mathbb{R}) \to N(\mathbb{R}) \xrightarrow{ad} W \xrightarrow{\delta} H^1(\mathbb{R},T)$$

where, for $\omega \in W$ lifting to $w \in N(\mathbb{C})$, $\delta(\omega)$ is represented by $w^{-1}.\iota w \in \mathrm{Ker}(1 + \iota\colon T(\mathbb{C}) \to T(\mathbb{C}))$.

<u>Proposition 4.2</u>. The class $\delta(\omega)$ is represented by $(\omega^{-1}\mu)(-1)/\mu(-1) \in T(\mathbb{C})$.

<u>Proof</u>: Note that $\delta(\omega_1\omega_2) = \omega_2^{-1}\,\delta(\omega_1).\delta(\omega_2)$ while $(\omega_1\omega_2)^{-1}\mu(-1)/\mu(-1) = \omega_2^{-1}\,(\omega_1^{-1}\mu(-1)/\mu(-1)).(\omega_2^{-1}\mu(-1)/\mu(-1))$ and so it suffices to prove the proposition for a generator s_α of W.

We make the identifications $T(\mathbb{C}) = X_*(T) \otimes \mathbb{C}^\times$, $\underline{t}_{\mathbb{C}} = X_*(T) \otimes \mathbb{C}$, and $\underline{t}^\vee_{\mathbb{C}} = X^*(T) \otimes \mathbb{C}$. If $\check{\alpha}$ is a coroot and H_α is the element of $\underline{t}_{\mathbb{C}}$ corresponding to α, then $\exp \pi i H_\alpha = \check{\alpha}(-1)$. Let $X_\alpha \in \underline{g}_\alpha$ and $X_{-\alpha} \in \underline{g}_{-\alpha}$ be such that $[X_\alpha, X_{-\alpha}] = H_\alpha$. As $\iota\alpha = -\alpha$, we have that $\iota H_\alpha = -H_\alpha$ and that $\iota X_\alpha = cX_{-\alpha}$ and $\iota X_{-\alpha} = dX_\alpha$ with $c,d \in \mathbb{C}$. The conditions $[X_\alpha, X_{-\alpha}] = H_\alpha$ and $\iota^2 = 1$ imply that $cd = 1$ and $\iota c.d = 1$, and so c is real and $d = c^{-1}$. If we replace X_α by aX_α then we must replace $X_{-\alpha}$ by $\frac{1}{a}X_{-\alpha}$ and c by a^2c. Thus, for a given α, there are two possibilities:

either X_α can be chosen so that $\iota X_\alpha = -X_{-\alpha}$ or X_α can be chosen so that $\iota X_\alpha = X_{-\alpha}$. In the first case α is compact and in the second it is noncompact.

Assume that α is compact; then the map $\underline{su}_2 \to \underline{g}$ such that $\begin{pmatrix}1 & 0\\ 0 & -1\end{pmatrix} \mapsto H_\alpha$, $\begin{pmatrix}0 & i\\ 0 & 0\end{pmatrix} \mapsto X_\alpha$, $\begin{pmatrix}0 & 0\\ -i & 0\end{pmatrix} \mapsto X_{-\alpha}$ lifts to a homomorphism $SU_2 \to G_{\mathbb{R}}$ (defined over $\mathbb{R}$). The image w of $\begin{pmatrix}0 & 1\\ -1 & 0\end{pmatrix}$ in $G(\mathbb{R})$ represents s_α. Thus $\delta(s_\alpha) = 1$ in this case. On the other hand, $s_\alpha(\mu)-\mu = -\langle\alpha,\mu\rangle\alpha^\vee$, and so $s_\alpha\mu(-1)/\mu(-1) = \alpha^\vee(-1)^{-\langle\alpha,\mu\rangle} = 1$ (by 4.1).

If α is noncompact, then the map $s\ell_2 \to \underline{g}$ such that $\begin{pmatrix}0 & -i\\ i & 0\end{pmatrix} \mapsto H_\alpha$, $\frac{1}{2}\begin{pmatrix}-i & 1\\ 1 & i\end{pmatrix} \mapsto X_\alpha$, $\frac{1}{2}\begin{pmatrix}i & 1\\ 1 & -i\end{pmatrix} \mapsto X_{-\alpha}$ lifts to a homomorphism $SL_2 \to G_{\mathbb{R}}$. The image w of $\begin{pmatrix}i & 0\\ 0 & -i\end{pmatrix}$ in $G(\mathbb{C})$ represents s_α. Then $w^{-1}.\iota w$ is the image of $\begin{pmatrix}-1 & 0\\ 0 & -1\end{pmatrix}$, which is $\exp \pi i H_\alpha = \alpha^\vee(-1)$. On the other hand $s_\alpha\mu(-1)/\mu(-1) = \alpha^\vee(-1)^{-\langle\alpha,\mu\rangle} = \alpha^\vee(-1)$ (by 4.1).

<u>Corollary 4.3</u>. If the reflex field $E(G,X)$ of (G,X) is real then there exists an $n \in N(\mathbb{R})$ such that $\underset{\sim}{ad}(n)\circ\mu = \iota\mu$.

<u>Proof</u>: Since ι fixes $E(G,X)$ there is an element w in $G(\mathbb{C})$, which we can choose to lie in $N(\mathbb{C})$, such that $\iota\mu = \underset{\sim}{ad}(w)\circ\mu$. The proposition shows that the image of $\underset{\sim}{ad}\, w$ in $H^1(\mathbb{C}/\mathbb{R}, T(\mathbb{C}))$ is represented by $(\iota-1)\mu(-1)$, and therefore is zero. Thus there is an $n \in N(\mathbb{R})$ representing $\underset{\sim}{ad}\, w$.

When the reflex field $E(G,X)$ is real and $Sh(G,X)$ has a canonical model over $E(G,X)$ then ι defines an antiholomorphic involution of $Sh(G,X)$. One of the conjectures of Langlands gives an explicit description of this involution.

Let h , as before, be special and let ${}^{\iota}h$ be the element of X corresponding to $\iota\mu$. If n is as in the corollary, then $ad(n)\circ h = {}^{\iota}h$. Since K_∞ is the centralizer of $h(i)$, and of ${}^{\iota}h(i)$, we see that n normalizes K_∞ . Thus $g \mapsto gn : G(\mathbb{R}) \to G(\mathbb{R})$ induces a map on the quotient $G(\mathbb{R})/K_\infty$, which we can transfer to X by means of the isomorphism $g \mapsto adg\circ h : G(\mathbb{R})/K_\infty \xrightarrow{\approx} X$. Thus we obtain an antiholomorphic isomorphism $\eta = (adg\circ h \mapsto ad(gn)\circ h : X \to X)$

Conjecture B. (Langlands [1, p. 418], [2, p. 2.7, Conjecture B], [3, p. 234]). The involution of $Sh(G,X)$ defined by ι is $[x,g] \mapsto [\eta(x),g]$.

Remark 4.4. The conjecture is true for all special h if it is true for one, and it follows from Deligne [1, 5.2] that to prove the conjecture it suffices to show $\iota[h, 1] = [\eta(h), 1]$ $(=[{}^{\iota}h, 1])$ for a single special h. Conjecture B is easy to prove if $Sh(G,X)$ is a moduli variety for abelian varieties over $E(G,X)$. (More generally, if it is a "moduli variety for motives", see (10. 7). It is proved for all Shimura varieties of abelian type in Milne-Shih [1].

The conjecture of Langlands concerning conjugates of Shimura varieties is expressed in terms of the Taniyama group; thus let

$$1 \to S \to T \xrightarrow{\pi} Gal(\bar{\mathbb{Q}}/\mathbb{Q}) \to 1$$

be the extension, and $sp : Gal(\bar{\mathbb{Q}}/\mathbb{Q}) \to \underset{\sim}{T}(\mathbb{A}^f)$ the splitting, defined in (III. 3). For any $\tau \in Gal(\bar{\mathbb{Q}}/\mathbb{Q})$, ${}^{\tau}S \overset{df}{=} \pi^{-1}(\tau)$ is a right S-torsor, and $sp(\tau) \in {}^{\tau}S(\mathbb{A}^f)$ defines a trivialization of ${}^{\tau}S$ over $\mathbb{A}^f$. (For any finite Galois extension L of $\mathbb{Q}$ and $\tau \in Gal(L^{ab}/\mathbb{Q})$ we can also define an S^L-torsor ${}^{\tau}S^L$; it corresponds to the cohomology class $\gamma(\tau) \in H^1(L/\mathbb{Q}, S^L)$; (see III. 2.9).)

Let G,X,h,μ,T be as at the start of this section. As $T^{ad}_{\mathbb{R}}$ is anisotropic, $\mu^{ad} \overset{df}{=} (\mathbb{G}_m \xrightarrow{\mu} T \to T^{ad})$ satisfies (III. 1.1) and so factors into $\mathbb{G}_m \xrightarrow{\mu_{can}} S \xrightarrow{\rho_\mu} T^{ad} \subset G^{ad}$. Thus S acts on G, and we can use ${}^{\tau}S$ to twist G: we define ${}^{\tau}G$ (or ${}^{\tau,\mu}G$) to be ${}^{\tau}S \times {}^{S}G$. (If $L \supset \mathbb{Q}$ splits T then there is an isomorphism $f : G_L \xrightarrow{\approx} {}^{\tau}G_L$ such that $\sigma f = f \circ \underset{\sim}{ad}\, \gamma_\sigma(\tau,\mu^{ad})$.) Note that the action of S on T is trivial, and so $T = {}^{\tau}S \times {}^{S}T \subset {}^{\tau}G$. Define ${}^{\tau}h$ to be the homomorphism $\mathbb{S} \to {}^{\tau}G_{\mathbb{R}}$ associated with $\mathbb{G}_m \xrightarrow{\tau\mu} T_{\mathbb{C}} \subset {}^{\tau}G_{\mathbb{C}}$, and ${}^{\tau}X$ (or ${}^{\tau,\mu}X$) to be the $G(\mathbb{R})$-conjugacy class of maps $\mathbb{S} \to {}^{\tau}G_{\mathbb{R}}$ containing ${}^{\tau}h$. The element $sp(\tau) \in {}^{\tau}S(\mathbb{A}^f)$ provides a canonical isomorphism $g \mapsto sp(\tau).g : G(\mathbb{A}^f) \to {}^{\tau}G(\mathbb{A}^f)$, which we write as $g \mapsto {}^{\tau}g$ (or $g \mapsto {}^{\tau,\mu}g$). Langlands has shown [3, p. 231] that $({}^{\tau}G, {}^{\tau}X)$ satisfies (1.1); he asserts [3, p. 233] that if h' is a second special point of X and $\mu' = \mu_{h'}$ then there is an isomorphism

$$\phi(\tau;\mu',\mu) : Sh({}^{\tau,\mu}G, {}^{\tau,\mu}X) \to Sh({}^{\tau,\mu'}G, {}^{\tau,\mu'}X)$$

such that

$$\phi(\tau;\mu',\mu) \circ \mathcal{J}({}^{\tau,\mu}g) = \mathcal{J}({}^{\tau,\mu'}g) \circ \phi(\tau;\mu',\mu) .$$

<u>Conjecture C</u>. (Langlands [3, p. 232-33]) (a) For any special $h \in X$ there is an isomorphism

$$\phi_\tau = \phi_{\tau,\mu_h} : \tau Sh(G,X) \xrightarrow{\approx} Sh({}^\tau G, {}^\tau X)$$

such that

$$\phi_\tau(\tau[h,1]) = [{}^\tau h, 1]$$
$$\phi_\tau \circ \tau \mathcal{J}(g) = \mathcal{J}({}^\tau g) \circ \phi_\tau, \text{ all } g \in G(\mathbb{A}^f).$$

(b) If h' is a second special element of X and $\mu = \mu_h$, $\mu' = \mu_{h'}$, then

$$\begin{array}{ccc} \tau Sh(G,X) & \xrightarrow{\phi_{\tau,\mu}} & Sh({}^{\tau,\mu}G, {}^{\tau,\mu}X) \\ & \searrow^{\phi_{\tau,\mu'}} & \downarrow \phi(\tau;\mu',\mu) \\ & & Sh({}^{\tau,\mu'}G, {}^{\tau,\mu}X) \end{array}$$

commutes.

<u>Remark 4.5</u>. For a given h there is at most one map $\phi_{\tau,\mu}$ satisfying the conditions in part (a) of the conjecture (this follows from Deligne [1, 5.2]).

We note one consequence of conjecture C. Assume that $Sh(G,X)$ has a canonical model $(M(G,X), f : M(G,X)_{\mathbb{C}} \xrightarrow{\approx} Sh(G,X))$, and let $h \in X$ be special with associated cocharacter μ. Then for any automorphism τ of $\mathbb{C}$, $\tau M(G,X)$ is defined over $\tau E(G,X)$, and obviously $\tau E(G,X) = E({}^{\tau,\mu}G, {}^{\tau,\mu}X)$. Moreover, if we make

${}^{\tau}g \in {}^{\tau}G(\mathbb{A}^f)$ act on $\tau M(G,X)$ as $\tau\mathcal{J}(g)$, then $(\tau M(G,X), \tau M(G,X)_{\mathbb{C}} \xrightarrow{\phi_{\tau,\mu}\circ\tau f} Sh({}^{\tau,\mu}G, {}^{\tau,\mu}X)$ satisfies the condition, relative to h, to be a canonical model. Part (b) of conjecture C shows that everything is essentially independent of h, and so $\tau M(G,X)$ is a canonical model for $Sh({}^{\tau,\mu}G, {}^{\tau,\mu}X)$. For the sake of reference, and because it is the original form of conjecture C, we state another conjecture which is a weak form of this consequence.

Conjecture A. (Langlands [1, p. 417], [2, p. 2.5]) Assume that $Sh(G,X)$ has a canonical model $(M(G,X),f)$, and let h be some special point of X with associated cocharacter μ. Then there exists an isomorphism $g \mapsto g' : G(\mathbb{A}^f) \to {}^{\tau,\mu}G(\mathbb{A}^f)$ such that, if $g' \in {}^{\tau}G(\mathbb{A}^f)$ is made to act on $\tau M(G,X)$ as $\tau(\mathcal{J}(g))$, then $\tau M(G,X)$ is a canonical model for $Sh({}^{\tau,\mu}G, {}^{\tau,\mu}X)$.

Remark 4.6. Conjecture A appears to depend on the choice of h. One can, however, use the maps $\phi(\tau;\mu',\mu)$ to show that if the conjecture is true with one special point h then it is true with any special h.

We shall need to use several properties of the maps $\phi(\tau;\mu',\mu)$. Thus we prove them for the Shimura varieties of interest to us, namely those of abelian type. We begin by defining the maps in an easy case.

Let (G,X) satisfy (1.1). Assume:

(4.7a) for all special $h \in X$ and all $\tau \in \mathrm{Aut}(\mathbb{C})$,

$$(\tau-1)(\iota+1)\mu_h = 0 = (\iota+1)(\tau-1)\mu_h ;$$

(4.7b) if h is special and $\rho_h : S \longrightarrow G$ is the map defined by μ_h (see III.1) then the element $\gamma(\tau,\mu) \overset{df}{=} \rho_h(\gamma(\tau))$ of $H^1(\mathbb{Q},G)$ is independent of h.

Now fix two special points h and h' of X and let $\mu = \mu_h$ and $\mu' = \mu_{h'}$. We write $'G$ for ${}^{\tau,\mu}G$, $''G$ for ${}^{\tau,\mu'}G$, etc.. Let L be some large finite Galois extension of $\mathbb{Q}$ and let $a(\tau)$ be a section to $\underset{\sim}{T}^L \to \mathrm{Gal}(L^{ab}/\mathbb{Q})$. Then there are defined $\beta(\tau) \in S(\mathbb{A}_L^f)$, $\beta(\tau,\mu) \overset{df}{=} \rho_h(\beta(\tau)) \in G(\mathbb{A}_L^f)$, and $\beta(\tau,\mu') \overset{df}{=} \rho_{h'}(\beta(\tau)) \in G(\mathbb{A}_L^f)$, and cocycles $\gamma_\sigma(\tau)$, $\gamma_\sigma(\tau,\mu) \overset{df}{=} \rho_h(\gamma_\sigma(\tau))$, and $\gamma_\sigma(\tau,\mu')$. Moreover there are maps $f' = (g \mapsto a(\tau).g) : G_L \xrightarrow{\approx} 'G_L$, $f'' = (g \mapsto a(\tau).g) : G_L \xrightarrow{\approx} ''G_L$, and $f = f'' \circ f'^{-1} : 'G_L \to ''G_L$.

According to (4.7b) there is a $v \in G(L)$ such that $\gamma_\sigma(\tau,\mu') = v^{-1}.\gamma_\sigma(\tau,\mu).\sigma v$. The map $f_1 = f \circ \underset{\sim}{ad}\, f'(v^{-1}) : 'G_L \xrightarrow{\approx} ''G_L$ is defined over $\mathbb{Q}$ and sends $'X$ into $''X$. It therefore defines an isomorphism $\mathrm{Sh}(f_1) : \mathrm{Sh}('G,'X) \xrightarrow{\approx} \mathrm{Sh}(''G,''X)$.

As $B \overset{df}{=} \beta(\tau,\mu')\, v^{-1}\, \beta(\tau,\mu)^{-1}$ is fixed by $\mathrm{Gal}(L/\mathbb{Q})$ it lies in $G(\mathbb{A}^f)$, and hence $'B \overset{df}{=} {}^{\tau,\mu}B = f'(\beta(\tau,\mu)^{-1}\, \beta(\tau,\mu')\, v^{-1})$ lies in $'G(A^f)$. We define $\phi(\tau;\mu',\mu)$ to be the composite $\mathrm{Sh}(f_1) \circ \mathfrak{J}('B)$. Thus

$$\phi(\tau;\mu',\mu)\,[x,'g] = [f_1 \circ x,\ ''(Bg)]$$

Evidently,

$$\phi(\tau;\mu',\mu) \circ \mathfrak{J}('g) = \mathfrak{J}(''g) \circ \phi(\tau;\mu',\mu)\ .$$

Replace $a(\tau)$ by $a(\tau)u$ with $u \in S^L(L)$, and let $u_1 = \rho_h(u)$ and $u_2 = \rho_{h'}(u)$. This forces the following changes:

f'	f	$\gamma_\sigma(\tau,\mu)$	v^{-1}	$\beta(\tau)$
$f' \circ \underset{\sim}{ad}\, u_1$	$f \circ \underset{\sim}{ad}(f'(u_2 u_1^{-1}))$	$u_1^{-1}\, \gamma_\sigma(\tau,\mu)\sigma u_1$	$u_2^{-1} v^{-1} u_1$	$\beta(\tau) u$.

Thus f_1 and B are unchanged, and so also is $\phi(\tau;\mu',\mu)$. If v^{-1} is replaced by $v^{-1}\, u^{-1}$ where $u \in G(L)$ satisfies $u = \gamma_\sigma(\tau,\mu).\sigma u$, then $[\underset{\sim}{ad}\, f'(u^{-1}) \circ x\ ,\ f'(u^{-1})g] = [\ x,g]$ for any $[x,g] \in Sh('G,'X)$ because $f'(u) \in {'G}(\mathbb{Q})$. Again $\phi(\tau;\mu',\mu)$ is unchanged, and is therefore well-defined.

Example 4.8. Let $(G,X) = (CSp(V),\ S^{\pm})$. For $h \in X$ special, we can use $\rho_h : S \to CSp(V)$ to define an action of S on V . Let ${}^{\tau,\mu}V = {}^{\tau}S \times^{S} V$; clearly ${}^{\tau,\mu}G = CSp({}^{\tau,\mu}V)$. The element $sp(\tau) \in S(\mathbb{A}^f)$ defines an isomorphism $sp(\tau,\mu) : V(\mathbb{A}^f) \to {}^{\tau,\mu}V(\mathbb{A}^f)$. Under the bijections $Sh(G,X) \xrightarrow{\approx} \mathcal{A}(G,X,V)$ defined in (2.3), $\phi(\tau;\mu',\mu)$ corresponds to the map $[A,t,\ sp(\tau,\mu) \circ k] \longrightarrow [A,t,sp(\tau,\mu')\circ k]$.

Example 4.9. Suppose $h' = \underset{\sim}{ad}\, q \circ h$ with $q \in G(\mathbb{Q})$. Then $B = q$ and $v^{-1} = q$. Thus $\phi(\tau;\mu',\mu)$ is the map

$$[x,'q] \mapsto [f \circ \underset{\sim}{ad}\, f'(q) \circ x\ ,\ ''(qg)]\ .$$

Note that, even without the assumption (4.7), this expression gives a well-defined map.

To be able to apply the above discussion, we need to know when (4.7) holds. Clearly (4.7a) is valid if G is an adjoint group or if there is a map $(G,X) \longrightarrow (CSp(V),\ S^{\pm})$ such that the kernel of $G \to CSp(V)$ is finite.

Lemma 4.10. The pair (G,X) satisfies (4.7) if it is of abelian type and G is adjoint.

Proof. We can assume G to be $\mathbb{Q}$-simple. There is a diagram

$$(G,X) \leftarrow (G_1,X_1) \rightarrow (CSp(V),\ S^{\pm})$$

such that $G_1^{ad} = G$, $G_1^{der} = \tilde{G}$, and $G_1 \rightarrow CSp(V)$ has finite kernel (cf. 1.4). We shall prove (4.7b) holds for (G_1,X_1). To show that the two classes $\gamma(\tau,\mu)$ and $\gamma(\tau,\mu')$ are equal in $H^1(\mathbb{Q},G_1)$ it suffices to show they have the same images in $H^1(\mathbb{Q},G_1/G_1^{der})$ and in $H^1(\mathbb{R},G_1)$ (see 7.3). The first is obvious since μ and μ' map to the same element of $X_*(G_1/G_1^{der})$. For the second we use (III.3.14). Thus $\gamma = ((\tau-1)\mu)(-1)$ and $\gamma' = ((\tau-1)\mu')(-1)$ represent the images of $\gamma(\tau,\mu)$ and $\gamma(\tau,\mu')$ in $H^1(\mathbb{R},G_1)$. For any $z \in G(\mathbb{C})$ we write $z(\mu)$ for $\underset{\sim}{ad}\, z \circ \mu$. A direct calculation shows that if $\mu' = x(\mu)$, $x \in G(\mathbb{R})$, then

$$x^{-1}\gamma'\, x\, \gamma^{-1} = (x^{-1}.\ \tau x - \tau)\mu\ (-1)\ .$$

Let T be a maximal $\mathbb{Q}$-rational torus in G such that μ factors through $T(\mathbb{C})$, and let N be the normalizer of T. If $w \in N(\mathbb{C})$ then

$$\iota w.\gamma.\ w^{-1}.\gamma^{-1} = (\iota w.\ w^{-1})\ [(w-1)(\tau-1)\mu(-1)]$$

According to (4.2), $\iota w.\ w^{-1} = (\iota c.\ c^{-1})\ [(w-1)\mu(-1)]$ for some $c \in T(\mathbb{C})$.

Thus

$$\iota w.\gamma.\ w^{-1}.\ \gamma^{-1} = (\iota c.\ c^{-1})\ [(w-1)\tau\mu(-1)]\ .$$

If we choose w to act on the roots of (G,T) as $x^{-1}.\ \tau x.\ \tau^{-1}$, then $(w-1)\tau\mu(-1) = (x^{-1}\tau x-\tau)\ \mu\ (-1)$, and it follows that $x\ \gamma'\ x^{-1} = \iota(c^{-1}w).\gamma.(c^{-1}w)^{-1}$, which completes the proof.

<u>Lemma 4.11</u>. Let (G,G',X^{+}) define a connected Shimura variety and assume (G,X) is of abelian type. Then there exists a map $(G_0,X_0) \longrightarrow (G,X)$ such that $G_0^{ad} = G$, $G_0^{der} = G'$, $G_0(\mathbb{Q}) \longrightarrow G(\mathbb{Q})$ is surjective, and (G_0,X_0) satisfies (4.7).

<u>Proof</u>. Clearly the lemma is true for a product if it is true for each factor, and is true for (G,G',X^{+}) if it is true for $(G,\tilde{G},X^{+})$. Thus we can assume G is $\mathbb{Q}$-simple and $G' = \tilde{G}$. Choose (G_1,X_1) as in the proof of (7.10). Let L be a finite Galois extension of $\mathbb{Q}$ that splits $Z(G_1)$. There exists a subjective map $M \longrightarrow X^{*}(Z(G_1))$ with M a finitely-generated free $\mathbb{Z}[\mathrm{Gal}(L/\mathbb{Q})]$-module. Let $Z(G_1) \hookrightarrow Z$ be the corresponding map of tori, and define $G_0 = \tilde{G} *_{Z(\tilde{G})} Z$ (see Deligne [2,2.0.1]). The map $Z(G_1) \hookrightarrow Z$ induces an inclusion $G_1 \hookrightarrow G_0$, and we define X_0 to be the composite of X_1 with this inclusion. Then (G_0,X_0) satisfies (4.7) because (G_1,X_1) does, and $G_0(\mathbb{Q}) \longrightarrow G(\mathbb{Q})$ is surjective because $Z(G_0) = Z$ and $H^1(\mathbb{Q},Z) = 0$.

Let (G,G',X^+) and (G_o,X_0) be as in (4.11), and let h and h' be special elements of X^+. Write $\mu = \mu_h$ and $\mu' = \mu_{h'}$. The map

$$\phi(\tau;\mu',\mu) : Sh('G_0,'X_0) \xrightarrow{\approx} Sh("G_0,"X_0)$$

induces an isomorphism

$$\phi^0(\tau;\mu',\mu) : Sh^0('G,'G','X) \longrightarrow Sh^0("G,"G',"X).$$

(As before, we have substituted ' and " for the superscripts τ,μ and τ,μ'.) The usual argument shows that ϕ^0 is independent of (G_0,X_0). Moreover, the surjectivity of $G_0(\mathbb{Q}) \longrightarrow G(\mathbb{Q})$ shows that

$$\phi^0(\tau;\mu',\mu) \circ '\gamma. = "\gamma. \circ \phi^0(\tau;\mu',\mu)$$

for all $\gamma \in G(\mathbb{Q})^{+\wedge}$ (rel G') where $\gamma.$ denotes the canonical left action on Sh^0. (For the fact that $\gamma \mapsto \gamma' = {}^{\tau,\mu}\gamma$ maps $G(\mathbb{Q})^{+\wedge}$ into $'G(\mathbb{Q})^{+\wedge}$, see 8.1.)

Proposition 4.12. Let (G,X) be such that (G^{ad},X) is of abelian type. Then there is a unique family of isomorphisms

$$\phi(\tau;\mu',\mu) : Sh({}^{\tau,\mu}G,{}^{\tau,\mu}X) \longrightarrow Sh({}^{\tau,\mu'}G,{}^{\tau,\mu'}X) ,$$

$\tau \in Aut(\mathbb{C})$, $\mu = \mu_h$, $\mu' = \mu_{h'}$ with h and h' special, such that:

(a) $\phi(\tau;\mu',\mu) \circ \mathcal{J}('g) = \mathcal{J}(''g) \circ \phi(\tau;\mu',\mu)$, $g \in G(\mathbb{A}^f)$;

(b) $\phi(\tau;\mu'',\mu') \circ \phi(\tau;\mu',\mu) = \phi(\tau;\mu'',\mu)$;

(c) if h and h' belong to the same connected component X^+ of X , then $\phi(\tau;\mu',\mu)$ restricted to the connected component of $Sh({}^{\tau,\mu}G, {}^{\tau,\mu}X)$ is the map $\phi^0(\tau;\mu',\mu)$ defined above;

(d) if $h' = \underset{\sim}{ad}(q) \circ h$ with $q \in G(\mathbb{Q})$, then $\phi(\tau;\mu',\mu)$ is the map defined in (4.9).

Proof. There is clearly at most one family of maps with these properties. To show the existence one uses the standard technique for extending a map from the connected component of a variety to the whole variety (see Deligne [2, 2.7], or § 9).

Remark 4.13. In the case that τ fixes $E(G,X)$, we define in (7.8) below a map $\phi(\tau;\mu) : Sh(G,X) \longrightarrow Sh({}^{\tau,\mu}G, {}^{\tau,\mu}X)$. On comparing the two definitions one finds that

$$\phi(\tau;\mu',\mu) = \phi(\tau;\mu') \circ \phi(\tau;\mu)^{-1} .$$

Remark 4.14. Let $h' = \underset{\sim}{ad}(q) \circ h$ with $q \in G(\mathbb{Q})$, and assume part (a) of conjecture C holds. One checks directly that $\phi = \phi(\tau;\mu',\mu) \circ \phi_{\tau,\mu}$ has the following properties:

$$\phi(\tau[h',1]) = \phi(\tau[h,q^{-1}]) = [{}^{\tau,\mu'}h',1]$$
$$\phi \circ \tau\mathcal{J}(g) = \mathcal{J}({}^{\tau,\mu'}g) \circ \phi .$$

Thus $\phi = \phi_{\tau,\mu'}$, and part (b) of the conjecture holds (for μ and μ').

§5. A cocycle calculation. (cf. IV C).

Let A be an abelian variety over $\mathbb{C}$ of CM-type, so that there is a product F of CM-fields acting on A in such a way that $H_1(A,\mathbb{Q})$ is a free F-module of rank 1. Assume that there is a homogeneous polarization $[\psi]$ on A whose Rosati involution stabilizes $F \subset \mathrm{End}(A)$ and induces ι on it. Let $F_0 = \{f \in F \mid \iota f = f\}$; thus F_0 is a product of totally real fields. Note that the Hodge structure h on $V = H_1(A,\mathbb{Q})$ is compatible with the action of F. Let $\psi \in [\psi]$ be a polarization of A (or (V,h)) ; for any choice of an element $f \in F^\times$ with $\iota f = -f$ there exists a unique F-Hermitian form ϕ on V such that $\psi(x,y) = \mathrm{Tr}_{F/\mathbb{Q}}(f\phi(x,y))$ (see I.4.6).

Let Σ be the set of embeddings $F_0 \hookrightarrow \mathbb{C}$; then

$$H_1(A,\mathbb{C}) = V \otimes_{\mathbb{Q}} \mathbb{C} = \oplus_{\sigma\in\Sigma} V_\sigma , \text{ where } V_\sigma = V \otimes_{F_0,\sigma} \mathbb{C} .$$

Moreover:

V_σ is a free $F \otimes_{F_0,\sigma} \mathbb{C}$-module of rank 1;

$V_\sigma = V_\sigma^+ \oplus V_\sigma^-$, where $h(z)$ acts as z on V_σ^+ and ιz on V_σ^- ;

ϕ defines a Hermitian form ϕ_σ on V_σ such that $\phi_\sigma > 0$ on V_σ^+ and $\phi_\sigma < 0$ on V_σ^- .

Let τ be an automorphism of $\mathbb{C}$ and let $V' = H_1(\tau A,\mathbb{Q})$. The action of F on A induces an action of F on τA , and $[\psi]$ gives rise to a homogeneous polarization $[\tau\psi]$ on τA . Thus there is a decomposition $H_1(\tau A,\mathbb{C}) = \oplus_{\sigma\in\Sigma} V'_\sigma$ where the V'_σ have similar structures to the V_σ .

Our purpose is to construct an isomorphism $\theta : H_1(A,\mathbb{C}) \longrightarrow H_1(\tau A,\mathbb{C})$ that is $F \otimes \mathbb{C}$-linear and takes $[\psi]$ to $[\tau\psi]$. It will suffice to define θ on each component V_σ of $H_1(A,\mathbb{C})$.

As $H_1(A,\mathbb{C})$ is canonically dual to the de Rham cohomology group $H^1_{dR}(A)$, and $H^1_{dR}(\tau A) = H^1_{dR}(A) \otimes_{\mathbb{C},\tau} \mathbb{C}$, we see that $H_1(\tau A,\mathbb{C}) = H_1(A,\mathbb{C}) \otimes_{\mathbb{C},\tau} \mathbb{C}$. Under this identification, the two actions of F correspond, and $\tau\psi$ corresponds to ψ.

Fix a $\sigma \in \Sigma$, and consider V_σ and V'_σ. Since F_0 acts on $V_\sigma \otimes_{\mathbb{C},\tau} \mathbb{C}$ through $\tau\sigma$, we see that we must have $V'_\sigma = V_{\tau^{-1}\sigma} \otimes_{\mathbb{C},\tau} \mathbb{C}$. There is an $F_0 \otimes \mathbb{C}$-linear isomorphism $\theta_1 : V_\sigma \xrightarrow{\approx} V'_\sigma$ and, since F acts on V^+_σ and V^-_σ through distinct enbeddings $F \hookrightarrow \mathbb{C}$, exactly one of the following must hold:

$$(+) \quad \theta_1 : V^+_\sigma \xrightarrow{\approx} V'^+_\sigma, \quad \theta_1 : V^-_\sigma \xrightarrow{\approx} V'^-_\sigma ;$$

$$(-) \quad \theta_1 : V^+_\sigma \xrightarrow{\approx} V'^-_\sigma, \quad \theta_1 : V^-_\sigma \xrightarrow{\approx} V'^+_\sigma .$$

Choose a basis for V_σ compatible with the decomposition $V_\sigma = V^+_\sigma \oplus V^-_\sigma$ and define θ_σ to be θ_1 in case $(+)$ and to be the composite of θ_1 with $\begin{pmatrix} 0 & i \\ i & 0 \end{pmatrix}$ in case $(-)$. Then θ_σ is an $F \otimes_{F_0,\sigma} \mathbb{C}$-linear isomorphism $V_\sigma \to V'_\sigma$ taking ϕ_σ to a multiple of ϕ'_σ.

<u>Lemma 5.1</u>. With the above notations, there exists an isomorphism $\theta : H_1(A,\mathbb{C}) \longrightarrow H_1(\tau A,\mathbb{C})$ such that:

(a) $\theta \circ f = f \circ \theta$ for all $f \in F$;

(b) $\theta(\tau[\psi]) = [\psi]$;

(c) $\iota\theta = \theta.(\tau\mu(-1)/\mu(-1))$

Proof. Define $\theta = \oplus_{\sigma}\theta_{\sigma}$ and note that $\iota\theta_{\sigma} = \theta_{\sigma}$ in case (+) while $\iota\theta_{\sigma} = -\theta_{\sigma}$ in case (-) . On the other hand, $\mu(-1)$ acts as $\begin{pmatrix} 0 & -i \\ i & 0 \end{pmatrix}$ on $V_{\sigma} = V_{\sigma}^{+} \oplus V_{\sigma}^{-}$ and $\tau\mu(-1) = \mu(-1)$ in case (+) while $\tau\mu(-1) = \begin{pmatrix} 0 & i \\ -i & 0 \end{pmatrix}$ in case (-) .

We shall need a slightly more precise result.

Proposition 5.2. Let A be an abelian variety over $\mathbb{C}$ that is of CM-type, and let τ be an automorphism of $\mathbb{C}$. There exists an isomorphism $\theta : H_1(A,\mathbb{C}) \longrightarrow H_1(\tau A,\mathbb{C})$ such that:

(a) $\theta(s) = \tau s$ for all Hodge cycles s on A ;

(b) $\iota\theta = \theta.\gamma$ where γ is the class in $H^1(\mathbb{R},MT(A))$ represented by $\tau\mu(-1)/\mu(-1)$. (MT(A) = Mumford-Tate group of A.)

Proof. Note that, if we let

$$P(R) = \{\theta : H_1(A,R) \xrightarrow{\approx} H_1(\tau A,R) \mid \theta \text{ satisfies } (5.2a)\}$$

for any $\mathbb{Q}$-algebra R , then P is a right MT(A)-torsor. Proposition (5.2) describes the class of $P_{\mathbb{R}}$ in $H^1(\mathbb{R},MT(A))$. The lemma shows that image of the class in $H^1(\mathbb{R},T)$ is correct, where T is the subtorus of $F^{\times}$ of elements whose norm to F_0 lies in $\mathbb{Q}^{\times}$.

We shall complete the proof of the proposition by showing that $H^1(\mathbb{R}, MT(A)) \to H^1(\mathbb{R},T)$ is injective.

The norm map N_{F/F_0} defines a surjection $T \to \mathbb{G}_m$, and we define ST and SMT to make the rows in the following diagram exact:

$$\begin{array}{ccccccccc} 1 & \to & SMT & \to & MT & \to & \mathbb{G}_m & \to & 1 \\ & & \cup & & \cup & & \| & & \\ 1 & \to & ST & \to & T & \to & \mathbb{G}_m & \to & 1 \end{array} .$$

This diagram gives rise to an exact commutative diagram

$$\begin{array}{ccccccc} \mathbb{R}^\times & \to & H^1(\mathbb{R},SMT) & \to & H^1(\mathbb{R},MT) & \to & 0 \\ \| & & \downarrow & & \downarrow & & \\ \mathbb{R}^\times & \to & H^1(\mathbb{R},ST) & \to & H^1(\mathbb{R},T) & \to & 0 \end{array} .$$

Note that ST (and hence SMT) is anisotropic over $\mathbb{R}$, and that for an anisotropic torus S', $H^1(\mathbb{R},S') = \mathrm{Ker}(S'(\mathbb{C}) \xrightarrow{2} S'(\mathbb{C}))$. Thus $H^1(\mathbb{R},SMT) \to H^1(\mathbb{R},ST)$ is injective, and the five-lemma shows that $H^1(\mathbb{R},MT) \to H^1(\mathbb{R},T)$ is injective.

<u>Remark 5.3</u>. Let A, F, and $V = H_1(A,\mathbb{Q})$ be as in the first paragraph. Then h can be regarded as a map $h : \mathbb{S} \to F^\times(\mathbb{R})$ (thinking of $F^\times$ as a $\mathbb{Q}$-rational torus). It is clear from the discussion preceeding (5.1) that τA is the abelian variety corresponding to $(V, {}^\tau h)$, where ${}^\tau h$ is the map $\mathbb{S} \to F^\times(\mathbb{R})$ with associated cocharacter $\tau\mu_h \in X_*(F^\times)$.

§6. Conjugates of abelian varieties of CM-type.

Let A be an abelian variety of CM-type over $\mathbb{C}$, let $V = H_1(A,\mathbb{Q})$, and let h be the (natural) Hodge structure on V. Fix some family $(s_\alpha)_{\alpha\in J}$ of tensors such that the Mumford-Tate group $MT(A)$ of A is $\mathrm{Aut}(V,(s_\alpha))$ (see I.3). The canonical map $S \xrightarrow{\rho} MT(A)$ induces an action of S on $(V,(s_\alpha))$ and, for any automorphism τ of $\mathbb{C}$, we define $({}^{\tau}V,({}^{\tau}s_\alpha)) = {}^{\tau}S\times^{S}(V,(s_\alpha))$. The element $sp(\tau)\in{}^{\tau}S(\mathbb{A}^f)$ defines an isomorphism

$$v \mapsto sp(\tau).v\colon\ (V(\mathbb{A}^f),(s_\alpha)) \xrightarrow{\approx} ({}^{\tau}V(\mathbb{A}^f),({}^{\tau}s_\alpha)),$$

which we shall again denote by $sp(\tau)$.

Lemma 6.1. There is an isomorphism $f : (H_1(\tau A,\mathbb{Q}),(\tau s_\alpha)) \xrightarrow{\approx} ({}^{\tau}V,({}^{\tau}s_\alpha))$

Proof. Let P_A be the functor such that, for any $\mathbb{Q}$-algebra R, $P_A(R)$ is the set of isomorphisms $(H_1(A,R),(s_\alpha)) \xrightarrow{\approx} (H_1(\tau A,R),(\tau s_\alpha))$. Clearly P_A is representable, and is a right $MT(A)$-torsor. Since $P_A\times^{MT(A)}(H_1(A,\mathbb{Q}),(s_\alpha)) = (H_1(\tau A,\mathbb{Q}),(\tau s_\alpha))$, to prove the lemma it suffices to show that P_A is isomorphic to the $MT(A)$-torsor $\rho_*({}^{\tau}S)$. We shall show this simultaneously for all abelian varieties (over $\mathbb{C}$, of CM-type) whose Mumford-Tate groups are split by a fixed finite Galois extension L of $\mathbb{Q}$.

According to (III.1.7), $S^L=\varprojlim MT(A)$, and it will suffice to show that the two S^L-torsors $P = \varprojlim P_A$ and ${}^{\tau}S^L$ are isomorphic. As $H^1(\mathbb{Q},S^L)$ satisfies the Hasse principle (III.1.5) this only has to be shown locally. The isomorphisms

$H_1(A,\mathbb{Q}_\ell) = V^f(A) \xrightarrow{\tau} V^f(\tau A) = H_1(\tau A,\mathbb{Q}_\ell)$ show P to be trivial over $\mathbb{Q}_\ell$, while $sp(\tau) \in {}^{\tau}S^L(\mathbb{Q}_\ell)$ shows ${}^{\tau}S^L_{\mathbb{Q}_\ell}$ to be trivial. Finally (III. 3.14) and (5.2) show ${}^{\tau}S^L_{\mathbb{R}}$ and $P_{\mathbb{R}}$ define the same cohomology class in $H^1(\mathbb{R},S^L)$ and are therefore isomorphic.

Note that f is uniquely determined up to right multiplication by an element of $MT(A)(\mathbb{Q})$.

<u>Conjecture CM (first form)</u>. The isomorphism f of (6.1) can be chosen to make the following diagram commute:

$$\begin{array}{ccc} V^f(A) & \xrightarrow{\tau} & V^f(\tau A) \\ \| & & \downarrow f\otimes 1 \\ V(\mathbb{A}^f) & \xrightarrow{sp(\tau)} & {}^{\tau}V(\mathbb{A}^f) \end{array}$$

We next restate the conjecture in a form that is closer to the usual statements of the main theorem of complex multiplication. Let $T = MT(A)$, and choose a polarization ψ for (V,h) which we shall assume to be one of the s_α . From the inclusion $(T,\{h\}) \hookrightarrow (CSp(V),S^{\pm})$ we obtain, as in §2, a bijection

$$Sh(T,\{h\}) \xrightarrow{\sim} \mathcal{A}(T,\{h\},V)$$

where $\mathcal{A}(T,\{h\},V)$ consists of certain isomorphism classes of triples $(A',(t_\alpha),k)$.

The torus T continues to act on ${}^{\tau}V$, and in fact $T = Aut({}^{\tau}V,({}^{\tau}s_\alpha))$. One of the ${}^{\tau}s_\alpha$ is ${}^{\tau}\psi$, which is a polarization for $({}^{\tau}V,{}^{\tau}h)$, where ${}^{\tau}h$ is the homomorphism $\mathbb{S} \to T$ corresponding to $\tau\mu_h$. Thus we have an inclusion

$(T,\{{}^{\tau}h\}) \hookrightarrow (CSp({}^{\tau}V), S^{\pm})$ and, as before, a bijection

$$Sh(T,\{{}^{\tau}h\}) \xrightarrow{\approx} \mathcal{A}(T,\{{}^{\tau}h\},{}^{\tau}V) .$$

We define $\chi_{\tau} : \mathcal{A}(T,\{h\},V) \longrightarrow \mathcal{A}(T,\{{}^{\tau}h\},{}^{\tau}V)$ to be the mapping that sends $[A',(t_{\alpha}),k]$ to the class $[\tau A',(\tau t_{\alpha}),{}^{\tau}k]$ where ${}^{\tau}k$ is the composite $V^{f}(\tau A) \xrightarrow{\tau^{-1}} V^{f}(A) \xrightarrow{k} V(\mathbb{A}^{f}) \xrightarrow{sp(\tau)} {}^{\tau}V(\mathbb{A}^{f})$. Lemma (6.1) shows that $[\tau A',(\tau t_{\alpha}),{}^{\tau}k]$ satisfies condition (2.1a) to lie in $\mathcal{A}(T,\{{}^{\tau}h\},{}^{\tau}V)$ and (5.3) shows that it satisfies (2.1b).

<u>Conjecture CM (second form)</u>. The following diagram commutes:

$$\begin{array}{ccccc}
[h,g] & & Sh(T,\{h\}) & \xrightarrow{\approx} & \mathcal{A}(T,\{h\},V) \\
\big\downarrow & & \big\downarrow{\scriptstyle\approx} & & \chi_{\tau}\big\downarrow{\scriptstyle\approx} \\
[{}^{\tau}h,g] & & Sh(T,\{{}^{\tau}h\}) & \xrightarrow{\approx} & \mathcal{A}(T,\{{}^{\tau}h\},{}^{\tau}V)
\end{array}$$

It is easy to check that the two forms of the conjecture are equivalent.

<u>Remark 6.2</u>. When τ fixes the reflex field $E(T,\{h\})$, then conjecture CM becomes the main theorem of complex multiplication (see Milne-Shih [1, 2.6]).

<u>Example 6.3</u>. Let F be a CM-field and Σ a CM-type for F . Let A be an abelian variety (an actual abelian variety - not an isogeny class of abelian varieties!) of type (F,Σ) . Then $H_1(A,\mathbb{Z})$ is a locally free module of rank one over the ring of integers O_F in F , and hence defines an element $I(A)$ of $Pic(O_F)$. Consider

$$
\begin{array}{c}
(S^L(\mathbb{A}_L^f)/S^L(L))^{Gal(L/\mathbb{Q})} \\
\downarrow \\
1 \longrightarrow T(\mathbb{A}^f)/T(\mathbb{Q}) \longrightarrow (T(\mathbb{A}_L^f)/T(L))^{Gal(L/\mathbb{Q})} \longrightarrow H^1(L/\mathbb{Q},T)
\end{array}
$$

where $T = Res_{F/\mathbb{Q}}\, \mathbb{G}_m$, L is a (sufficiently large) finite Galois extension of $\mathbb{Q}$, and the vertical map is induced by the canonical map $\rho : S^L \rightarrow T$. As $H^1(L/\mathbb{Q},T) = 0$, the image of $\bar{\beta}(\tau)$ in $(T(\mathbb{A}_L^f)/T(L))^{Gal(L/\mathbb{Q})}$ arises from an element $\beta'(\tau) \in T(\mathbb{A}^f)/T(\mathbb{Q})$. This defines an ideal class $I(\tau) \in Pic(O_F)$, and the conjecture predicts that $I(\tau A) = I(\tau)\, I(A)$.

Remark 6.4. Let A be an abelian variety of potential CM-type defined over a number field k . Conjecture CM would imply that the zeta function of A is an alternating product of L-series associated to complex representations of the Weil group of k . Deligne has proved this result without, however, proving the conjecture (cf. IV.).

§ 7. Conjecture C, conjecture CM, and canonical models.

Let $(M(G,X),\ f\colon M(G,X)_{\mathbb{C}} \xrightarrow{\approx} Sh(G,X))$ be a canonical model for $Sh(G,X)$ (Deligne [2, 2.2.5]) and, for each automorphism τ of $\mathbb{C}$ fixing $E(G,X)$, set $\psi_\tau = f \circ (\tau f)^{-1}$. These isomorphisms $\psi_\tau\colon \tau Sh(G,X) \to Sh(G,X)$ satisfy the following conditions:

(7.1a) $\psi_{\tau_1\tau_2} = \psi_{\tau_1} \circ (\tau_1\psi_{\tau_2}),\ \tau_1,\tau_2 \in \mathrm{Aut}(\mathbb{C}/E(G,X))$;

(7.1b) $\psi_\tau \circ \tau(\mathcal{J}(g)) = \mathcal{J}(g) \circ \psi_\tau,\ \tau \in \mathrm{Aut}(\mathbb{C}/E(G,X)),\ g \in G(\mathbb{A}^f)$;

(7.1c) let $h \in X$ be special and assume that τ fixes the reflex field $E(h)$ of h; then $\psi_\tau(\tau[h,1]) = [h, \tilde{r}(\tau)]$. (Here $\tilde{r}(\tau) \in G(\mathbb{A}^f)$ represents $r_E(T,h)(\tau) \in T(\mathbb{A}^f)/T(\mathbb{Q})^\wedge$ where T is some $\mathbb{Q}$-rational torus such that h factors through $T_{\mathbb{R}}$ and $r_E(T,h)$ is the reciprocity morphism (Deligne [2, 2.2.3]).) Note that the family (ψ_τ) is uniquely determined by (G,X): if $(M(G,X)', f')$ is a second canonical model, there is an isomorphism $q\colon M(G,X)' \to M(G,X)$ such that $f' = f \circ q_{\mathbb{C}}$, and so $f' \circ (\tau f')^{-1} = f \circ (\tau f)^{-1} = \psi_\tau$. Moreover, descent theory shows that every family (ψ_τ) satisfying (7.1) arises from a canonical model for $Sh(G,X)$.

If τ fixes $E(G,X)$ and $M(G,X)$ is a canonical model for $Sh(G,X)$, then $\tau M(G,X) = M(G,X)$ is again canonical model for $Sh(G,X)$, and so conjecture A suggests that we should have $Sh(G,X) \xrightarrow{\approx} Sh(^\tau G, {}^\tau X)$. We shall prove this. Thus let (G,X) be any pair satisfying (1.1) and let $h \in X$ be special. Choose a

$\mathbb{Q}$-rational maximal torus T in G such that h factors through $T_{\mathbb{R}}$, and let $\mu = \mu_h$. If τ is an automorphism of $\mathbb{C}$ that fixes $E(G,X)$ then $\tau\mu$ and μ have the same weight; thus $(1 + \iota)\tau\mu = (1 + \iota)\mu$, and (see III. 3.18) there is a well-defined cohomology class $\gamma(\tau,\mu) \in H^1(\mathbb{Q}, T)$.

Lemma 7.2. The image of $\gamma(\tau,\mu)$ in $H^1(\mathbb{Q},G)$ is trivial.

Proof: After replacing (G,X) with the pair (G_1,X_1) constructed in (3.4), we can assume G^{der} is simply connected. Let $H = G/G^{der}$ and let μ' be the composite of μ with $G \to H$. As $\tau\mu$ is conjugate to μ, $\tau\mu' = \mu'$ and (III. 3.10) shows that $\gamma(\tau,\mu')$ is trivial.

Let $w \in G(\mathbb{C})$ normalize $T(\mathbb{C})$ and be such that $\tau\mu = \underset{\sim}{ad}w \circ \mu$. According to (III. 3.14), the image of $\gamma(\tau,\mu)$ in $H^1(\mathbb{R},G)$ is represented by $\tau\mu(-1)/\mu(-1) = (\underset{\sim}{ad}w \circ \mu)(-1)/\mu(-1)$ which (see 4.2) is also represented by $w \cdot \iota w^{-1}$; it is therefore trivial. The lemma is now a consequence of the following easy result.

Sublemma 7.3. Let G be a reductive group over $\mathbb{Q}$ such that G^{der} is simply connected. An element γ of $H^1(\mathbb{Q},G)$ is trivial if its images in $H^1(\mathbb{Q}, G/G^{der})$ and $H^1(\mathbb{R}, G)$ are trivial.

We continue with the notations of the second paragraph of this section; thus $h \in X$ is special, $\mu = \mu_h$, and τ fixes $E(G,X)$. Choose an element $a(\tau) \in {}^{\tau}S(\overline{\mathbb{Q}})$ and let $f\colon G_{\overline{\mathbb{Q}}} \to {}^{\tau}G_{\overline{\mathbb{Q}}}$

be the isomorphism $g \longmapsto a(\tau).g$. It will often be convenient to regard f as being defined over L where L is some sufficiently large finite Galois extension of $\mathbb{Q}$ contained in $\overline{\mathbb{Q}}$. Let $\beta(\tau) = sp(\tau)^{-1}\, a(\tau) \in S(\mathbb{A}_L^f)$ and let $\beta(\tau, \overline{\mu})$ be the image of $\beta(\tau)$ in $T^{ad}(\mathbb{A}_L^f)$ under the map $\rho_{\overline{\mu}} : S \to T^{ad}$ defined by $\mu^{ad} \overset{df}{=} \overline{\mu} \overset{df}{=} (\mathbb{G}_m \overset{\mu}{\to} T \to T^{ad})$. Recall (III.3.18) that we have also defined an element $\overline{\beta}(\tau,\mu) \in T(\mathbb{A}_L^f)/T(L)T(\mathbb{Q})^\wedge$. Since $\beta(\tau,\overline{\mu})$ and $\overline{\beta}(\tau,\mu)$ have the same image in $T(\mathbb{A}_L^f)/Z(\mathbb{A}_L^f)\,T(L)\,T(\mathbb{Q})^\wedge$ we can choose an element $\tilde{\beta}(\tau,\mu) \in T(\mathbb{A}_L^f)$ that lifts both $\beta(\tau,\overline{\mu})$ and $\overline{\beta}(\tau,\mu)$; it is determined up to multiplication by an element of $Z(\mathbb{A}_L^f) \cap T(L)\,T(\mathbb{Q})^\wedge = Z(L)\,Z(\mathbb{Q})^\wedge$. (Note that $T(\mathbb{Q})Z(\mathbb{Q})^\wedge = T(\mathbb{Q})^\wedge$ because $T^{ad}(\mathbb{Q})$ is a discrete subgroup of $T^{ad}(\mathbb{A}^f)$.) Let $\sigma\tilde{\beta}(\tau,\mu) = \tilde{\beta}(\tau,\mu)\gamma_\sigma$; then (γ_σ) is a 1-cocycle representing $\gamma(\tau,\mu) \in H^1(\mathbb{Q},T)$. We have $\sigma f = f \circ \underset{\sim}{ad}\,\gamma_\sigma$. The lemma shows that there is an element $v \in G(\overline{\mathbb{Q}})$ such that $\gamma_\sigma = v^{-1}.\sigma v$ for all $\sigma \in Gal(\overline{\mathbb{Q}}/\mathbb{Q})$. We define an isomorphism $f_1 \colon G \to {}^{\tau}G$ and an element $\beta_1(\tau,\mu) \in G(\mathbb{A}^f)$ by the formulas:

$$f_1 = f \circ \underset{\sim}{ad}\ v^{-1} \tag{7.4a}$$

$$\beta_1(\tau,\mu) = \tilde{\beta}(\tau,\mu)v^{-1} \tag{7.4b}$$

<u>Remark</u> 7.5. In the above we have had to choose an $a(\tau)$, $\tilde{\beta}(\tau,\mu)$, and v. For example, if $a(\tau)$ is replaced by $a(\tau)u$, $u \in S(L)$, then $\beta(\tau,\overline{\mu})$ is replaced by $\beta(\tau,\overline{\mu})\,\rho_{\overline{\mu}}(u)$. We show that the cosets defined by $\beta_1(\tau,\mu)$ and $\beta_1(\tau,\mu)^{-1}$ in $G(\mathbb{Q}) \setminus G(\mathbb{A}^f)/Z(\mathbb{Q})^\wedge$ are independent of all choices.

Consider the exact commutative diagram

$$\begin{array}{ccccc} 1 \to T(\mathbb{Q})\backslash T(\mathbb{A}^f) & \longrightarrow & (T(L)\backslash T(\mathbb{A}^f_L))^{\mathrm{Gal}(L/\mathbb{Q})} & \longrightarrow & H^1(L/\mathbb{Q},T(L)) \\ \downarrow & & \downarrow & & \downarrow \\ 1 \to G(\mathbb{Q})\backslash G(\mathbb{A}^f) & \longrightarrow & (G(L)\backslash G(\mathbb{A}^f_L))^{\mathrm{Gal}(L/\mathbb{Q})} & \longrightarrow & H^1(L/\mathbb{Q},G(L)) \end{array}$$

in which the vertical arrows are induced by the inclusion $T \hookrightarrow G$. On dividing by $Z(\mathbb{Q})^\wedge$ we obtain

$$\begin{array}{ccccc} 1 \longrightarrow T(\mathbb{A}^f)/T(\mathbb{Q})^\wedge & \longrightarrow & (T(L)\backslash T(\mathbb{A}^f_L)/T(\mathbb{Q})^\wedge)^{\mathrm{Gal}(L/\mathbb{Q})} & \longrightarrow & H^1(L/\mathbb{Q},T(L)) \\ \downarrow & & \downarrow & & \downarrow \\ 1 \longrightarrow G(\mathbb{Q})\backslash G(\mathbb{A}^f)/Z(\mathbb{Q})^\wedge & \longrightarrow & (G(L)\backslash G(\mathbb{A}^f_L)/Z(\mathbb{Q})^\wedge)^{\mathrm{Gal}(L/\mathbb{Q})} & \longrightarrow & H^1(L/\mathbb{Q},G(L)). \end{array}$$

Lemma 7.2. shows that the image of $\overline{\beta}(\tau,\mu)$ (or $\overline{\beta}(\tau,\mu)^{-1}$) under the middle vertical arrow lies in $G(\mathbb{Q})\backslash G(\mathbb{A}^f)/Z(\mathbb{Q})^\wedge$; it is represented by $\beta_1(\tau,\mu)$.

<u>Remark</u> 7.6. Everything is much simpler when μ satisfies (III. 1.1). Then there is a map $\rho_\mu : S \to T$ and we can choose $\tilde{\beta}(\tau,\mu) = \beta(\tau,\mu) \overset{\mathrm{df}}{=} \rho_\mu(\beta(\tau))$. A change in the choices of $a(\tau)$ and v forces the following changes:

$a(\tau)$	$\beta(\tau,\mu)$	γ_σ	v	f	$\beta_1(\tau,\mu)$
$a(\tau)u_0$	$\beta(\tau,\mu)u_2$	$u_2^{-1}\gamma_\sigma{}^\sigma u_2$	$u_3 v u_2$	$f \circ \mathrm{ad}\,u_2$	$\beta_1(\tau,\mu)u_3^{-1}$

$u_0 \in S(L)$, $u_2 = \rho_\mu(u_0) \in T(L)$, $u_3 \in G(\mathbb{Q})$.

We shall abuse notation by writing ${}^\tau h$ also for the map $\mathbb{S} \to G_{\mathbb{R}}$ associated with $\tau\mu\colon \mathbb{G}_m \to G_{\mathbb{C}}$; thus ${}^\tau h$ (in the sense of § 4) $= f \circ {}^\tau h$(this sense).

Lemma 7.7. Regard v as an element of $G(\mathbb{C})$; then $\operatorname{ad} v \circ {}^{\tau}h \in X$.

Proof. Let $w \in G(\mathbb{C})$ normalize $T(\mathbb{C})$ and be such that $\tau\mu = \operatorname{ad} w \circ \mu$. Then (see the proof of 7.2) v^{-1} and w represent the same cocycle, and so $vw \in G(\mathbb{R})$. Hence $\operatorname{ad} v \circ {}^{\tau}h = \operatorname{ad} v \circ \operatorname{ad} w \circ h \in X$.

Since $\operatorname{ad} v \circ {}^{\tau}h \in X$, and $f_1 \circ \operatorname{ad} v \circ {}^{\tau}h = {}^{\tau}h \in {}^{\tau}X$, we see that $f_1 : G \xrightarrow{\approx} {}^{\tau}G$ defines an isomorphism $Sh(f_1) : Sh(G,X) \xrightarrow{\approx} Sh({}^{\tau}G, {}^{\tau}X)$.

Proposition 7.8. Let $\phi(\tau;\mu)$ be the map

$$Sh(f_1) \circ \mathcal{J}(\beta_1(\tau,\mu)) : Sh(G,X) \xrightarrow{\approx} Sh({}^{\tau,\mu}G, {}^{\tau,\mu}X).$$

Then $\phi(\tau;\mu)$ is independent of the choices of $a(\tau)$, $\tilde{\beta}(\tau,\mu)$, and v; moreover

$$\phi(\tau;\mu)[\operatorname{ad} v \circ {}^{\tau}h, \beta_1(\tau,\mu)^{-1}] = [{}^{\tau}h, 1]$$

$$\phi(\tau;\mu) \circ \mathcal{J}(g) = \mathcal{J}({}^{\tau,\mu}g) \circ \phi(\tau;\mu).$$

Proof. The formula $\phi(\tau;\mu)[x,g] = [f_1 \circ x, f_1(g\beta_1(\tau,\mu))]$ shows immediately that $\phi(\tau;\mu)$ maps $[\operatorname{ad} v \circ {}^{\tau}h, \beta_1(\tau,\mu)^{-1}]$ to $[{}^{\tau}h, 1]$ and that $\phi(\tau;\mu)\, \mathcal{J}(g) = \mathcal{J}(g')\, \phi(\tau;\mu)$ with $g' = f_1(\beta_1^{-1} g \beta_1) = f \circ \operatorname{ad} \beta(\tau,\bar{\mu})(g) = {}^{\tau}g$. The independence assertion is a consequence of this, the following lemma, and Deligne [1,5.2].

<u>Lemma 7.9.</u> The element $[\underset{\sim}{ad}v \circ {}^{\tau}h, \beta_1(\tau,\mu)^{-1}] \in Sh(G,X)$ is independent of the choices of $a(\tau)$, $\tilde{\beta}(\tau,\mu)$, and v.

<u>Proof</u>. Suppose that, after a change in the choices of $a(\tau)$, $\tilde{\beta}(\tau,\mu)$ and v, the elements β_1 and v are replaced by β_1' and v'. Remark (7.5) shows that $(\beta_1')^{-1} = u\beta_1^{-1}z$ with $u \in G(\mathbb{Q})$ and $z \in Z(\mathbb{Q})^\wedge$; moreover $\underset{\sim}{ad}(\beta_1' v') = \underset{\sim}{ad}(\beta(\tau,\bar{\mu})u_1) = \underset{\sim}{ad}(\beta_1 v u_1)$ with $u_1 \in T^{ad}(L)$. Thus $\underset{\sim}{ad}(z^{-1} \beta_1 u^{-1} v') = \underset{\sim}{ad}(\beta_1 v u_1)$ and, on cancelling the β_1, we find $\underset{\sim}{ad}(u^{-1}v') = \underset{\sim}{ad}(vu_1)$. Hence $[\underset{\sim}{ad}v' \circ {}^{\tau}h, (\beta_1')^{-1}] = [\underset{\sim}{ad}v' \circ {}^{\tau}h, u\beta_1^{-1}z] = [\underset{\sim}{ad}u^{-1}v' \circ {}^{\tau}h, \beta_1^{-1}] = [\underset{\sim}{ad}v \circ \underset{\sim}{ad}u_1 \circ {}^{\tau}h, \beta_1^{-1}] = [\underset{\sim}{ad}v \circ {}^{\tau}h, \beta_1^{-1}]$ because ${}^{\tau}h$ maps into $T(\mathbb{R})$ and $u_1 \in T^{ad}(L)$.

<u>Remark 7.10.</u> Under the hypothesis of (7.6), the map $\phi(\tau;\mu)$ becomes $[x,g] \longmapsto [f \circ \underset{\sim}{ad}v^{-1} \circ x, f(v^{-1} g\beta(\tau,\mu))]$ and the element in (7.9) becomes $[\underset{\sim}{ad}v \circ {}^{\tau}h, v\beta(\tau,\mu)^{-1}]$. Both can be directly shown to be independent of all choices.

<u>Proposition 7.11.</u> Assume that $Sh(G,X)$ has a canonical model and let (ψ_τ), $\tau \in Aut(\mathbb{C}/E(G,X))$, be the corresponding family of maps as in (7.1) above. Conjecture C is true for $Sh(G,X)$ and a particular $\tau \in Aut(\mathbb{C}/E(G,X))$ if

$$\psi_\tau(\tau[h,1]) = [\underset{\sim}{ad}v \circ {}^{\tau}h, \beta_1(\tau,\mu)^{-1}] \tag{7.12}$$

holds for all special $h \in X$.

Proof: Note that Lemma 7.9 shows (7.12) makes sense. Define $\phi_{\tau,\mu} = \phi(\tau;\mu) \circ \psi_\tau$. Then

$$\phi_{\tau,\mu}(\tau[h,1]) = \phi(\tau;\mu)[\underset{\sim}{\mathrm{ad}}v \circ {}^{\tau}h,\ \beta_1(\tau,\mu)^{-1}] \quad \text{by } (7.12)$$

$$= [{}^{\tau}h,1] \quad \text{by } (7.8)$$

Moreover,

$$\phi_{\tau,\mu} \circ (\tau\mathfrak{J}(g)) = \phi(\tau;\mu) \circ \mathfrak{J}(g) \circ \psi_\tau$$

$$= \mathfrak{J}({}^{\tau,\mu}g) \circ \phi_{\tau,\mu} \quad \text{by } (7.8)$$

Thus $\phi_{\tau,\mu}$ satisfies condition (a) of conjecture C. Let h' be a second special point and let $\mu' = \mu_{h'}$. Then

$$\phi(\tau;\mu',\mu) \circ \phi_{\tau,\mu} = \phi(\tau;\mu',\mu) \circ \phi(\tau;\mu) \circ \psi_\tau$$

$$= \phi_{\tau,\mu'}$$

because $\phi(\tau;\mu',\mu) = \phi(\tau;\mu') \circ \phi(\tau;\mu)^{-1}$ (4.13).

Remark 7.13 In certain situations, (7.12) simplifies. For example, under the hypothesis of (7.6) it becomes

$$\psi_\tau(\tau[h,1]) = [\underset{\sim}{\mathrm{ad}}v \circ {}^{\tau}h,\ v\,\beta(\tau,\mu)^{-1}] \qquad (7.13a)$$

(see 7.10). On the other hand, if we identify $Sh(G,X)$ with $M(G,X)_{\mathbb{C}}$, then (7.12) becomes

$$\tau[h,1] = [\underset{\sim}{\mathrm{ad}}v \circ {}^{\tau}h,\ \beta_1(\tau,\mu)^{-1}] \qquad (7.13b)$$

If $\gamma(\tau,\bar{\mu}) \overset{df}{=} \rho_{\bar{\mu}}(\gamma(\tau))$ is trivial in $H^1(\mathbb{Q}, \rho_{\bar{\mu}}(S))$ then there exists a $u \in S(L)$ such that $\rho_{\bar{\mu}}(u)^{-1}(\sigma\rho_{\bar{\mu}}(u)) = \gamma_\sigma \pmod{Z(G)}$.

After replacing $a(\tau)$ with $a(\tau)u$ one finds that f is defined over $\mathbb{Q}$, that $\beta(\tau,\mu)$ can be chosen to lie in $T(\mathbb{A}^f)$, and consequently that $v = 1$. Thus (7.12) becomes

$$\psi_\tau(\tau[h,1]) = [{}^\tau h,\ \tilde{\beta}(\tau,\mu)^{-1}] \tag{7.13c}$$

Finally, if τ fixes $E(h)$ then the hypothesis of (7.6) is satisfied, $\gamma(\tau)$ is trivial in $H^1(\mathbb{Q}, S^{E(h)})$, and (7.12) can be written

$$\psi_\tau(\tau[h,1]) = [h,\ \beta(\tau,\mu)^{-1}] = [h,\ \tilde{r}(\tau)] \tag{7.13d}$$

(see III.3.10), which is one of the defining conditions for $M(G,X)$ to be canonical model (see 7.1c).

<u>Proposition 7.14.</u> Assume that conjecture C is true for $Sh(G,X)$ and all $\tau \in Aut(\mathbb{C}/E(G,X))$; then $Sh(G,X)$ has a canonical model and the maps ψ_τ (as in 7.1) satisfy $\psi_\tau = \phi(\tau;\mu_h)^{-1} \circ \phi_{\tau,\mu_h}$ for any special $h \in X$; equation (7.12) is true for all τ fixing $E(G,X)$ and all special $h \in X$.

<u>Proof:</u> Choose a special h and set $\psi_\tau = \phi(\tau;\mu)^{-1} \circ \phi_{\tau,\mu}$ with $\mu = \mu_h$. Arguments reverse to those in the proof of (7.11) show that ψ_τ is independent of h, that $\psi_\tau \circ \tau\mathcal{J}(g) = \mathcal{J}(g) \circ \psi_\tau$, and that $\psi_\tau(\tau[h,1]) = [\mathrm{ad}\, v \circ {}^\tau h,\ \beta_1(\tau,\mu)^{-1}]$. To complete the proof it must be shown that $\psi_{\tau_1\tau_2} = \psi_{\tau_1} \circ (\tau_1\psi_{\tau_2})$, but it can be checked directly that the two maps agree at the point $\tau_1\tau_2[h,1]$, and this implies they agree everywhere.

Corollary 7.15. In addition to the assumption of (7.14) suppose that $E(G,X) \subset \mathbb{R}$. Then conjecture B is true for $Sh(G,X)$.

Proof. If we identify $Sh(G,X)$ with $M(G,X)_{\mathbb{C}}$ then (7.14) and (7.13) show that $\iota[h,1] = [{}^{\iota}h, \tilde{\beta}(\iota,\mu)^{-1}]$ for any special h, where $\tilde{\beta}(\iota,\mu)$ has been chosen to be in $T(\mathbb{A}^f)$. But, according to (III. 3.9), $\tilde{\beta}(\iota,\mu) = 1$ and so $\beta(\iota,\mu) \in T(\mathbb{Q})$. Thus $\iota[h,1] = [{}^{\iota}h, \tilde{\beta}(\iota,\mu)^{-1}] = [{}^{\iota}h, 1]$, which implies conjecture B (4.4).

We come now to the relation between conjectures C and CM. Let A be an abelian variety of CM-type, let $V = H_1(A,\mathbb{Q})$, let h be the natural Hodge structure on V, and let ψ be a Riemann form for A. If T is the Mumford-Tate group of A then we have an embedding $(T,h) \hookrightarrow (CSp(V), S^{\pm})$.

Proposition 7.16. Conjecture CM is true for A and a given $\tau \in Aut(\mathbb{C})$ if and only if (7.12) holds for $Sh(CSp(V),S^{\pm})$, h, and τ.

Proof. Write (G,X) for $(CSp(V), S^{\pm})$. Recall (2.3) that there is a bijection $Sh(G,X) \xrightarrow{\approx} \mathcal{A}(G,X,V)$ where $\mathcal{A}(G,X,V)$ consists of certain isomorphism classes of triples (A',t,k). Let $\mu = \mu_h$; we define $\chi_{\tau,\mu} : \mathcal{A}(G,X,V) \longrightarrow \mathcal{A}({}^{\tau,\mu}G, {}^{\tau,\mu}X, {}^{\tau,\mu}V)$ to be the map $[A',t,k] \longmapsto [\tau A', \tau t, {}^{\tau}k]$ where ${}^{\tau}k$ is the composite $V^f(\tau A') \xrightarrow{\tau^{-1}} V^f(A') \xrightarrow{k} V(\mathbb{A}^f) \xrightarrow{sp(\tau)} {}^{\tau,\mu}V(\mathbb{A}^f)$. Clearly there is a commutative diagram

$$\begin{array}{ccc} \mathcal{A}(T,\ h,\ V) & \hookrightarrow & \mathcal{A}(G,\ X,\ V). \\ \downarrow \chi_\tau & & \downarrow \chi_{\tau,\mu} \\ \mathcal{A}(T,{}^\tau h,{}^{\tau,\mu}V) & \hookrightarrow & \mathcal{A}({}^{\tau,\mu}G,{}^{\tau,\mu}X,{}^{\tau,\mu}V) \end{array}$$

where χ_τ is as defined in §6. On the other hand, as the canonical model for $Sh(G,X)$ is the moduli variety, $\tau: Sh(G,X) \to Sh(G,X)$ corresponds to the map $\tau : \mathcal{A}(G,X,V) \to \mathcal{A}(G,X,V)$ such that $[A',t,k] \mapsto [\tau A',\tau t, \tau k]$ (where $\tau k = k \circ \tau^{-1}$). It is easily verified that $\phi(\tau;\mu)$ corresponds to the map $[A',t,k] \longrightarrow [A',t,sp(\tau) \circ k]$; thus $\phi(\tau;\mu) \circ \tau$ corresponds to $\chi_{\tau,\mu}$. Since $\phi(\tau;\mu)$ is an isomorphism, (7.12) is equivalent to the equation $\phi(\tau;\mu)(\tau[h,1]) = [{}^\tau h,1]$, or, to the assertion that $\chi_{\tau,\mu}$ maps the triple corresponding to $[h,1]$ to the triple corresponding to $[{}^\tau h,1]$. But this is precisely the second form of conjecture CM.

Corollary 7.17. Conjecture CM is true if and only if conjecture C is true for all Shimura varieties of the form $Sh(CSp(V), S^\pm)$.

Proof. Combine (7.16) with (7.11) and (7.14).

Remark 7.18. The same arguments as above show that conjecture CM implies conjecture C for Shimura varieties of the form $Sh(G,X)$ when (G,X) embeds into $(CSp(V), S^\pm)$. We shall show, however, that conjecture C for Shimura varieties of the form $Sh(CSp(V), S^\pm)$ implies conjecture C for all Shimura varieties of abelian type, see Theorem 9.8. Thus, at least for these varieties, conjecture C is equivalent to a statement involving nothing more than abelian varieties of CM-type.

Remark 7.19. It is easy to verify conjecture CM in the case that $\tau = \iota$ (cf. 4.4). On combining this remark with (6.2), we find that conjecture CM is true whenever τ fixes the maximal real subfield of $E(G,X)$. In particular, conjecture CM is true for elliptic curves. Now (7.16) shows that conjecture C is true for $Sh(GL_2, S^{\pm})$. (Cf. Shimura [1, 6.9]).

Even if τ does not fix the maximal totally real subfield of $E(G,X)$, conjecture CM still holds in some cases. Such examples are given in Milne-Shih [1, 2.7]. They arise naturally when one analyzes conjecture B, for details see (ibid, §6).

§ 8. Statement of conjecture $C°$.

Let (G,X) satisfy (1.1), let $h \in X$ be special, and let $\mu = \mu_h$. Recall that there is a unique homomorphism $\rho_{\bar{\mu}} : S \to G^{ad}$ such that $\rho_{\bar{\mu}} \circ \mu_{can} = \bar{\mu} \overset{df}{=} \mu^{ad}$; then $\rho_{\bar{\mu}}$ defines an action of S on G, and we write ${}^{\tau}G$ for ${}^{\tau}S \times^{S} G$ and $g \mapsto {}^{\tau}g : G(\mathbb{A}^f) \to {}^{\tau}G(\mathbb{A}^f)$ for $g \mapsto sp(\tau).g$.

<u>Lemma 8.1</u>. The isomorphism $g \mapsto {}^{\tau}g : G(\mathbb{A}^f) \to {}^{\tau}G(\mathbb{A}^f)$ maps the subgroup $G(\mathbb{Q})^{+\wedge}$ of $G(\mathbb{A}^f)$ into ${}^{\tau}G(\mathbb{Q})^{+\wedge}$ and $G(\mathbb{Q})^{\wedge}_{+}$ into ${}^{\tau}G(\mathbb{Q})^{\wedge}_{+}$.

<u>Proof</u>. Choose an element $a(\tau) \in {}^{\tau}S(L)$ for some finite Galois extension L of $\mathbb{Q}$, and let $f : G_L \to {}^{\tau}G_L$ be the isomorphism $g \to a(\tau).g$. In (3.6) we have defined an isomorphism $\pi_0\pi(f) : \pi_0\pi(G) \to \pi_0\pi({}^{\tau}G)$ and it is easily checked that the following diagram commutes:

$$\begin{array}{ccc} g \mapsto {}^{\tau}g : G(\mathbb{A}^f) & \to & {}^{\tau}G(\mathbb{A}^f) \\ \downarrow & & \downarrow \\ \pi_0\pi(f) : \pi_0\pi(G) & \to & \pi_0\pi({}^{\tau}G) \end{array}$$

Since the kernels of the two vertical arrows in this diagram are $G(\mathbb{Q})^{+\wedge}$ and ${}^{\tau}G(\mathbb{Q})^{+\wedge}$ (Deligne [2, 2.5.1]), $g \mapsto {}^{\tau}g$ maps the first group into the second. Clearly $g \mapsto {}^{\tau}g$ maps $Z(G)(\mathbb{Q})$ into $Z({}^{\tau}G)(\mathbb{Q})$ and so it maps $G(\mathbb{Q})^{\wedge}_{+} = G(\mathbb{Q})^{+\wedge} Z(\mathbb{Q})$ into ${}^{\tau}G(\mathbb{Q})^{\wedge}_{+} = {}^{\tau}G(\mathbb{Q})^{+\wedge}({}^{\tau}Z(\mathbb{Q}))$.

Lemma 8.2. Let (G,G',X^+) define a connected Shimura variety, let $h \in X$ be special, and let $\mu = \mu_h$. Then there exists a unique isomorphism $g \mapsto {}^\tau g : G(\mathbb{Q})^{+\wedge}(\mathrm{rel}\ G') \longrightarrow {}^\tau G(\mathbb{Q})^{+\wedge}(\mathrm{rel}\ {}^\tau G')$ with the following property: for any map $(G_1,X_1) \longrightarrow (G,X)$ such that $G_1^{ad} = G$ and G_1^{der} is a covering of G', the diagram

$$\begin{array}{ccc} g \mapsto {}^\tau g : G_1(\mathbb{Q})^\wedge_+ & \longrightarrow & {}^\tau G_1(\mathbb{Q})^\wedge_+ \\ \downarrow & & \downarrow \\ g \mapsto {}^\tau g : G(\mathbb{Q})^{+\wedge}(\mathrm{rel}\ G') & \longrightarrow & {}^\tau G(\mathbb{Q})^{+\wedge}(\mathrm{rel}\ G') \end{array}$$

commutes.

Proof. According to (3.4) we can choose a (G_1,X_1), as in the statement of the lemma, such that $Z(G_1)$ is a torus having trivial cohomology. Then $G_1(\mathbb{Q}) \longrightarrow G(\mathbb{Q})$ is surjective, and the equality

$$G(\mathbb{Q})^{+\wedge}(\mathrm{rel}\ G') = G_1(\mathbb{Q})^\wedge_+ *_{G_1(\mathbb{Q})_+} G(\mathbb{Q})^+$$

(Deligne [2, 2.1.6.2]) shows that $G_1(\mathbb{Q})^\wedge_+ \longrightarrow G(\mathbb{Q})^{+\wedge}(\mathrm{rel}\ G')$ is surjective. Thus we can define $g \mapsto {}^\tau g$ to be the map induced by its namesake on $G_1(\mathbb{Q})^\wedge_+$.

Let $(G_2,X_2) \longrightarrow (G,X)$ be a second map as in statement of the lemma and define G_3 to be the identity component of $G_2 \times_G G_1$. There is an X_3 for which there are maps $(G_3,X_3) \longrightarrow (G_1,X_1)$ and $(G_3,X_3) \longrightarrow (G_2,X_2)$. Since $\mathrm{Ker}(G_3 \longrightarrow G_2) = \mathrm{Ker}(G_1 \longrightarrow G)$,

$G_3(\mathbb{Q}) \to G_2(\mathbb{Q})$ is surjective and the image of $G_3(\mathbb{Q})^{\wedge}_+$ is dense in $G_2(\mathbb{Q})^{\wedge}_+$. Clearly the maps $g \mapsto {}^{\tau}g$ for G_3, G_1, and G are compatible, as are the same maps for G_3 and G_2. This forces the maps $g \mapsto {}^{\tau}g$ for G_2 and G to be compatible.

When necessary, we shall denote the map defined in the lemma by $\gamma \mapsto {}^{\tau,\mu}\gamma$.

Recall that any $\gamma \in G(\mathbb{Q})^{+\wedge}$ (rel G') defines an automorphism $\gamma.$ of $Sh^{\circ}(G,G',X^+)$ which, when $\gamma \in G(\mathbb{Q})^+$, is equal to the family of maps $\underset{\sim}{ad}\,\gamma : \Gamma\backslash X^+ \to \gamma\Gamma\gamma^{-1}\backslash X^+$.

<u>Conjecture C°</u>. Let (G,G',X^+) define a connected Shimura variety and let τ be an automorphism of $\mathbb{C}$.

a) For any special $h \in X^+$, with $\mu = \mu_h$, there is an isomorphism

$$\phi^{\circ}_{\tau} = \phi^{\circ}_{\tau,\mu} : \tau Sh^{\circ}(G,G',X^+) \to Sh^{\circ}({}^{\tau}G, {}^{\tau}G', {}^{\tau}X^+)$$

such that

$$\phi^{\circ}_{\tau}(\tau[h]) = [{}^{\tau}h]$$

$$\phi^{\circ}_{\tau} \circ \tau(\gamma.) = {}^{\tau}\gamma. \circ \phi^{\circ}_{\tau}, \quad \gamma \in G(\mathbb{Q})^{+\wedge} \text{ (rel } G')\,.$$

b) If $h' \in X^+$ is a second special element and $\mu' = \mu_{h'}$, then

$$\begin{array}{ccc} \tau Sh^{\circ}(G,G',X^{+}) & \xrightarrow{\phi^{\circ}_{\tau,\mu}} & Sh^{\circ}({}^{\tau,\mu}G, {}^{\tau,\mu}G', {}^{\tau,\mu}X^{+}) \\ & \searrow^{\phi^{\circ}_{\tau,\mu'}} & \downarrow \phi^{\circ}(\tau;\mu',\mu) \\ & & Sh^{\circ}({}^{\tau,\mu'}G, {}^{\tau,\mu'}G', {}^{\tau,\mu'}X^{+}) \end{array}$$

commutes.

(For $\phi^{\circ}(\tau;\mu',\mu)$, see §4.)

§ 9. Reduction of the proof of conjecture C to the case of the symplectic group.

Let (G,X) satisfy (1.1), let $\gamma \in G^{ad}(\mathbb{Q})$, and let $h \in X$ be special. If the image of γ in $G^{ad}(\mathbb{R})$ lifts to an element of $G(\mathbb{R})$, then $h' = \underset{\sim}{ad}\, \gamma \circ h$ is also a special point of X. Write $\mu = \mu_h$, $\mu' = \mu_{h'}$, and choose an $a(\tau) \in {}^{\tau}S^{L}(L)$ for some finite Galois extension of $\mathbb{Q}$. Then $f_1 = (a(\tau).g \mapsto a(\tau).\,gg\gamma^{-1})$ is $\mathbb{Q}$-rational isomorphism ${}^{\tau,\mu}G \to {}^{\tau,\mu'}G$ which is independent of the choice of $a(\tau)$ and maps ${}^{\tau,\mu}X$ into ${}^{\tau,\mu'}X$.

<u>Lemma 9.1.</u> With the above notations, the composite

$$Sh({}^{\tau,\mu}G, {}^{\tau,\mu}X) \xrightarrow{\tau\gamma} Sh({}^{\tau,\mu}G, {}^{\tau,\mu}X) \xrightarrow{\phi(\tau;\mu',\mu)} Sh({}^{\tau,\mu'}G, {}^{\tau,\mu'}X)$$

is equal to $Sh(f_1)$.

<u>Proof</u>. If γ lifts to an element of $G(\mathbb{Q})$, this is immediate from the definition of $\phi(\tau;\mu',\mu)$ (see 4.12d). Since we can always find a group with the same adjoint and derived groups as G, but with cohomologically trivial centre, this shows that the two maps agree on a connected component of $Sh({}^{\tau,\mu}G, {}^{\tau,\mu}X)$. To complete the proof we only have to note that both maps transfer the action of $\mathfrak{J}(g)$ on $Sh({}^{\tau,\mu}G, {}^{\tau,\mu}X)$ into the action of $\mathfrak{J}(f_1(g))$ on $Sh({}^{\tau,\mu'}G, {}^{\tau,\mu'}X)$.

Lemma 9.2. Suppose conjecture C is true for (G,X) and let $h \in X$ be special with $\mu = \mu_h$. Then for any $\gamma \in G^{ad}(\mathbb{Q})^+$,

$$\phi_{\tau,\mu} \circ \tau(\gamma.) = {}^{\tau}\gamma. \circ \phi_{\tau,\mu} .$$

Proof. Note that the image of γ in $G^{ad}(\mathbb{R})$, being in $G(\mathbb{R})^+$, lifts to $G(\mathbb{R})$. Let $h' = \underset{\sim}{ad}\, \gamma \circ h$ and $\mu' = \mu_{h'}$, and consider the diagram

$$\begin{array}{ccccc}
\tau Sh(G,X) & \xrightarrow{\phi_{\tau,\mu}} & Sh({}^{\tau,\mu}G, {}^{\tau,\mu}X) & & \\
\downarrow{\scriptstyle \tau(\gamma.)} & & \downarrow{\scriptstyle {}^{\tau}\gamma.} & & \\
\tau Sh(G,X) & \xrightarrow{\phi_{\tau,\mu}} & Sh({}^{\tau,\mu}G, {}^{\tau,\mu}X) & & \downarrow{\scriptstyle Sh(f_1)} \\
 & \searrow{\scriptstyle \phi_{\tau,\mu'}} & \downarrow{\scriptstyle \phi(\tau;\mu',\mu)} & & \\
 & & Sh({}^{\tau,\mu'}G, {}^{\tau,\mu'}X) & &
\end{array}$$

Since we are assuming that the bottom triangle commutes, it suffices to show that the diagram commutes with the lower $\phi_{\tau,\mu}$ removed. But clearly

$$Sh(f_1) \circ \phi_{\tau,\mu}(\tau[h,1]) = [{}^{\tau}h',1] = \phi_{\tau,\mu'} \circ \tau(\gamma.) \quad (\tau[h,1]),$$

$$Sh(f_1) \circ \phi_{\tau,\mu} \circ \tau\mathcal{J}(g) = \mathcal{J}({}^{\tau,\mu'}ad\ \gamma(g)) \circ Sh(f_1) \circ \phi_{\tau,\mu},$$

$$\phi_{\tau,\mu'} \circ \tau(\gamma.) \circ \tau\mathcal{J}(g) = \mathcal{J}({}^{\tau,\mu'}ad\ \gamma(g)) \circ \phi_{\tau,\mu'} \circ \tau(\gamma.),$$

which completes the proof.

Remark 9.3. If, in (9.2), γ lifts to $\delta \in G(\mathbb{Q})$, then the statement of the lemma becomes $\phi_{\tau,\mu} \circ \mathcal{J}(\delta^{-1}) = \mathcal{J}({}^{\tau}\delta^{-1}) \circ \phi_{\tau,\mu}$, which is part of (a) of conjecture C.

Proposition 9.4. Let (G,X) satisfy (1.1) and let X^+ be one connected component of X. Then conjecture C is true for $Sh(G,X)$ if and only if conjecture C° is true for $Sh^\circ(G^{ad},G^{der},X^+)$.

Proof. Assume conjecture C and let $h \in X^+$ be special. Then $\phi_{\tau,\mu}$, with $\mu = \mu_h$, maps $\tau[h,1]$ to $[{}^{\tau}h,1]$ and therefore it maps $Sh^\circ(G,G',X^+)$ into $Sh^\circ({}^{\tau}G,{}^{\tau}G',{}^{\tau}X^+)$. We can therefore define $\phi^\circ_{\tau,\mu}$ to be the restriction of $\phi_{\tau,\mu}$ to $Sh^\circ(G,G',X^+)$. Part (a) of conjecture C° follows from part (a) of conjecture C and (9.2), while part (b) of conjecture C° follows from part (b) of conjecture C.

Next assume conjecture C° holds for $Sh^\circ(G^{ad},G^{der},X^+)$. Suppose that, for special $h \in X^+$, we have extended $\phi^\circ_{\tau,\mu}$, $\mu = \mu_h$, to a map $\phi_{\tau,\mu} : \tau Sh(G,X) \to Sh({}^{\tau}G,{}^{\tau}X)$ satisfying $\phi_{\tau,\mu} \circ \tau\mathcal{J}(g) = \mathcal{J}({}^{\tau}g) \circ \phi_{\tau,\mu}$. Then $\phi_{\tau,\mu}(\tau[h,1]) = [{}^{\tau}h,1]$ and, for $\mu' = \mu_{h'}$ with $h' \in X^+$, $\phi_{\tau,\mu'} = \phi(\tau;\mu',\mu) \circ \phi_{\tau,\mu}$, because the maps ϕ° have the corresponding properties. If h' is a special element of X, but $h' \notin X^+$, we write $h' = \mathrm{ad}\, q \circ h$ with $h \in X^+$ and $q \in G(\mathbb{Q})$, and define $\phi_{\tau,\mu'}$ to be $\phi(\tau;\mu',\mu) \circ \phi_{\tau,\mu}$. We have already noted in (4.14) that this map automatically satisfies part (a) of conjecture C. That the entire family, (ϕ_{τ,μ_h}), $h \in X$ special, satisfies part b of conjecture C follows easily from the definitions and from (4.12b).

It remains to see how to extend $\phi^{\circ}_{\tau,\mu}$. For this we use Deligne [2, 2.7.3]. Write τSh for $\tau Sh(G,X)$ and ${}^{\tau}Sh$ for $Sh({}^{\tau}G,{}^{\tau}X)$. Recall (Deligne [2, 2.1.16]) that $G(\mathbb{A}^f)$ acts transitively on $\pi_0(\tau Sh)$ $(= \tau\,\pi_0(Sh))$ and that the stabilizer of $\tau e \overset{df}{=} \tau Sh^{\circ}(G^{ad},G^{der},X^+)$ is $G(\mathbb{Q})^{\wedge}_{+}$. Similarly ${}^{\tau}G(\mathbb{A}^f)$ acts transitively on $\pi_0({}^{\tau}Sh)$ and the stabilizer of ${}^{\tau}e$ is ${}^{\tau}G(\mathbb{Q})^{\wedge}_{+}$. We have compatible isomorphisms $G(\mathbb{A}^f) \longrightarrow {}^{\tau}G(\mathbb{A}^f)$ and $\pi_0(\tau Sh) \longrightarrow \pi_0({}^{\tau}Sh)$ (see the proof of 8.1). Thus giving a morphism $\tau Sh \longrightarrow {}^{\tau}Sh$ that is compatible with these two morphisms is equivalent to giving a morphism $\tau e \longrightarrow {}^{\tau}e$ that is equivariant for the actions of the stabilizers of τe and ${}^{\tau}e$. But $\phi^{\circ}_{\tau,\mu}$ is such a morphism.

<u>Lemma 9.5</u>. Suppose that (G,X) and (G',X') satisfy (1.1) and that there is a map $(G,X) \longrightarrow (G',X')$ with $G \longrightarrow G'$ injective. If conjecture C is true for $Sh(G',X')$ then it is also true for $Sh(G,X)$.

<u>Proof</u>. According to Deligne [1, 1.15.1] the map $Sh(G,X) \longrightarrow Sh(G',X')$ is injective. A special point h of X maps to a special point h' of X', and the map $\phi_{\tau,\mu_{h'}}$ sends $\tau[h,1]$ to $[{}^{\tau}h,1] \in Sh({}^{\tau}G,{}^{\tau}X) \subset Sh({}^{\tau}G',{}^{\tau}X')$. It therefore sends $\tau[h,g]$ to $[{}^{\tau}h,{}^{\tau}g] \in Sh({}^{\tau}G,{}^{\tau}X)$ for any $g \in G(\mathbb{A}^f)$, which implies that it maps $\tau Sh(G,X)$ into $Sh({}^{\tau}G,{}^{\tau}X)$. We define ϕ_{τ,μ_h} to be the restriction of $\phi_{\tau,\mu_{h'}}$ to $\tau Sh(G,X)$.

Lemma 9.6. If conjecture C° is true for $Sh^\circ(G,G',X^+)$, and G" is a quotient of G' , then conjecture C° is true for $Sh^\circ(G,G'',X^+)$.

Proof. This follows immediately from the general fact that $Sh^\circ(G,G'',X^+)$ is the quotient of $Sh^\circ(G,G',X^+)$ by the kernel of the surjective map

$$G(\mathbb{Q})^{+\wedge} \text{ (rel } G') \longrightarrow G(\mathbb{Q})^{+\wedge} \text{ (rel } G'') .$$

Lemma 9.7. If conjecture C° is true for $Sh^\circ(G_i,G_i',X_i^+)$, $i = 1,\ldots,n$, then the conjecture is true for $Sh^\circ(\Pi G_i,\Pi G_i',\Pi X_i^+)$.

Proof. Easy.

Theorem 9.8. If conjecture C is true for all varieties of the form $Sh(CSp(V), S^\pm)$ then it is true for all Shimura varieties of abelian type.

Proof. If conjecture C is true for varieties of the form $Sh(CSp(V), S^\pm)$ then (1.4), (9.5), and (9.4) show that conjecture C° is true for all connected Shimura varieties of primitive abelian type. Then (9.6) and (9.7) show that conjecture C° is true for all connected Shimura varieties of abelian type. Finally (9.4) then implies that conjecture C is true for all Shimura varieties of abelian type.

Corollary 9.9. Conjecture CM implies that conjecture C is true for all Shimura varieties of abelian type.

Proof. Combine (7.17) with (9.8).

§ 10. Application of the motivic Galois group.

Let M be an extension of $\mathrm{Gal}(\overline{\mathbb{Q}}/\mathbb{Q})$ by the Serre group S,

$$1 \longrightarrow S \longrightarrow M \xrightarrow{\tilde{\pi}} \mathrm{Gal}(\overline{\mathbb{Q}}/\mathbb{Q}) \longrightarrow 1 \qquad (10.1)$$

together with a splitting $\widetilde{sp}$ over $\mathbb{A}^f$, in the sense defined in (III.2). This means, in particular, that the action of $\mathrm{Gal}(\overline{\mathbb{Q}}/\mathbb{Q})$ on S by inner automorphisms in M is the algebraic action described in (III 1.8), and that (10.1) is the projective limit of a system of extensions

$$1 \longrightarrow S^L \longrightarrow M^L \longrightarrow \mathrm{Gal}(L^{ab}/\mathbb{Q}) \longrightarrow 1 \qquad (10.2)$$

over fields L finite and Galois over $\mathbb{Q}$ where $\mathbb{Q} \subset L \subset L^{ab} \subset \overline{\mathbb{Q}} \subset \mathbb{C}$. We assume in addition that the right S-torsor ${}^{\tau}\tilde{S} \overset{df}{=} \tilde{\pi}^{-1}(\tau)$ is isomorphic to the S-torsor ${}^{\tau}S$ arising from the Taniyama group (see §4). Since the existence of $\widetilde{sp}(\tau)$ implies that ${}^{\tau}\tilde{S}_{\mathbb{Q}_\ell}$ is trivial for all ℓ, (III 1.5) shows that the assumption holds if ${}^{\tau}S_{\mathbb{R}} \simeq {}^{\tau}\tilde{S}_{\mathbb{R}}$ as $S_{\mathbb{R}}$-torsors.

Let A be an abelian variety over $\mathbb{Q}$ of potential CM-type and let $V = H_1(A(\mathbb{C}),\mathbb{Q})$. We identify the Mumford-Tate group $MT(A)$ of A with the algebraic group $\mathrm{Aut}(V,(s_\alpha))$, where (s_α) is the family of all Hodge cycles on A (and its powers etc). There is a canonical map $\rho : S \to MT(A)$ (and $S = \varprojlim MT(A)$). As in (6.1) we let P_A be the $MT(A)$-torsor such that, for any $\mathbb{Q}$-algebra R,

$$P_A(R) = \{a : H_1(A,R) \xrightarrow{\approx} H_1(\tau A,R) \mid a(s_\alpha) = \tau s_\alpha \text{ , all } \alpha\} .$$

The proof of (6.1) shows that there exists an S-equivariant morphism ${}^{\tau}S \longrightarrow P_A$. Note that, as $H_1(A,\mathbb{A}^f) = V^f(A)$ and $H_1(\tau A,\mathbb{A}^f) = V^f(\tau A)$, $P_A(\mathbb{A}^f)$ contains a canonical element, namely τ , the map induced by letting τ act on the points of finite order of A . As in § 6, we let $({}^{\tau}V,({}^{\tau}s_\alpha)) = {}^{\tau}S \times {}^{S}(V,(s_\alpha))$.

<u>Lemma 10.3</u>. The following are equivalent:

(a) there exists an S-equivariant morphism $p : {}^{\tau}\tilde{S} \to P_A$ such that $p(\widetilde{sp}(\tau)) = \tau$;

(b) there exists an isomorphism $f : (H_1(\tau A,\mathbb{Q}), (\tau s_\alpha)) \to ({}^{\tau}V,({}^{\tau}s_\alpha))$ such that

$$\begin{array}{ccc} V^f(A) & \xrightarrow{\quad\tau\quad} & V^f(\tau A) \\ \Big\| & & \Big\downarrow f\otimes 1 \\ V(\mathbb{A}^f) & \xrightarrow{\ \widetilde{sp}(\tau)\ } & {}^{\tau}V(\mathbb{A}^f) \end{array}$$

is commutative.

<u>Proof</u>. First note that if p_o is one S-equivariant morphism ${}^{\tau}S \to P_A$ then the other such morphisms are of the form $m\circ p_o$, $m \in MT(A)(\mathbb{Q})$, and that if f_o is one isomorphism $(H_1(\tau A,\mathbb{Q}),\tau s_\alpha) \to ({}^{\tau}V,({}^{\tau}s_\alpha))$ then the others are of the form $f_o \circ m$, $m \in MT(A)(\mathbb{Q})$. Choose a p_o and define f_o to be the inverse of

$$s.v \longmapsto p_o(s)(V) \ : \ {}^{\tau}V \longrightarrow H_1(\tau A,\mathbb{Q}) .$$

Then $(f_o \otimes 1)^{-1} \circ \widetilde{sp}(\tau)$ is the map $v \longmapsto p_o(\widetilde{sp}(\tau))(v)$. The equivalence of (a) and (b) now obvious.

Note that (b) of the lemma says that, if in the statement of conjecture CM, T is replaced by M , then the conjecture becomes true for A .

By the motivic Galois group we shall mean the group associated with the Tannakian category of (absolute Hodge) motives generated by the abelian varieties over $\mathbb{Q}$ of potential CM-type; ie. the group called the Serre group in (II.6). It is an extension of $Gal(\bar{\mathbb{Q}}/\mathbb{Q})$ by S in the sense of the first paragraph of this section (see II.6, and IV, especially B).

Proposition 10.4. (a) If, in the statement of conjecture CM, T is replaced by the motivic Galois group, then the conjecture becomes true for all abelian varieties over $\mathbb{Q}$ of potential CM-type.

(b) Let M be an extension of $Gal(\bar{\mathbb{Q}}/\mathbb{Q})$ by S as in the first paragraph of this section. If conjecture CM becomes true for all abelian varieties over $\mathbb{Q}$ of potential CM-type when T is replaced by M , then M is isomorphic to the motivic Galois group (as extensions of $Gal(\bar{\mathbb{Q}}/\mathbb{Q})$ by S with splittings over $\mathbb{A}^f$).

Proof. (a) It is shown in (II.6) that the motivic Galois group satisfies (a) of (10.3).

(b) Fix a finite Galois extension L of $\mathbb{Q}$ such that $L \subset \bar{\mathbb{Q}}$. For each abelian variety A over $\mathbb{Q}$ of potential CM-type whose Mumford-Tate group is split by L , (10.3) gives us an

S-equivariant morphism $p_A : {}^{\tau}\tilde{S}^L \longrightarrow P_A$ such that $p_A(\widetilde{sp}(\tau)) = \tau$ (here ${}^{\tau}\tilde{S}^L$ is the inverse image of τ in (10.2)). On passing to the inverse limit over A, we obtain an isomorphism $p : {}^{\tau}\tilde{S}^L \longrightarrow {}^{\tau}\tilde{\tilde{S}}^L$ such that $p(\widetilde{sp}(\tau)) = \widetilde{\widetilde{sp}}(\tau)$, where ${}^{\tau}\tilde{\tilde{S}}$ and $\widetilde{\widetilde{sp}}(\tau)$ refer to all motivic Galois group. Choose sections $\breve{a}$ and $\breve{\breve{a}}$ to π for M and the motivic Galois group, and define $\tilde{\beta}(\tau)$ and $\tilde{\tilde{\beta}}(\tau)$ in $S^L(\mathbb{A}^f_L)$ by the formulas (see III.2.9)

$$\widetilde{sp}(\tau)\,\breve{\beta}(\tau) = \breve{a}(\tau)$$

$$\widetilde{\widetilde{sp}}(\tau)\,\breve{\breve{\beta}}(\tau) = \breve{\breve{a}}(\tau)$$

Since $p(\breve{a}(\tau)) = \breve{\breve{a}}(\tau)a$, $a \in S^L(L)$, it follows that

$$\tilde{\beta}(\tau) \equiv \tilde{\tilde{\beta}}(\tau) \mod S^L(L) .$$

This implies (b) (by (III.2.7, 2.9)).

For the remainder of this article, M will denote the motivic Galois group. We retain the notations of the first paragraph of this section.

Much of sections §4 - §9 of this paper remains valid when the Taniyama group T is replaced by the motivic Galois group M. In particular, there are isomorphisms

$$\tilde{\Phi}(\tau;\mu',\mu") : Sh({}^{\tau,\mu}G, {}^{\tau,\mu}X) \longrightarrow Sh({}^{\tau,\mu'}G, {}^{\tau,\mu'}X)$$

as in (4.12) except now defined relative to M. The isomorphisms

$$\tilde{\phi}(\tau;\mu) : Sh(G,X) \longrightarrow Sh({}^{\tau,\mu}G, {}^{\tau,\mu}X)$$

of (7.8) are not defined in the same generality because, in their definition, we have used that $b(\tau,\mu)$ is defined whenever μ satisfies (III.3.3) (rather than III.1.1). The alternative definition (see 7.6, 7.10) is, however, valid and provides a map $\tilde{\phi}(\tau;\mu)$ when (G,X) satisfies (0.1) and (0.2).

Theorem 10.5. If, in the statement of conjecture C, the Taniyama group is replaced with the motivic Galois group, then the conjecture becomes true for all Shimura varieties of abelian type.

Proof. As in (7.17) one proves that conjecture CM implies that conjecture C is true for Shimura varieties of the form $Sh(CSp(V),S^{\pm})$, and as in (9.8) that this implies that conjecture C is true for all Shimura varieties of abelian type.

Corollary 10.6. Conjecture A is true for all Shimura varieties of abelian type.

Proof. We remark that, because the S-torsors $\pi^{-1}(\tau)$ and $\tilde{\pi}^{-1}(\tau)$ defined by T and M are isomorphic, so also are the pairs $({}^{\tau,\mu}G, {}^{\tau,\mu}X)$ defined by the two groups. (The maps $g \longmapsto {}^{\tau,\mu}g : G(\mathbb{A}^f) \longrightarrow {}^{\tau,\mu}G(\mathbb{A}^f)$ could however, differ.) The discussion preceding the statement of conjecture A in § 4 therefore shows that the corollary follows from (10.5).

Corollary 10.7. If (G,X) is of abelian type and satisfies (0.1) and (0.2), then conjecture B is true for $Sh(G,X)$.

Proof. The torsor ${}^{\iota}\tilde{S}$ is trivial because $\iota : A(\mathbb{C}) \longrightarrow \iota A(\mathbb{C})$ is a homeomorphism and therefore induces a map $H_1(A,\mathbb{Q}) \to H_1(\iota A,\mathbb{Q})$ which can be shown to map the Hodge cycle s_α to ιs_α . Thus in (0.5) (whose relevant part is implied by (7.14)) we can take $a(\iota) \in {}^{\iota}S(\mathbb{Q})$, $c(\iota) = \rho_\mu(a(\iota))$, $v = 1$, and $\alpha = \rho_\mu(\tilde{\beta}(\iota)^{-1})$ where $\tilde{\beta}(\iota)$ is defined by $\widetilde{sp}(\tau)\,\tilde{\beta}(\iota) = a(\iota)$. Hence (10.5) implies $\iota[h,1] = [{}^{\iota}h, \rho_\mu(\tilde{\beta}(\iota)^{-1})]$. But conjecture CM in its original form is true when $\tau = \iota$ (see 7.19) and this implies $\tilde{\beta}(\iota) \equiv \beta(\iota) \bmod S^L(L)$. Since $\beta(\iota) \equiv 1 \bmod S^L(L)$ by (III.3.9), we have $\iota[h,1] = [{}^{\iota}h,1]$.

Remark 10.8. In (Milne-Shih [1]) conjecture B is proved for all Shimura varieties of abelian type. There is a good reason why it is easy to prove conjecture B under the assumption of (0.1) and (0.2): these conditions should imply $Sh(G,X)$ is a moduli variety for motives.

Remark 10.9. Theorem 10.5 together with the proof of (7.14) show that $Sh(G,X)$ has a canonical model whenever (G,X) is of abelian type and satisfies (0.1) and (0.2). Presumably if the maps $\tilde{\phi}(\tau,\mu)$ were defined (using M) for all Shimura varieties of abelian type, then one would recover the main theorem of Deligne [2], but there seems little point in this (except that it would give a proof not involving $E_E(G,G', X^+)$) .

Deligne has conjectured the following:

<u>Conjecture D.</u> The Taniyama group and the motivic Galois group are isomorphic (as extensions of $Gal(\overline{\mathbb{Q}}/\mathbb{Q})$ by S together with a splitting over $\mathbb{A}^f$) .

See IV, where the two groups are shown to be isomorphic as extensions of $Gal(\overline{\mathbb{Q}}/\mathbb{Q})$ by S . (It therefore remains to show that the isomorphism can be chosen to carry sp into $\widetilde{sp}$.)

Deligne also suggested that his conjecture D should be equivalent to Langlands's conjecture C. We prove:

<u>Proposition 10.10.</u> Conjecture D is true if and only if conjecture C is true for all Shimura varieties of abelian type (equivalently, for all Shimura varieties of the form $Sh(CSp(V),S^{\pm}))$.

<u>Proof.</u> If conjecture D is true, then (10.5) shows that conjecture C is true for Shimura varieties of abelian type. Conversely, if conjecture C is true for varieties of the form $Sh(CSp(V),S^{\pm})$ then (7.17) shows that conjecture CM is true, and (10.4b) that conjecture D is true.

<u>Remark 10.11.</u> Let $L \subset \overline{\mathbb{Q}}$ be a finite Galois extension of $\mathbb{Q}$ and let K be subfield of L . We write ${}_K T^L$ and ${}_K M^L$ for the pull-backs of T^L and M^L relative $Gal(L^{ab}/K) \hookrightarrow Gal(L^{ab}/\mathbb{Q})$. Assume that L is a CM-field. If A is an abelian variety of CM-type whose Mumford-Tate group is split by L , then the reflex field of A is contained in L , and the main theorem of complex

multiplication shows that conjecture CM is true for all such A and all τ fixing L (cf. 6.2). Thus an obvious variant of (10.4b) shows that ${}_L T^L \approx {}_L M^L$ as extensions of $\mathrm{Gal}(L^{ab}/L)$ by S^L with splittings over $\mathbb{A}^f$. Since conjecture CM is known to be true for $\tau = \iota$, it is therefore also true for any τ fixing the maximal totally real subfield K of L; thus ${}_K T^L \approx {}_K M^L$ (as extensions ...). The results of Shih[1] (see also Milne-Shih [1]) often allow one to replace K in this isomorphism by a subfield of L over which L has degree 4.

For example, let F_o be a totally real field of finite degree over $\mathbb{Q}$ and let F_1 and F_2 be distinct quadratic totally imaginary extensions of F_o. Let $F_3 = F_1 \otimes_{F_o} F_2$ and choose a subset Σ_o of $I_o \overset{df}{=} \mathrm{Hom}(F_o, \mathbb{R})$. For each $\sigma \in I_o$ choose extensions σ_1 and σ_2 of σ_o to F_1 and F_2. Let

$$I_1 = \mathrm{Hom}(F_1,\mathbb{C}),\ \Sigma_1 = \{\sigma_1 \mid \sigma \in \Sigma_o\}$$

$$I_2 = \mathrm{Hom}(F_2,\mathbb{C}),\ \Sigma_2 = \{\sigma_2 \mid \sigma \notin \Sigma_o\}$$

$$I_3 = \mathrm{Hom}(F_3,\mathbb{C}),\ I_1 \times_{I_o} I_2$$

$$\Sigma_3 = \{(\sigma_1,\sigma_2) \mid \sigma \in I_o\} \cup \{(\iota\sigma_1,\sigma_2) \mid \sigma \in \Sigma_o\} \cup \{(\sigma_1,\iota\sigma_2) \mid \sigma \notin \Sigma_o\}.$$

Define E_i to be the subfield of $\mathbb{C}$ of elements fixed by $\{\tau \in \mathrm{Aut}(\mathbb{C}) \mid \tau\,\Sigma_i \subset \Sigma_i\}$, $i = 0, 1, 2, 3$. Then E_o is totally real and $E_3 = E_1 E_2$ is a CM-field. In general, $[E_3 : E_o] = 4$. When E_3 is Galois over $\mathbb{Q}$, ${}_{E_o} T^{E_3} \approx {}_{E_o} M^{E_3}$ (as extensions ...).

Remark 10.12 Our original approach to the results of this section was a little more elementary. We showed directly that there exists a compatible family of maps

$$e^{-L} : \mathrm{Gal}(L/\mathbb{Q}) \longrightarrow S^L(\mathbb{A}^f)/S^L(\mathbb{Q})$$

such that if M' is the extension and splitting defined by

$$\tau \longmapsto \bar{b}(\tau)\bar{e}(\tau) \ : \mathrm{Gal}(L^{ab}/\mathbb{Q}) \longrightarrow S^L(\mathbb{A}_L^f)/S^L(L)$$

with $\bar{b}$ as in (III.3.11) (cf. III.2.7) then conjecture CM holds for M' . Thus, in all of the above, the motivic Galois group can be replaced by M' . Of course (10.4b) shows that M' is isomorphic to the motivic Galois group (as extensions ...).

REFERENCES

Deligne, P.
1. Travaux de Shimura, Sém. Bourbaki Février 71, Exposé 389, Lecture Notes in Math., 244, Springer, Berlin, 1971.

2. Variétés de Shimura: interpretation modulaire, et techniques de construction de modéles canoniques. Proc. Symp. Pure Math., A.M.S. 33 (1979) part 2, 247-290.

3. Valeurs de fonctions L et périodes d´intégrales. Proc. Symp. Pure Math., A.M.S., 33 (1979) part 2, 313-346.

Doi, K. and Naganuma, H.
1. On the algebraic curves uniformized by arithmetical automorphic functions. Ann. Math. 86 (1967) 449-460.

Kazhdan, D.
1. On arithmetic varieties. Lie Groups and their representations, Halsted, New York, 1975. 158-217.

Langlands, R.
1. Some contemporary problems with origins in the Jugendtraum. Proc. Symp. Pure Math., A.M.S. 28(1976) 401-418.

2. Conjugation of Shimura varieties (preliminary version of [3]).

3. Automorphic representations, Shimura varieties, and motives. Ein Märchen. Proc. Symp. Pure Math., A.M.S., 33 (1979) part 2, 205-246.

Milne, J. and Shih, K.-y.
1. The action of complex conjugation on a Shimura variety, Annals of Math, 113 (1981) 569-599.

Serre, J.-P.
1. Cohomologie Galoisienne, Lecture Notes in Math. 5, Springer, Berlin, 1964.

Shih, K.-y.
1. Anti-holomorphic automorphisms of arithmetic automorphic function fields, Ann. of Math. 103 (1976) 81-102.

2. Conjugations of arithmetic automorphic function fields, Invent. Math. 44 (1978) 87-102.

Shimura, G.
1. Introduction to the arithmetic theory of automorphic functions. Princeton Univ. Press 1971.

VI. Hodge Cycles and Crystalline Cohomology

Arthur Ogus
Mathematics Department
University of California
Berkeley, California

0. Introduction

This paper is a collection of musings about several questions related to crystalline cohomology that have plagued me for the past few years. It contains many more conjectures than proofs, and my justification for publishing is the hope that others will find the problems as intriguing as I did but perhaps have more success in solving them.

The first section is the oldest, and is essentially contained in a preprint ("Differentials of the second kind and the conjugate sequence") nearly five years old. It gives a geometric interpretation of the second spectral sequence of De Rham hypercohomology in positive characteristics, and in particular of the associated filtration of the abutment. Surprisingly, the result is the same as in characteristic zero [2]: it is the filtration by codimension of support. Even more surprising is the trivial nature of the proof, especially compared with the rather complicated proof in characteristic zero. This section also contains some of the strange consequences this result has for the De Rham cohomology of nonprojective

Partially supported by NSF Grant MCS 80-23848

varieties of finite type over $\mathbb{Z}$.

The second section, which dates from about the same time as the first, stems from an attempt to further link the mod p and characteristic zero theories by asking the following question: If ω is a De Rham cohomology class which lies in r^{th} level of the filtration above mod p for almost all primes p, does it in fact lie in this level of the corresponding filtration in characteristic zero? This question was inspired by the special case which asks: if a meromorphic differential form is exact mod p for almost all primes p, is it exact? This question is an analogue of the problem beautifully addressed by Katz in [9]. I am not able to say much about it except in some very special cases, in which it turns out to amount to other famous conjectures -- such as the Hodge conjecture.

The third section introduces a new invariant $\in \{\pm 1\}$ of a variety in characteristic p, derived from its crystalline cohomology. I came across this invariant (called the "crystalline discriminant") while working on supersingular K3 surfaces [18 §3], and only later realized that it is interesting more generally. There is a conjectural formula for this invariant in terms of the Betti numbers -- a sort of p-adic Hodge index formula -- which I verify in many cases. This conjectural formula turns out to be a very special case (for liftable varieties) of the philosophy of the last section, that Hodge cycles should behave well under the action of Frobenius in crystalline cohomology.

Let me take the time now to describe the essential idea of the fourth section in a simplified setting. This section was inspired by Deligne's seminar at IHES (1978-1979) on his notion of absolutely Hodge cycles, and is in fact an attempt to substitute crystalline for étale cohomology in his theory. The idea is the following: If R is an étale $\mathbb{Z}$-algebra and X is a smooth projective R-scheme, choose a $\mathbb{C}$-valued point σ of R and let X_σ be the associated scheme over $\mathbb{C}$. If the Hodge conjecture is true for X_σ, then any Hodge class z in $H^{2r}_{DR}(X/\mathbb{C})$ is the cohomology class of an algebraic cycle. It is easy to see that this class will contain an algebraic cycle defined over $\overline{\mathbb{Q}}$, and hence also over some other étale $\mathbb{Z}$-algebra $R' \supseteq R$. It follows that z in fact lies in $H^{2r}_{DR}(X/R) \otimes R'$. Moreover, if s is a closed point of Spec R' and W is the completion of R' at s, then via the canonical isomorphism: $H_{DR}(X/R) \otimes W \cong H_{cris}(X(s)/W)$, we get an action of absolute Frobenius Φ_s on $H_{DR}(X/R) \otimes W$, and necessarily z will be a Tate class with respect to this action, i.e. will satisfy $\Phi_s(z) = p^r z$. Any z which satisfies this for every s is called an "absolutely Tate class." (Actually in §4 we consider a slightly twisted version of this.) Thus, it is natural to try to prove directly that any Hodge class defines an absolutely Tate class, which we are in fact able to do in some cases by adopting Deligne's methods. It is also possible to reverse the process: The set of absolutely Tate classes forms a finite dimensional vector space over $\mathbb{Q}$, and it is reasonable to expect that they are all given by algebraic cycles, or at least Hodge cycles, as we are able to verify for abelian varieties of CM type.

It is a great pleasure to thank N. Katz and P. Deligne for the overwhelming influence they have had on this research, which was inspired throughout by one or the other of them.

1. The conjugate spectral sequence in positive characteristics

If X is a smooth scheme over a field k, its de Rham cohomology $H_{DR}(X/k)$ comes with two spectral sequences, often called "Hodge" ($E_1^{pq} = H^q(X,\Omega^p_{X/k}) \Rightarrow F^{\cdot}_{Hodge}H^i(X,\Omega^{\cdot}_{X/k})$) and "conjugate" ($E_2^{pq} = H^p(X,\underline{H}^q(\Omega^{\cdot}_{X/k})) \Rightarrow F^{\cdot}_{con}H^i(X,\Omega^{\cdot}_{X/k})$). In this section we give a geometric interpretation of the conjugate spectral sequence when k has positive characteristic: it agrees, from E_2 on, with the spectral sequence which begins with $E_1^{pq} = H^{q+p}_{Z^p/Z^{p+1}}(X,\Omega^{\cdot}_{X/k})$, attached to X and its filtration $Z^{\cdot}$ by codimension of support [12]. In characteristic zero this result is also true [2], but for a different (deeper) reason. We begin with a preliminary result that emphasizes the difference between the two cases.

<u>(1.1) Theorem</u>: <u>With the above hypotheses (and when char</u> $k > 0$), <u>let</u> Y <u>be any closed subset of</u> X <u>of codimension</u> r <u>such that</u> $X-Y$ <u>has cohomological dimension</u> $<r$ (i.e., <u>such that</u> $H^i(X-Y,\underline{F})$ <u>vanishes for every coherent sheaf</u> $\underline{F}$ <u>and every</u> $i \geq r$). <u>Then</u>:

$$F^r_{con}H^i_{DR}(X/k) = \mathrm{Im}(H^i_Y(X,\Omega^{\cdot}_{X/k}) \to H^i(X,\Omega^{\cdot}_{X/k})) = F^r_Z.H^i_{DR}(X/k) .$$

<u>Proof</u>: If $Y \subseteq X$ is any closed subset, let Im_Y denote the image of the canonical map: $H_Y(X/k) \to H_{DR}(X/k)$. By definition, $F^r_Z.$ is the union of all the Im_Y such that $\mathrm{codim}\, Y \geq r$, so it suffices to prove the following two statements:

(a) For any Y of codimension $\geq r$, $\mathrm{Im}_Y \subseteq F^r_{con}$.

(b) For any Y with $\mathrm{cd}(X-Y) < r$, $F^r_{con} \subseteq \mathrm{Im}_Y$.

To prove (a) observe that the map: $H^i_Y(X,\Omega^{\cdot}_{X/k}) \to H^i(X,\Omega^{\cdot}_{X/k})$ is compatible with the filtrations arising from the second spectral sequence of hypercohomology, which I will denote by $F^{\cdot}_{con}$ in both cases. Thus it suffices to show that $F^r_{con}H^i_Y(X,\Omega^{\cdot}_{X/k}) = H^i_Y(X,\Omega^{\cdot}_{X/k})$.

Let $F_{X/k}: X \to X'$ be the relative Frobenius map, and recall that we have the inverse Cartier isomorphism: $C^{-1}: \Omega^q_{X'/k} \to F_*\underline{H}^q(\Omega^{\cdot}_{X/k})$. Since $F_{X/k}$ is a homeomorphism mapping Y to Y', we find that $E_2^{pq} = H^p_Y(X,\underline{H}^q(\Omega^{\cdot}_{X/k})) \cong H^p_{Y'}(X',F_*\underline{H}^q(\Omega^{\cdot}_{X/k})) \cong H^p_{Y'}(X',\Omega^q_{X'/p})$. Since $\Omega^q_{X'/k}$ is locally free and X' is Cohen-Macaulay, this vanishes if $p < r$, and the result follows.

It is of course statement (b) which is special to positive characteristics. To prove it, consider the conjugate spectral sequence on X-Y. Again by Cartier, we get $E_2^{pq} = H^p(X\text{-}Y,\underline{H}^q(\Omega^{\cdot}_{X/k})) \cong H^p(X'\text{-}Y', F_*\underline{H}^q(\Omega^{\cdot}_{X/k})) \cong H^p(X'\text{-}Y',\Omega^q_{X'/k})$. Since the base change map $\pi: X'\text{-}Y' \to X\text{-}Y$ is affine, this is $H^p(X\text{-}Y,\pi_*\Omega^q_{X'/k})$, which vanishes if $p \geq r$. This shows that $F^r_{\text{con}}H^i_{DR}(X\text{-}Y/k) = 0$, and it follows that $F^r_{\text{con}}H^i_{DR}(X/k)$ maps to zero in $H^i_{DR}(X\text{-}Y/k)$, **hence** is contained in Im_Y. □

(1.2) <u>Corollary</u>: <u>If</u> X/k <u>is smooth and quasi-projective, then</u> $F^r_{\text{con}}H^i_{DR}(X/k) = F^r_Z.H^i_{DR}(X/k)$.

<u>Proof</u>: If k has characteristic zero, this is proved in [2]. In positive characteristics, choose a $Y \subseteq X$ which is of codimension r and the intersection of r ample divisors. Then X-Y has cohomological dimension less than r, so (1.1) applies. □

To prove that in fact the conjugate spectral sequence agrees with the spectral sequence associated with the filtration $Z^{\cdot}$, and to eliminate the hypothesis of quasi-projectivity in (1.2), we follow the method of [2]. The main step is to obtain a sort of Cousin resolution of the de Rham cohomology sheaves $\underline{H}^q(\Omega^{\cdot}_{X/k})$. Again, this task is immensely simplified in positive characteristics by the Cartier isomorphism.

(1.3) Theorem: If X is a smooth k-scheme:

(1.3.1) There is a resolution of the sheaf $\underline{H}^q(\Omega^{\cdot}_{X/k})$ by flasque sheaves:

$$0 \to \underline{H}^q(\Omega^{\cdot}_{X/k}) \to \underline{H}^q_{Z^0/Z^1}(\Omega^{\cdot}_{X/k}) \to \underline{H}^{q+1}_{Z^1/Z^2}(\Omega^{\cdot}_{X/k}) \to \dots \underline{H}^{q+p}_{Z^p/Z^{p+1}}(\Omega^{\cdot}_{X/k}) \dots$$

(1.3.2) From E_2 on, the conjugate spectral sequence and the spectral sequence associated with the filtration $Z^{\cdot}$ agree.

(1.3.3) The filtrations $F_Z\cdot$ (by codimension of support) and $F^{\cdot}_{con}$ on $H^i_{DR}(X/k)$ agree.

Proof: Let me first remark that in positive characteristics the sequence (1.3.1) does not stop after q steps, as it does in characteristic zero. Aside from this, the situations are the same. The existence of the complex in (1.3.1) and the flasqueness of the sheaves $\underline{H}^{q+p}_{Z^p/Z^{p+1}}(\Omega^{\cdot}_{X/k})$ follow from general nonsense about filtered spaces [2]. If we prove that the sequence is exact, it will follow that the complex of global sections of this complex, which can be identified with (E_1^{pq}, d_1), calculates $H^p(X,\underline{H}^q(\Omega^{\cdot}_{X/k}))$. A slightly technical argument explained in [2] then shows that (1.3.2) and (hence) (1.3.1) are exact. By a trick used in [2, 4.2], we find that it suffices to prove that the maps: $\underline{H}^q_{Z^{p+1}} \to \underline{H}^q_{Z^p}$ are zero. This is also true in positive characteristics, but so is much more:

(1.4) Proposition: Assume char $k > 0$, and let $F : X \to X'$ be the relative Frobenius map. Then we have isomorphisms:

$$C^{-1}_{Z^p} : \underline{H}^p_{Z^p}(\Omega^q_{X'/k}) \to F_*\underline{H}^{q+p}_{Z^p}(\Omega^{\cdot}_{X/k}) \quad \text{and}$$

$$C^{-1}_{Z^p/Z^{p+1}} : \underline{H}^p_{Z^p/Z^{p+1}}(\Omega^q_{X'/k}) \to F_*\underline{H}^{q+p}_{Z^p/Z^{p+1}}(\Omega^{\cdot}_{X/k}).$$

(1.4.2) <u>The maps:</u> $\underline{H}^i_{Z^{p+1}}(\Omega^{\cdot}_{X/k}) \to \underline{H}^i_{Z^p}(\Omega^{\cdot}_{X/k})$ <u>vanish</u>.

<u>Proof</u>: Consider the second spectral sequence of the **hyperderived** functor Γ_{Z^ν}: $E_2^{pq} = \underline{H}^p_{Z^\nu}(\underline{H}^q(\Omega^{\cdot}_{X/k})) \Longrightarrow F^{\cdot}_{con}\underline{H}^i_{Z^\nu}(\Omega^{\cdot}_{X/k})$. Because **F** is a homeomorphism, $F_*E^{\cdot\cdot}$ is still a spectral sequence and $F_*\underline{H}^p_{Z^\nu} \cong \underline{H}^p_{Z^\nu} F_*$. Thus, we have $F_*E_2^{pq} \cong \underline{H}^p_{Z^\nu}(F_*\underline{H}^q(\Omega^{\cdot}_{X/k})) \Longrightarrow F_*F^{\cdot}_{con}\underline{H}^i_{Z^\nu}(\Omega^{\cdot}_{X/k})$. But on X', $F_*\underline{H}^q(\Omega^{\cdot}_{X/k}) \cong \Omega^q_{X'/k}$, and $\underline{H}^p_{Z^\nu}(\Omega^q_{X'/k}) = 0$ if $p \neq \nu$. Thus, the spectral sequence degenerates: $F_*E_2^{pq} = 0$ if $p \neq \nu$ and $F_*E_2^{\nu q} = \underline{H}^\nu_{Z^\nu}(\Omega^q_{X'/k})$. This provides the first isomorphism of (1.4.1); the second can be proved similarly.

One can deduce from (1.4.1) that the sequence of (1.3.1) is isomorphic to the Cousin resolution of $\Omega^q_{X'/k}$ and hence is exact, but it is just as easy to prove (1.4.2). From the degeneration of the spectral sequence in the previous paragraph we see that the induced filtration $F^{\cdot}_{con}$ of $\underline{H}^i_{Z^\nu}(\Omega^{\cdot}_{X/k})$ is all in level ν, i.e. $F^p_{con} = 0$ if $p > \nu$ and $F^\nu_{con} = \underline{H}^i_{Z^\nu}(\Omega^{\cdot}_{X/k})$. The fact that the map $\underline{H}^i_{Z^{\nu+1}}(\Omega^{\cdot}_{X/k}) \to \underline{H}^i_{Z^\nu}(\Omega^{\cdot}_{X/k})$ is compatible with the filtration $F^{\cdot}_{con}$ now makes it vanish. □

(1.5) <u>Remark</u>: Because the sheaves $F_*\underline{H}(\Omega^{\cdot}_{X/k})$ are coherent, one has that for all affines in characteristic p, the sequence:

$$0 \to H^q_{DR}(X/k) \to H^q_{Z^0/Z^1}(X/k) \to H^{q+1}_{Z^1/Z^2}(X/k) \to \cdots$$

is exact. In particular, if X is affine and $U \subseteq X$ is open, the map $H^q_{DR}(X/k) \to H^q_{DR}(U/k)$ is injective. Similarly one can see that for affine X, the maps $H^i_{Z^{p+1}}(X/k) \to H^i_{Z^p}(X/k)$ vanish. In characteristic zero, this is only true if X is the spectrum of a local ring. A consequence of the disparate behavior of de Rham cohomology in characteristics zero and p is the unpleasant nature of de Rham cohomology for noncomplete varieties over **Z**, as evidenced by the following examples:

(1.6) <u>Example</u>: There exists a smooth variety U over $\mathbb{Z}$ and a nontorsion cohomology class $\xi \in H^2_{DR}(U/\mathbb{Z})$ which is divisible by <u>every</u> prime p, i.e. such that $0 = \xi(p) \in H^2_{DR}(U_p/\mathbb{F}_p)$ for every prime p. In fact, Corollary (1.2) makes this appear quite common. Suppose $X/\mathbb{Z}$ is smooth, projective, geometrically connected, and $F^1_Z.H^2_{DR}(X_{\mathbb{Q}}/\mathbb{Q})$ has rank 2. If $Y \subseteq X$ is smooth and ample, $Y_{\mathbb{Q}}$ is geometrically irreducible, so $F^1_Z.H^2_{DR}(X_{\mathbb{Q}}/\mathbb{Q})/H^2_Y(X_{\mathbb{Q}}/\mathbb{Q}) \neq 0$. In characteristic p, however, (1.2) shows that this group vanishes. Thus, if η is an element of $F^1_Z.H^2_{DR}(X/\mathbb{Z})$ whose image in $H^2_{DR}(X_{\mathbb{Q}}/\mathbb{Q})$ is not in Im_Y, $\eta|_{X-Y}$ will do for ξ. For a specific example, take $X = \mathbb{P}^1 \times \mathbb{P}^1$, Y the diagonal, and η the class of a ruling. As explained in the next section, I believe that this phenomenon cannot occur in an H^1.

(1.7) <u>Example</u>: Let $R = \mathbb{Z}[1/N]$, and let X/R be an elliptic curve with complex multiplication. Then if U is an affine open subset of X, the image ξ in $H^1_{DR}(U/R)$ of a nonzero differential ω of the first kind on X is nontorsion, but is divisible by <u>infinitely many</u> primes p. To see this, observe that according to (1.2), ξ will map to zero in the reduction $U(p)$ of U to $\mathbb{F}_p$ iff $\omega(p)$ lies in $F^1_{con}H^1_{DR}(X(p)/\mathbb{F}_p)$, i.e. iff $F^1_{con}(p) = F^1_{Hodge}(p)$, i.e. iff $X(p)$ is supersingular [9, 2.3.4.1.5]. But this holds for "half" of the primes p, as is well-known.

Finally, let me warn the reader of two points likely to cause confusion when working with $F^{\cdot}_Z.$ in the above contexts. First of all, over a field k which is not algebraically closed, $F^r_Z.H^{2r}_{DR}(X/k)$ consists of more than just the cohomology classes of cycles of codimension r, even if the characteristic of k is zero. In fact, since de Rham cohomology with supports commutes with flat base change and since every subset of codimension r defined over a finite extension is contained in a subset of codimention r defined over k, the dimension of this vector space does not change when we enlarge k. For instance, suppose that k has characteristic zero and $Y \subset X$ is smooth and irreducible but not geometrically irreducible. Then $H^{2r}_Y(X/k) = H^0_{DR}(Y/k)$, which is a k-vector space whose dimension is equal to the number of geometric components of Y, and which contains a distinguished element, the cohomology class of Y. On the other hand, in characteristic $p > 0$, even if k is algebraically closed, each $H^{2r}_Y(X/k)$ is "too big," in general, so of course $F^r_{con}H^{2r}_{DR}(X/k)$ is "too big." But it can also be too small. For example, if X is any abelian surface over k, $F^1_{con}H^2_{DR}(X/k)$ has dimension 5. If X is supersingular, NS(X) has rank 6, and so the map: $NS \otimes k \to H^2_{DR}(X/k)$ cannot be injective. (For a more thorough investigation of this phenomenon, the reader is referred to [18], where it is proved that, nonetheless, $NS \otimes \mathbb{Z}/p\mathbb{Z} \to H^2_{DR}(X/k)$ is injective, and that the kernel of $NS \otimes k \to H^2_{DR}(X/k)$ classifies X, up to isomorphism.)

2. Problems -- Reduction of the conjugate filtration mod p.

Despite the disparate behavior of $F^{\cdot}_{con}$ in characteristics zero and p, and despite examples (1.6) and (1.7), it is tempting to try to understand characteristic zero by reducing mod p. One way (which I have not seriously considered) would be by investigating the conjugate filtration in "characteristic p^n." The idea I want to consider here is to reduce modulo all primes, along the lines of [9].

(2.1) Problem: *Suppose R is an integral domain of finite type over $\mathbb{Z}$ with fraction field K, and X is a smooth proper R-scheme. Let $\xi \in H^i_{DR}(X/R)$ be such that for every closed point s of S = Spec R, the image $\xi(s)$ of ξ in $H^i_{DR}(X(s)/k(s))$ lies in $F^r_{con}H^i_{DR}(X(s)/k(s)))$. Then does ξ_K lie in $F^r_{con}H^i_{DR}(X_K/K)$?*

Let me first remark that one can always replace S by an open subset without changing the problem. Also notice that the problem is trivial if K has positive characteristic. To see this, recall that the inverse Cartier isomorphism implies that ${}_{con}E_2^{pq} = R^pf_*\underline{H}^q(\Omega^{\cdot}_{X/S}) \cong R^pf'_*\Omega^q_{X'/S}$, where $f: X \to S$ and $f': X' = X \times_{F_S} S \to S$ are the structure maps. It follows that ${}_{con}E_r^{pq}$ is coherent for every $r \geq 2$, and since S is reduced, we may shrink it until every ${}_{con}E_r^{pq}$ is locally free. Now since $\Omega^q_{X'/S}$ is flat over S, the theorem of exchange implies that formation of ${}_{con}E_2^{pq}$ commutes with arbitrary change of base, and, by induction, so does every ${}_{con}E_r^{pq}$. Applying Nakayama's lemma to the sheaves E_∞^{pq}, one immediately deduces that (2.1) is true.

If K has characteristic zero, as we shall henceforth assume, the problem looks very deep, and I can prove almost nothing about it. The rest of this section consists of some reformulations and investigations of special cases.

For example, notice that since K has characteristic zero, $F^r_{con}H^i_{DR}(X/K)$ vanishes if $p > i/2$, and in particular $F^i_{con}H^i_{DR}(X/K) = 0$ if $i > 0$. In positive characteristics, however, since each $k(s)$ is perfect, $F^i_{con}H^i_{DR}(X(s)/k(s))$ is the image of the absolute Frobenius endomorphism of de Rham cohomology. Thus, (2.1) predicts that these mod p vector spaces should have no common interpolation, i.e.:

(2.2) Special Case: Suppose $\xi \in H^i_{DR}(X/S)$ is such that $\xi(s)$ lies in the image of the Frobenius endomorphism of $H^i_{DR}(X(s)/k(s))$ for all closed points s. Then does $\xi_K \in H^i_{DR}(X/K)$ vanish?

Also extremely interesting is what (2.1) says at the other end of the spectrum, i.e. when $r = 1$. One way to formulate it is the following:

(2.3) Special Case: Suppose U/S is smooth and affine and $\omega \in \Omega^i_{U/S}$ is exact when restricted to $U(s)/k(s)$ for all closed points s of S. Is there an open subset of U on which ω becomes exact?

In other words, (2.3) says that a meromorphic differential form which becomes exact on each $U(s)$ should be exact as a meromorphic form. To see the relation with (2.1), begin by proving or assuming that ω (which is necessarily closed) is a differential form of the second kind, i.e., that its cohomology class in $H^i_{DR}(U/k)$ extends to some smooth compactification $\overline{U}$ of U. Then (after shrinking S) this class will map to $F^1_{con}H^i_{DR}(\overline{U}(s)/k(s))$ for all s, and so by (2.1) should map to $F^1_{con}H^i_{DR}(\overline{U}/K)$, i.e., should vanish in some open set.

It is now clear that (2.1) is really a problem in the construction of algebraic cycles. Indeed, using Mazur's interpretation of

$F^{\cdot}_{con}$ in terms of the action of Frobenius on crystalline cohomology, we shall see that an affirmative answer to (2.1) would prove the following analogue of Tate's conjecture:

(2.4) Problem: Suppose that K is a numberfield and that S is smooth over $\mathbb{Z}$. For each closed point s of S, let $W(s)$ be the Witt ring of $k(s)$, and let $F(s)$ be the semi-linear endomorphism of $H^{2r}_{DR}(X/S) \otimes W(s) \cong H^{2r}_{cris}(X(s)/W(s))$ induced by the absolute Frobenius endomorphism of $X(s)$. Suppose $\xi \in H^{2r}_{DR}(X/S)$ is such that $F(s)(\xi) = p^r\xi$ for every s. Then is ξ_K the cohomology class of an algebraic cycle of codimension r in $X_{\overline{K}}$?

To see that (2.1) implies (2.4), assume (as we may, by shrinking S) that the Hodge cohomology of X/S is locally free. Then it is clear from Mazur's description [15] of the Hodge and conjugate filtrations of de Rham cohomology in characteristic p that if $F(s)(\xi) = p^r\xi(s)$, then $\xi(s)$ lies in $F^r_{con}(s) \cap F^r_{Hodge}(s) \subseteq H^{2r}_{DR}(X(s)/k(s))$. Of course, this implies that ξ lies in $F^r_{Hodge}H^{2r}_{DR}(X/S)$. Thus, if (2.1) is true, even with the additional hypothesis that ξ lies in F^r_{Hodge}, we see that ξ will lie in $F^r_{con}H^{2r}_{DR}(X/K)$, which is just the $\overline{K}$-span of the set of cohomology classes of algebraic cycles of codimension r [2]. This set forms a $\mathbb{Q}$-vector space V, and it remains to prove that ξ lies in V, not just in $\overline{K} \otimes V$. But $F(s)$ acts on $W(s) \otimes_k K \otimes V = W(s) \otimes V$ as $p^rF_{W(s)} \otimes id_V$, and it is clear that the set of all elements ζ of this space such that $F(s)(\zeta) = p^r\zeta$ is just $\mathbb{Q}_p \otimes V$. It follows that any ξ as in (2.4) has degree one over every prime p, henee is rational.

Even for $i = 1$, (2.1) looks formidable. Indeed, this case amounts to a special case of Grothendieck's conjecture on algebraic solutions to differential equations [9]. To explain why this is so, let us note the following:

(2.5) Remark: Let $\xi \in H^1_{DR}(X/T)$ correspond to the extension class of $0 \to (\mathcal{O}_X, d) \to (E, \nabla) \to (\mathcal{O}_X, d) \to 0$, under the identification $H^1_{DR}(X/T) = \mathrm{Ext}^1_{MIC}(\mathcal{O}_X, \mathcal{O}_X)$. Then the following are equivalent:

(a) The meromorphic cohomology class defined by ξ vanishes, i.e. there is a nonvoid open set U on which ξ vanishes.

(b) There is a nonvoid open set V on which (E, ∇) becomes constant.

Proof: Trivially, (a) implies (b). For the converse, assume that V is an affine open subset of U, and that ξ is represented by a closed 1-form ω. Then one sees easily that (E, ∇) is the connection on $\mathcal{O}_X \oplus \mathcal{O}_X$ given by $\nabla e_1 = 0$ and $\nabla e_2 = \omega \otimes e_1$. If (E, ∇) is constant, it does not follow, a priori, that the extension splits, i.e. that ξ is zero on V, but there must be a horizontal section $e = ae_1 + be_2$ of E with b not identically zero. Then $da + b\omega = 0$ and $db = 0$, so on the nonvoid open set $V_b \subseteq X$, $\omega = -d(a/b)$, so $\xi|_{V_b} = 0$.

If S is the spectrum of a field of characteristic zero, (a) and (b) occur only if $\xi = 0$, of course, and if S has characteristic $p > 0$, (a) and (b) occur iff for any (equivalently every) affine open set V in X, $\xi|_V = 0$, or equivalently $(E, \nabla)_V$, is constant. In particular we could answer (2.3) (when $i = 1$) from Grothendieck's conjecture as follows: If ω is a closed 1-form on U which is exact on $U(s)$ for all s, then (E, ∇) is constant on $U(s)$ for all s. According to Grothendieck's conjecture, this should imply the existence of a finite étale cover $W_{\mathbb{C}}$ of $U_{\mathbb{C}}$ on which (E, ∇) becomes constant, and since the characteristic is now zero, this implies that ξ already maps to zero in $H^1_{DR}(U_K/K)$.

(2.6) Proposition: (Katz) Suppose $R/\mathbb{Z}$ is étale and X is an elliptic curve over R. Then (2.1) is true for any ξ in $H^0(X,\Omega^1_{X/S})$. If, moreover, $X(s)$ is supersingular for infinitely many values of s (as conjectured by Serre), then (2.1) is true for any $\xi \in H^1_{DR}(X/R)$.

Proof: We have to show that for infinitely many $s \in S$, $\xi(s) \notin F^1_{con}H^1_{DR}(X(s)/k(s))$, if $\xi \neq 0$. But since $\xi \in F^1_{Hodge}$, this is the same as finding infinitely many points s such that $F^1_{con}(s) \cap F^1_{Hodge}(s) = 0$, i.e. such that $X(s)$ is ordinary, and this has been done by Serre [20]. For the general case, it suffices to show that if $\xi(s) \in F^1_{con}H^1_{DR}(X(s)/k(s))$ for almost all s, then in fact $\xi \in F^1_{Hodge}$, and for that it suffices to find infinitely many values of s such that $\xi(s) \in F^1_{Hodge}(s)$. But this will be true at the supersingular values of s. □

Using a simple version of Serre's argument [20] as explained by Katz, we can extend the above result to some abelian varieties of higher genus. The argument is based on a study of the ℓ-adic representations associated to abelian varieties over numberfields, and it is instructive to discuss these abstractly. Let $K \subseteq \overline{\mathbb{Q}}$ be a numberfield and let $G_K \subseteq \mathrm{Gal}(\overline{\mathbb{Q}}/\mathbb{Q})$ be its Galois group, $\mathrm{Gal}(\overline{\mathbb{Q}}/K)$. Suppose that ρ is a continuous representation of G_K ina $\mathbb{Q}_\ell$-vector space V of dimension d, unramified outside a finite set Σ of places of K. If $\overline{p}$ is a place of $\overline{\mathbb{Q}}$ lying over an unramified place p of K, then the conjugacy class of geometric Frobenius $\rho(F_{\overline{p}})$ depends only on p. The representations we will be considering, since they arise from algebraic geometry, are integral and pure of some weight β, that is, for all unramified ρ, the characteristic polynomial of $\rho(F_{\overline{p}})$ lies in $\mathbb{Z}[T] \subseteq \mathbb{Q}_\ell[T]$, and all its roots are of absolute value $\mathrm{Nm}(p)^{\beta/2}$. Recall that if we use Dirichlet density with respect to the field K, then the set of p such that $k(p)$ is a prime field (say of characteristic p) has density one; for such p, $Nm(p) = p$.

(2.7) Proposition: If ρ is as above, then after repeatedly replacing K by some finite extension:

(2.7.1) (Katz) If $\beta = 1$, then for almost every prime p of degree one, $Tr\rho(F_{\bar{p}})$ is not divisible by p.

(2.7.2) If $\beta = 2$, and if p is an unramified prime of degree one such that p divides $Tr\rho(F_{\bar{p}})$, then $\rho(F_{\bar{p}})p^{-1}$ is unipotent. If this holds for almost all such primes p, then the semi-simplification of ρ is isomorphic to $\oplus\mathbb{Q}_\ell(-1)$.

Proof: Choose an invariant $\mathbb{Z}_\ell$-lattice $E \subseteq V$, choose n so that $\ell^n > 2^d$, and let $K' \supseteq K$ be a finite extension such that $\rho|_{G_{K'}}$ is trivial mod ℓ^n. Then if p is a degree one prime of K' (and unramified in ρ), the trace A_p of $\rho(F_{\bar{p}})$ is an integer congruent to d mod ℓ^n. Now if $\beta = 1$ and if $A_p = pB_p$, then $|B| \leq dp^{-1/2}$, so that if $p > d^2$, $B_p = 0$, a contradiction of the fact that $A \equiv d \bmod \ell^n$. This proves (2.7.1).

To prove (2.7.2), enlarge K' so that it contains all the $\ell^{n\text{th}}$ roots of unity. Then if $A_p = pB_p$, B_p is an integer of absolute value less than or equal to d. Moreover, since p splits completely in K' and K' contains all $\ell^{n\text{th}}$ roots of unity, $p \equiv 1 \bmod \ell^n$, and hence $B_p \equiv d \bmod \ell^n$. This implies that $B_p = d$, and so $A_p = pd$. But A_p is the sum of d complex numbers of absolute value p, and this is only possible if each of the numbers actually equals p, i.e. all the eigenvalues of (F_p) are equal to p. If this holds for a set of primes of density one, then by Chebataroff, $tr(\rho) = tr[\oplus Q_\ell(-1)]$, and since a semi-simple representation is determined by its trace, this proves the claim.□

(2.8) Corollary: If X is an abelian variety over a numberfield K, then (after enlarging K), for almost every prime p of K of degree one, the Hasse-Witt matrix of $X(p)$ is not nilpotent.

Proof: Indeed, it is well-known that if the Hasse-Witt matrix of $X(p)$ is nilpotent, the trace of the action of F_p on the Tate module of X is divisible by p.□

(2.9) Corollary: If X is an elliptic curve or an abelian surface, then (after enlarging K), $X(p)$ is ordinary for almost every p with $\deg(p) = 1$.

Proof: Of course, the case of curves is covered by the previous corollary. In the case of a surface, if $X(p)$ is not ordinary, the action of $F(p)$ on $H^1(X(p),\mathcal{O}_{X(p)})$ has rank one and hence is zero on $H^2(X(p),\mathcal{O}_{X(p)}) = \Lambda^2 H^1(X(p),\mathcal{O}_{X(p)})$. This implies that the trace of $F(p)$ acting on $H^2_{cris}(X(p)/W(p))$ is divisible by p. This is the same, of course, as the trace of F_p on $H^2_{et}(X \times \overline{K})$. If we have enlarged K so that (2.7.2) applies, then all the eigenvalues of F_p and $F(p)$ are p. But on an abelian surface, the action of $F(p)$ on crystalline cohomology is semi-simple, by the lemma below, and hence $F(p)$ is just multiplication by p. This is impossible, because $F^1_{Hodge}(p)$ is precisely the kernel of $F(p)$ mod p, so $F(p)$ cannot be divisible by p. □

(2.10) Lemma: If X is an abelian variety over a finite field $\mathbb{F}_{p^d}$, the action of $(F_{abs})^d$ on $H^i_{cris}(X/W) \times_{\mathbb{Z}} \mathbb{Q}$. is semi-simple.

Proof: This is well-known, but I don't know of a reference, so here is a proof. It is clear that we may replace X by any X' isogenous to X and that we may assume that X is simple. Now the characteristic polynomial of $(F_{abs})^d$ has coefficients in $\mathbb{Q}$, and it follows that there exist polynomials $p(T)$ and $q(T)$ with rational coefficients which, when evaluated at $(F_{abs})^d$, yield its semi-simple and nilpotent parts. In particular the nilpotent part N of $(F_{abs})^d$ belongs to $End(X) \otimes \mathbb{Q}$, and since X is simple, the kernel of N cannot be a proper subscheme, so N vanishes. □

(2.11) Corollary: If X is an abelian variety of genus g defined over $\mathbb{Q}$ and if E is a totally real $\mathbb{Q}$-algebra of degree g acting faithfully via rational isogenies of X, then (2.1) is true for any differential of the first kind on X.

Proof: The hypotheses on E mean that it is a product of totally real fields E_i, and that we have an injective homomorphism $E \to End(X) \otimes \mathbb{Q}$. If $e_i \in E$ is the idempotent corresponding to the unit element of E_i and if $X_i = \{x \in X: e_j x = 0 \forall j \neq i\}$, then X is isogeneous to the product of the X_i's, each E_i acts faithfully on X_i, and since E_i is totally real $\deg E_i \leq \dim X_i$. Thus, we are reduced to the case in which E is a field.

Choose some $R = \mathbb{Z}[1/N]$ so that X is defined over R, and let $C_R = \{\omega \in H^0(X_R, \Omega^1_{X/R}): \omega(s) \in F^1_{con}(s)$ for all closed points s of Spec $R\}$. It is clear that if $R \subseteq R' \subseteq \mathbb{Q}$, we get natural inclusions $C_R \subseteq C_{R'} \subseteq H^0(X, \Omega^1_{X/\mathbb{Q}})$. For R sufficiently large, the $\mathbb{Q}$-vector space spanned by C_R is independent of R; call it C. Now it is clear that any automorphism of X_R/R acts on $H^1_{DR}(X/R)$ compatibly with F^1_{Hodge} and with each $F^1_{con}(s)$, and hence preserves C_R. Enlarging R if necessary, we may assume that all the endomorphisms of $X/\mathbb{Q}$ are defined over R, and hence that the action of E preserves $C \subseteq H^0(X,\Omega^1_{X/\mathbb{Q}})$. But the latter is a one dimensional E-module, so C either vanishes or fills up all of $H^0(X,\Omega^1_{X/\mathbb{Q}})$. In the latter case, C_R is of finite index in $H^0(X,\Omega^1_{X/R})$, and in fact we have equality after enlarging R. But then $F^1_{con}(s) = F^1_{Hodge}(s)$ for all closed points s of Spec R, which is impossible by (2.8). □

We can use the above arguments for K3 surfaces as well, to obtain the following result:

(2.12) <u>Proposition</u>: <u>If</u> X <u>is a K3 or abelian surface over an étale</u> $\mathbb{Z}$<u>-algebra</u> R, (2.2) <u>holds for any</u> $\omega \in F^1_{Hodge}H^2_{DR}(X/R)$. □

<u>Proof</u>: It suffices to prove that for infinitely many closed points s of Spec R, $F^2_{con}(s) \not\subseteq F^1_{Hodge}(s)$. Since $F^2_{con}(s)$ is the image of $F_{abs}(s)$ and $F^1_{Hodge}(s)$ is its kernel, we have containment iff $F_{abs}(s)$ is zero on de Rham cohomology, in which case its trace in crystalline or étale cohomology will be divisible by p. Therefore the following result proves the proposition:

(2.13) <u>Lemma</u>: <u>After enlarging</u> R, $tr(F(s))$ <u>is not divisible by</u> p <u>for any</u> s <u>of degree one in</u> Spec R.

<u>Proof</u>: According to (2.7.2), it suffices to show that (after enlarging R), $F(s)p^{-1}$ cannot be unipotent. We have already checked this for abelian surfaces, and for K3's we also have:

(2.14) <u>Lemma</u>: <u>After localizing</u> R, <u>the action of</u> $(F_{abs}(s))^{deg(s)}$ <u>on</u> $H^2_{cris}(X(s)/W(s)) \times \mathbb{Q}$ <u>is semi-simple, for each closed point</u> s <u>of</u> Spec R.

Proof: We only provide a sketch. The idea is to use Deligne's method [5] of comparing the cohomology of X with the cohomology of an associated abelian variety constructed by Kuga and Satake [14]. He shows that there exist an étale $\mathbb{Z}$-scheme T, a smooth T-scheme S, a versal family of polarized K3-surfaces X/S, a family Y/S of abelian schemes on which a certain ring C operates, and an isomorphism of variations of Hodge structures on the analytic space S_σ^{an} (obtained from S via a chosen $\sigma \in T(\mathbb{C})$): $C^+(X/S) \cong \mathrm{End}_C H^1(Y/S)$, where $C^+(X/S)$ is the even part of the Clifford algebra associated with the primitive cohomology of X/S. Using the regularity of the Gauss-Manin connection, one deduces that (after shrinking S and T) there is an isomorphism of modules with integrable connection: $(C^+(X/S),\nabla) \cong (\mathrm{End}_C H^1(Y/S),\nabla)$. We may also assume, after further shrinking, that the fibers of S/T are geometrically irreducible. Of course, this family is constructed so that our original X/R can be obtained by specializing X/S at some R-valued point of S. (It may be necessary to replace R by a quasi-finite extension, but since an invertible matrix is semi-simple iff its powers are, this is harmless.)

Now if t is a closed point of T, we obtain F-crystals $C^+(X(t)/S(t))$ and $\mathrm{End}_C H^1(Y(t)/S(t))$ on $(S(t)/W(t))_{cris}$. If we can prove that these are isomorphic, we will be able to conclude from (2.10) that $F(t)$ acting on $C^+(X(t))$, hence also on $H^2_{cris}(X(t)/W(t))$, is semi-simple. Now the above isomorphism is horizontal, and if it is compatible with the (horizontal) action of Frobenius at one point, it must be compatible at every point. Thus, it is enough to find one such point.

Since X/S is versal, the map from S^{an} to the moduli space of all polarized K3's is open, and hence there are some Kummer surfaces in its image [19, §6]. It is easy to see that these surfaces will remain Kummer over some (shrunken) T, so we are reduced to proving that our isomorphism is compatible with Frobenius for Kummer surfaces. Deligne has proved that the eigenvalues agree, and it is clear that Frobenius of a Kummer is semi-simple. Thus our two Frobenius matrices are at least conjugate, so that after correcting the isomorphism by some constant automorphism, it becomes compatible with the two Frobenii. □

For abelian varieties of CM type and for Fermat hypersurfaces we can go much further and prove that (2.1) is equivalent to the generalized Hodge conjecture. For simplicity, we give the proof only for simple abelian varieties of CM type.

(2.15) <u>Theorem</u>: <u>If</u> X <u>is an abelian scheme of CM type or a Fermat hypersurface over an étale $\mathbb{Z}$-algebra</u> R, <u>then (2.1) is equivalent to the generalized Hodge conjecture for</u> X. <u>More precisely, if</u> σ <u>is an embedding of</u> R <u>in</u> $\mathbb{C}$ <u>and if</u> $C^rH^i =: \{\omega \in H^i_{DR}(X/R) \colon \omega(s) \in F^r_{con}(s)$ <u>for all closed points</u> s <u>of</u> Spec $R\}$, <u>then the image of</u> $C^rH^i \otimes_\sigma \mathbb{C}$ <u>in</u> $H^i_{DR}(X/R) \otimes_\sigma \mathbb{C} \cong H^i(X^{an}_\sigma, \mathbb{Q}) \otimes \mathbb{C}$ <u>is the $\mathbb{C}$-span of the smallest sub-$\mathbb{Q}$-Hodge structure</u> N^rH^i <u>of</u> $H^i(X_\sigma)$ <u>such that</u> $N^rH^i \subseteq F^r_{Hodge}$.

<u>Proof</u>: First let us recall the basic set-up of simple abelian varieties of CM type. If K is the fraction field of R (which of course we assume to be integral), and if E is the ring of isogenies of $X \otimes \mathbb{C}$, we may assume that K is large enough so that the action of E is rational over K and so that E (which is necessarily a field) can be embedded in K. Furthermore we assume that K is Galois over $\mathbb{Q}$, with group G. The set S of all embeddings of E in K is a left G-set in an obvious way, and $E \otimes K \cong K^S$. Since E acts on X/K, $E \otimes K$ acts on $H^1_{DR}(X/K)$,

and $H^1_{DR}(X/K)$ is a free $E \otimes K$-module of rank one. Thus, $H^1_{DR}(X/K) = \oplus\{H_\chi : \chi \in S\}$, where $H_\chi = \{\omega \in H^1_{DR} : e^*(\omega) = \chi(e)\omega\}$ for all $e \in E$. Let $\Sigma \subseteq S$ be the set of all those χ such that $H_\chi \subseteq F^1_{Hodge}$. Fix an embedding σ of R and K in $\mathbb{C}$, and regard K as a subfield of $\mathbb{C}$. The action of E on $H^1_{DR}(X/K) \otimes \mathbb{C} \cong H^1(X_\sigma, \mathbb{Q}) \otimes \mathbb{C}$ is compatible with this isomorphism and preserves the rational subspace $H^1(X_\sigma, \mathbb{Q})$, hence is also compatible with the (semi-linear) action of $\mathrm{Aut}(\mathbb{C}/\mathbb{Q})$ on $H^1(X_\sigma, \mathbb{Q}) \otimes \mathbb{C}$. One sees immediately that if $g \in \mathrm{Aut}(\mathbb{C}/\mathbb{Q})$ acts on S via the canonical homomorphism $\mathrm{Aut}(\mathbb{C}/\mathbb{Q}) \to \mathrm{Aut}(K/\mathbb{Q}) = G$, then $gH_\chi = H_{g\chi}$. In particular, if $-$ or c denotes complex conjugation, $\overline{H}_\chi = H_{\overline{\chi}}$, and hence $\{\Sigma, \overline{\Sigma}\}$ is a partition of S.

Since $H^i(X) \cong \Lambda^i H^1(X)$, it is clear that $H^i(X)$ breaks up naturally into a direct sum $\oplus H_T$, indexed by the family $\binom{S}{i}$ of all subsets T of S with $\mathrm{card}(T) = i$. Moreover, $H_T \subseteq F^k H^i(X)$ iff $\mathrm{card}(T \cap \Sigma) \geq k$. Using this it is easy to calculate N^r:

(2.16) Lemma: $N^r H^i = \oplus\{H_T : \underline{\mathrm{card}}(gT \cap \Sigma) \geq r \text{ for all } g \in G\}$.

Proof: Let M denote the sum on the right. It is clear that $M \subseteq F^r_{Hodge}$, that M is $\mathrm{Aut}(\mathbb{C}/\mathbb{Q})$-invariant and hence is defined over $\mathbb{Q}$, and that M is the largest such subspace. Let us also check that it is a Hodge structure, i.e. that $M = F^p M + \overline{F^{i-p+1} M}$ for every p. If H_T occurs in M, and if $\mathrm{card}(T \cap \Sigma) \geq p$, then $H_T \subseteq F^p$, so there is nothing to prove. If $\mathrm{card}(T \cap \Sigma) < p$, then $\mathrm{card}(T \cap \overline{\Sigma}) \geq i-p+1$, so in fact $H_T \subseteq \overline{F^{i-p+1}}$. □

If s is a closed point of Spec R, the next lemma tells us how to compute $F^k_{con} H^i_{DR}(X(s)/k(s))$. Suppose we have enlarged R so that E acts on $C^r H^i \otimes K$, and so that $H^i_{DR}(X/R) = \oplus H_T(X/R)$, where $H_T(X/R) =: H_T \cap H^i_{DR}(X/R)$. We let $F_s \in G = \mathrm{Gal}(K/\mathbb{Q})$ denote the geometric Frobenius element, i.e. the inverse of the element of G stabilizing the closed point s and acting as $a \to a^p$ on $k(s)$.

(2.17) <u>Lemma</u>: $H_T(s) \subseteq F^k_{con}(s)$ <u>iff</u> $\mathrm{card}(F_s T \cap \overline{\Sigma}) \geq k$.

<u>Proof</u>: It suffices to prove this when $i = 1$. Now in this case, F_{abs} defines a p-linear isomorphism: $F^*: H^1(X(s),\mathcal{O}_{S(s)}) \to F^1_{con}(s)$, compatible with the action of some order $E' \subseteq E$. For $\chi \in S$, H_χ projects to a nonzero subspace of $H^1(X(s),\mathcal{O}_{X(s)})$ iff $\chi \in \overline{\Sigma}$, and if $\omega \in H_\chi$ and $e \in E'$, then $eF^*(\omega) = F^*(e\omega) = F^*(\chi(e)\omega) = \chi(e)^p F^*(\omega) = F_s^{-1}\chi(e)F^*(\omega)$. Thus, $F^* H_\chi = H_{F_s^{-1}\chi}$, and the lemma follows.□

To prove the theorem, suppose that $H_T \subseteq C^r H^i \otimes K$. Then for every s, $H_T(s) \subseteq F^r_{con}(s)$, so $\mathrm{card}(F_s T \cap \overline{\Sigma}) \geq r$. The Chebotaroff density theorem tells us that every element g of G is equal to some F_s, so that in fact $\mathrm{card}(gT \cap \overline{\Sigma}) \geq r$ for every g. Replacing g by cg, this condition becomes $\mathrm{card}(gT \cap \Sigma) \geq r$ for every g, and we see that $H_T \subseteq N^r H^i$. Conversely, if $H_T \subseteq N^r H^i$, it is clear that $H_T(s) \subseteq F^r_{con}(s)$ for all s, so the theorem is obvious.□

3. Discriminants of F-Crystals

Let k be an algebraically closed field of characteristic p and let (H,Φ) be a nondegenerate F-crystal on k, as defined for instance in [10]. Let $(\ ,\ ): H \times H \to W$ be a nondegenerate symmetric bilinear form, satisfying the comptability: $(\Phi x,\Phi y) = p^m F(x,y)$ for some nonnegative integer m. We wish to define the discriminant "disc$(H,\Phi,(\ ,\ ))$," a member of $\mathbb{Z}_p$ modulo the action of $\mathbb{Z}_p^{*2}$.

The special case in which (H,Φ) is the twist of a unit root crystal was already treated in [18]. In this case, there is a nonnegative integer r so that Φ is p^r times a bijection, and $T =: \{x: \Phi(x) = p^r x\}$ is a $\mathbb{Z}_p$-form of H, i.e., the natural map: $(T \times_{\mathbb{Z}_p} W, \mathrm{id} \otimes F) \to (H,\Phi)$ is an isomorphism. It is easy to see that necessarily $r = \frac{1}{2}m$ and that $(\ ,\ )$ induces a symmetric bilinear form: $T \times T \to \mathbb{Z}_p$, whose discriminant is taken to be the discriminant of H.

Before passing to the general case, recall that if T has rank n, its discriminant is computed as follows: $\Lambda^n T$ is a free $\mathbb{Z}_p$-module of rank one, and it inherits a canonical symmetric bilinear form determined by $(x_1 \wedge \ldots x_n, y_1 \wedge \ldots y_n) = \Sigma\, \mathrm{sgn}(\sigma) \prod_i (x_i, y_{\sigma(i)})$, the sum being taken over all elements of the permutation group S_n. Choose a basis e of $\Lambda^n T$; the image of (e,e) in $\mathbb{Z}_p/\mathbb{Z}_p^{*2}$ does not depend on the choice of e, and this image is the discriminant.

Since an F-crystal of rank one is always a twist of a unit root crystal, this suggests the following definition in the general case.

(3.1) Definition: The discriminant "disc$(H,\Phi,(\ ,\ ))$" $\in \mathbb{Z}_p/\mathbb{Z}_p^{*2}$ is the discriminant of the maximal exterior power $(\Lambda^n H,\Phi,(\ ,\ ))$ of H. Explicitly, it is the image of $\mathbb{Z}_p/\mathbb{Z}_p^{*2}$ of (e,e), where e is a basis of $\Lambda^n H$ satisfying $\Phi(e) = p^{mn/2} e$.

It is apparent that the above definition agrees with the old one in the case of twists of unit root crystals. It is also apparent that the image disc_0 of disc in $\mathbb{Q}_p/\mathbb{Q}_p^{*2}$ depends only on the isogeny class of $(H,\Phi,(\ ,\ ))$. Moreover, the p-adic ordinal of disc does not depend on Φ; we shall therefore be mostly interested in "$\mathrm{disco}(H,\Phi,(\ ,\ ))$" $=: (\mathrm{disc})p^{-\mathrm{ord}(\mathrm{disc})} \in \mathbb{Z}_p^*/\mathbb{Z}_p^{*2}$. If $p \neq 2$, this last group is isomorphic to $\mathbb{F}_p^*/\mathbb{F}_p^{*2}$; we denote the Legendre symbol of the image of disco by $(\frac{H}{p})$.

The next result shows that disco depends in essence only on the part of H with slope $d =: m/2$. It seems reasonable to call the integer m the "weight" of the form $(\ ,\ )$.

(3.2) <u>Proposition</u>: <u>The restriction of</u> $(\ ,\ )$ <u>to the subcrystal</u> H_d consisting of the part of slope d is nondegenerate, and $\mathrm{disco}(H) = \mathrm{disco}(H_d) = {}^{(1/2)\,\mathrm{rank}(H/H_d)} \in \mathbb{Z}_p^*/\mathbb{Z}_p^{*2}$.

<u>Proof</u>: Suppose all the slopes of H are in $\mathbb{Z}[1/n]$, and let $\mathbb{Q}_p(\pi) =: \mathbb{Q}_p[X]/(X^n-p)$, with π the image of X. Define Φ on $H \otimes \mathbb{Q}_p(\pi)$ to be $\Phi \otimes \mathrm{id}$, and let $T_\lambda =: \{x \in H \otimes \mathbb{Q}_p(\pi):$ $\Phi(x) = \pi^{n\lambda}x\}$ for each $\lambda \in \mathbb{Z}[1/n]$. Each T_λ is a $\mathbb{Q}_p(\pi)$-vector space, and there is a natural isomorphism:

$$\bigoplus_\lambda T_\lambda \otimes_{\mathbb{Q}_p(\pi)} (W(\pi) \otimes \mathbb{Q}_p) \xrightarrow{\sim} H \otimes_{\mathbb{Z}_p} \mathbb{Q}_p(\pi).$$

It is clear that if $x \in T_\lambda$ and $y \in T_\mu$ with $(x,y) \neq 0$, then necessarily $\lambda + \mu = 2d$. This implies that there is an orthogonal direct sum decomposition: $H = H_d \oplus H'$, where H' is the part of slope $\neq d$, and where we have tensored with $\mathbb{Q}_p(\pi)$. Moreover, T' admits a hyperbolic decomposition $T' = (\bigoplus_{\lambda<d} T_\lambda) \oplus (\bigoplus_{\mu>d} T_\mu)$, and hence has a basis $\{x_i,y_i\}$ with $(x_i,y_i) = 1$ and all other products zero. If $e' = (x_1 \wedge y_1 \wedge x_2 \ \ldots\ x_{n'} \wedge y_{n'})$, one sees immediately that $(e',e') = (-1)^{n'}$. Since $\mathbb{Z}_p[\pi]^*/\mathbb{Z}_p[\pi]^{*2} \cong \mathbb{Z}_p^*/\mathbb{Z}_p^{*2}$, $\mathrm{disco}(H') = (-1)^{n'}$ in this group, and since $\mathrm{disco}(H) = \mathrm{disco}(H_d)\mathrm{disco}(H')$, the proposition follows. □

(3.3) Remark: If $(H,\Phi,(\ ,\ ))$ is ordinary [16, 3.13] and unimodular and if $p \neq 2$, then $(\frac{H}{p})$ can be computed from purely mod p information. Indeed, in this case H is a sum of twists of unit root crystals, and it is clear from the proposition that it suffices to compute $(\frac{H_d}{p})$. The map $p^{-d}\Phi$ induces a p-linear bijection: $\mathrm{gr}_F^d(H \otimes k) \to \mathrm{gr}_F^d(H \otimes k)$, which is compatible with the induced bilinear form, hence by descent we get an $\mathbb{F}_p$-vector space with a nondegenerate pairing; clearly its discriminant gives $\mathrm{disc}(H_d)$. This shows in particular that if H arises as the crystalline cohomology of a surface with a degenerate Hodge spectral sequence and without torsion in crystalline cohomology, then disco(H) is just the discriminant of the quadratic form induced by cup-product on the part of $H^1(\Omega^1_{X/k})$ fixed by the absolute Cartier operator, i.e., on $H^2_{fl}(X,\mu_p)$.

From now on we assume that p is odd.

(3.4) Formula: If x is any nonzero element of $\Lambda^n(H \otimes \mathbb{Q})$, write $\Phi(x) = p^{mn/2}\lambda x$ and $(x,x) = p^b f$, with $f \in W^*$. Then $\lambda \in W^*$ and $(\frac{H}{p}) = f^{\frac{p-1}{2}}\lambda^{-1}$ mod p.

Proof: It is clear that $\lambda \in W^*$. Moreover, if x is chosen so that $\lambda = 1$, we know that $f \in \mathbb{Z}_p^*$, and its image in $\mathbb{Z}_p^*/\mathbb{Z}_p^{*2}$ is by definition the discriminant of H. Thus, the formula is true in this case, since $f^{\frac{p-1}{2}}$ mod p is the Legendre symbol of f mod p.

If now $x' = ax$, with $a \in W^*$, then $\lambda' = F^*(a)a^{-1}$ and $f' = a^2 f$, so that $(f')^{\frac{p-1}{2}} (\lambda')^{-1} = f^{\frac{p-1}{2}} \lambda^{-1} \bmod p$. Since this quantity is also obviously unchanged if we multiply x by a power of p, the formula holds in general.□

Formula (3.4) allows us to give a useful "geometric" interpretation of the discriminant if the rank n of H is even and if the form (,) is unimodular. Since k is algebraically closed there then exist totally isotropic direct summands K of H of rank $n/2$.

(3.5) <u>Corollary</u>: If K is a totally isotropic direct summand of $H' =: H \otimes \mathbb{Q}$ of rank $n/2$, then $\left(\frac{H}{p}\right) = \left(\frac{-1}{p}\right)^{n/2} (-1)^{\dim(K/K \cap \Phi(K))}$.

<u>Proof</u>: Recall that the set of such K falls into two connected families, corresponding to $\{e \in \Lambda^n H' : (e,e) = (-1)^{n/2}\}$. This correspondence can be described as follows: The filtration $K \subseteq H'$ induces an isomorphism $\Lambda^{n/2} K \otimes \Lambda^{n/2} H'/K \to \Lambda^n H'$, and since K is totally isotropic, (,) induces a perfect pairing $K \times H'/K \to W \otimes \mathbb{Q}$. This induces an isomorphism: $\Lambda^{n/2} K \otimes \Lambda^{n/2} H'/K \xrightarrow{\sim} W \otimes \mathbb{Q}$ and the inverse image of 1 therefore maps to an element e_K of $\Lambda^n H'$.. It can be checked that $(e_K, e_K) = (-1)^{n/2}$ and that if K' is another such subspace, $e_{K'} = (-1)^{\dim K/K \cap K'} e_K$.

The point is that these constructions are compatible with Φ; that is, the following diagram commutes:

$$\begin{array}{ccccc} \Lambda^n H & \leftarrow & \Lambda^{n/2} K \otimes \Lambda^{n/2} H/K & \rightarrow & W \otimes \mathbb{Q} \\ \downarrow \Lambda^n \Phi & & \downarrow & & \downarrow p^{mn/2} F \\ \Lambda^n H & \leftarrow & \Lambda^{n/2} \Phi K \otimes \Lambda^{n/2} H/\Phi K & \rightarrow & W \otimes \mathbb{Q} \end{array}$$

This implies that $p^{-nm/2}\,\overset{n}{(\Lambda\phi)}e_K = e_{\Phi(K)} = (-1)^{\dim K/K\cap\Phi(K)}e_K$, and so the corollary follows from the formula (3.4). □

The next result shows that these invariants are constant in a family.

(3.6) Proposition: Suppose S/k is smooth and $(H, ,(\ ))$ is an F-crystal on (S/W) with a symmetric bilinear form as above. If S is connected, then the value of $\mathrm{disc}(H(s),\phi(s), (\ ,\ ))$ is independent of the closed point s.

Proof: We may assume that H has rank one, and, after twisting, that H is a unit root crystal. We may also assume that S is affine and choose a lifting (T,F_T) of (S,F_S). Let x be a basis of H_T; then $\Phi(x) = \lambda x$, with $\lambda \in \mathcal{O}_T^*$. Necessarily $d = 1$, so $F_T^*(x,x) = (\Phi x,\Phi x) = \lambda^2(x,x)$. This implies that $(x,x) = p^b$ times a unit f in $\mathcal{O}_T$, for some constant b, which is of course the p-adic ordinal of $\mathrm{disc}(H(s))$ for every s. On the other hand, $(\frac{H(s)}{p})$ is $f(s)^{\frac{p-1}{2}}\lambda(s)^{-1}$ mod p. But the image δ of $f^{\frac{p-1}{2}}\lambda^{-1}$ in $\mathcal{O}_S$ satisfies $\delta^2 = 1$, and since S is connected, $\delta = \pm 1$. □

It is also interesting to consider crystals defined over a non-algebraically closed field k. In this case there is another, standard, discriminant attached to the quadratic form, lying in k^*/k^{*2}. Here is a useful formula relating the two discriminants when $k = \mathbf{F}_q$.

(3.7) Proposition: Suppose $(H,\Phi, (\ ,\ ))$ descends to an F-crystal $(H_0,\Phi, (\ ,\ ))$ over $\mathbf{F}_q$. Let $(\frac{H_0}{q})$ denote the Legendre symbol of the discriminant of $(H_0, (\ ,\ ))$ in $\mathbf{F}_q^*/\mathbf{F}_q^{*2} \cong \mu_2$, and let δ be the determinant of the W_q-linear map Φ^d, where $q = p^d$. Then:

$$\left(\frac{H_0}{q}\right) = \delta\left(\frac{H}{p}\right)^d .$$

Proof: We may assume that H is a unit root crystal of rank one. If e_0 is a basis for H_0, then $(e_0,e_0) = p^b f$, with $f \in \mathbf{Z}_q^*$, and $\left(\frac{H_0}{q}\right) = f^{(q-1)/2} \bmod p$. On the other hand, if $\Phi(e_0) = \lambda e_0$ and if $q = p^d$, then $\Phi^d(e_0) = \lambda^p \ldots \lambda^{p^{d-1}} e_0$, so $\delta = \lambda^{1+p+\ldots+p^{d-1}} \bmod p$. Finally, $(\frac{H}{p})\lambda = f^{(p-1)/2}$, and since $(\frac{H}{p})^p = (\frac{H}{p})$, we see that $(\frac{H}{p})\lambda^{p^i} = f^{(p^{i+1}-p^i)/2}$ for all i. Multiplying these together for $i = 0, \ldots p^{d-1}$, we find $(\frac{H}{p})^d = f^{(p^d-1)/2} = f^{(q-1)/2}$.

We can generalize the above formula as follows: Suppose S_0 is a smooth $\mathbf{F}_q$-scheme, and $(H, \ , (\ , \ ))$ is an F-crystal with quadratic form on $(S_0/W_q)_{cris}$. Replace H_0 by a twist of its maximal exterior power Λ_0, a unit root crystal of rank one. As is well known (c.f. for instance [10]), (Λ_0,Φ) defines a $\mathbf{Z}_p^*$-valued character δ of $\pi_1(S_0,\bar{s})$. The quadratic form defines an isomorphism of F-crystals $(\Lambda_0,\Phi) \otimes (\Lambda_0,\Phi) \xrightarrow{\sim} (\mathcal{O}_{X/W_q},F)$, so δ is of order 2 and hence has values in $\mu_2 \subseteq \mathbf{Z}_p^*$ and is determined by its reduction mod p. On the other hand, the quadratic form on Λ_0 also defines a double covering $\tilde{S}_0$ of S_0, with local equation $t^2 = (e_0,e_0)$ for any basis e_0 of Λ_{0S_0}; on $\tilde{S}_0$ we have sections $\tilde{e}_0 = \pm t^{-1}e_0$ with $(\tilde{e}_0,\tilde{e}_0) = 1$. This defines another character ε of $\pi_1(S_0,\bar{s})$ with values in μ_2, determined by $g\tilde{e}_0 = \varepsilon(g)\tilde{e}_0$. Assume that $S_0/\mathbf{F}_q$ is absolulteIy irreducible, so that $(\frac{H(s)}{p})$ is constant, by (3.6). Recall that there is a canonical homomorphism $\deg: \pi_1(S_0,\bar{s}) \to \hat{\mathbf{Z}}$ (which factors through $\mathbf{Z}$).

(3.8) <u>Proposition</u>: With the above notations and hypotheses,

$$\varepsilon(g) = \left(\frac{H}{p}\right)^{\deg(g)} \delta(g) \quad \text{for any} \quad g \in \pi_1(S_0,\bar{s}).$$

<u>Proof</u>: If F_s is a Frobenius element, it is easy to see that $\delta(F_s)$ is the determinant of $\Phi(s)^{\deg(s)}$ and that $\varepsilon(F_s) = \left(\frac{H_0(s)}{q}\right)$. Thus (3.7) shows that (3.8) is true for F_s and hence for any g by Chebotaroff. It is however more revealing to prove (3.8) directly (and simultaneously give a less computational proof of (3.7)). Recall that δ is defined as follows: For a suitable étale cover S' of S, we can find a basis e' of $\Lambda_{S'}$ with $\Phi e' = e'$; then $ge' = \delta(g)e'$ for $g \in \pi_1$. If we increase S' a bit, we can find a u such that $(ue',ue') = 1$, and then $g(ue') = \varepsilon(g)ue'$ and $\varepsilon(g) = u^g u^{-1}\delta(g)$. Now since $\Phi(e',e') = (e',e')$, (e',e') necessarily belongs to $\mathbf{F}_p^*$, and by definition $(\frac{H}{p}) = (\frac{(e',e')}{p})$. Since u^{-1} is a square root of (e',e'), $u^g u^{-1} = (\frac{H}{p})^{\deg(g)}$. □

Now let X_0 be a smooth proper $\mathbf{F}_q$-scheme, and suppose for simplicity that its crystalline cohomology is torsion free. If X_0 has even dimension N, $H^N_{cris}(X_0/W_q)$ comes equipped with a unimodular pairing compatible with the action Φ of absolute Frobenius: $(\Phi x,\Phi y) = p^N(x,y)$. We denote by $(\frac{X}{p})$ the Legendre symbol of the reduction mod p of the corresponding crystalline discriminant (which depends only on the pull-back X of X_0 to the algebraic closure of $\mathbf{F}_q$). Set $d = \log_p(q)$.

(3.9) <u>Proposition</u>: <u>If</u> X_0 <u>is as above and if</u> $(\frac{X_0}{q})$ <u>denotes the Legendre symbol of the discriminant of the cup-product pairing on</u> $H^N_{DR}(X_0/\mathbf{F}_q)$, <u>one has</u>:

$$\left(\frac{X_0}{q}\right) = \left(\frac{X}{p}\right)^d \delta ,$$

where δ is the determinant of $(F_{abs})^d$ on $H^N(X_{et}, \mathbb{Q}_\ell)$.

Proof: It is known that F^d acting on $H^N(X_{et}, \mathbb{Q}_\ell)$ has the same characteristic polynomial as $(F_{abs})^d$ acting on $H^N_{cris}(X_0/W_q)$, so this follows from (3.7).□

(3.10) Corollary: Let S be a scheme of finite type over $\mathbb{Z}[1/2]$ and absolutely irreducible, and let $f: X \to S$ be a smooth proper family without torsion in crystalline cohomology. Then there exists a character $\varepsilon: \pi_1(S, \bar{s}) \to \mu_2$ such that $\varepsilon(F_s) = (\frac{X(s)}{p(s)})^{\deg(s)}$ for every closed point s of S.

Proof: Since the form on $H^N_{DR}(X/S)$ defines a perfect pairing, we get an unramified double covering $\tilde{S} \to S$, Galois with group μ_2. It is clear that if s is a closed point, the Frobenius element F_s in this group is the Legendre symbol $\left(\frac{H_{DR}(X(s)/k(s))}{Nm(s)}\right)$. On the other hand, the maximal exterior power of the corresponding étale cohomology group defines another character of $\pi_1(S, \bar{s})$, with values in $\mu_2 \subseteq \mathbb{Q}_\ell^*$. The previous proposition tells us that their product ε is $\left(\frac{X(s)}{p(s)}\right)^{\deg(s)}$ at F_s. □

(3.11) Conjecture: If X is a smooth proper scheme of even dimension N, let $\gamma(X) = \beta_0 + \beta_2 + \ldots \beta_{N-2} + \frac{1}{2}(\beta_N - (-1)^{N/2}\beta_N)$. Then $(\frac{X}{p}) = (\frac{-1}{p})^{\gamma(X)}$, where $\beta_i =: \dim H^i(X, \mathbb{Q}_\ell)$.

Let me explain the source of this conjecture. Suppose that X can be lifted to a smooth scheme Y over W, choose an embedding σ of $K = W \otimes \mathbb{Q}$ in $\mathbb{C}$, and let e be a basis for $\Lambda^{\beta_N} H^N(Y_\sigma(\mathbb{C}), \mathbb{Z}) \otimes \mathbb{Z}(\beta N/2)$. Clearly e is a Hodge cycle. If it is in fact algebraic, it should lie in $H^N_{DR}(Y/W)$ and be fixed by Φ, so by definition, $(\frac{X}{p}) = (\frac{(e,e)}{p})$. Since the pairings on De Rham and integral cohomologies are compatible, we can compute (e,e) in integral cohomology, using the Hodge index theorem to determine the sign. Unless I have made an error in applying the formula of [21], the sign is as shown.

(3.12) Remark: If X is smooth and proper over $\mathbb{Z}[1/2M]$, then by (3.10) we see that the function $p \to (\frac{X(p)}{p})$ is given by a μ_2-valued character of Galois, unramified outside 2M. Of course the same is true for its conjectured value $(\frac{-1}{p})^{\gamma(X)}$, and hence if the conjecture holds for a set of primes of density greater than 1/2, it holds for all primes not dividing 2M.

(3.13) Remark: Still assuming that X is smooth and proper over $\mathbb{Z}(1/2\ M]$, let us remark that the truth of (3.11) for all $p \nmid 2M$ is equivalent to $\Lambda =: \overset{\beta}{\Lambda} H^N(X)(N\beta/2)$ being spanned by absolutely Hodge cycles, in the sense of Deligne [6]. To recall what this means, we choose an algebraic closure $\overline{\mathbb{Q}}$ of $\mathbb{Q}$ and an embedding $\overline{\sigma}\colon \overline{\mathbb{Q}} \hookrightarrow \mathbb{C}$. Then there are natural embeddings of $\Lambda_{\mathbb{Q}} =: \Lambda_{\mathbb{Z}} \otimes \mathbb{Q}$ in $\Lambda_\ell =: \overset{\beta}{\Lambda} H^N_{\text{ét}}(X_{\overline{\mathbb{Q}}}, \mathbb{Q}_\ell)(N\beta/2)$ and in $\Lambda_{DR} \otimes_{\mathbb{Q}} \mathbb{C}$, where $\Lambda_{DR} =: \overset{\beta}{\Lambda} H^N_{DR}(X/\mathbb{Q})\ (N\beta/2)$. An element of the product $\Lambda_{\mathbb{Q}} =: \pi_\ell \Lambda_\ell \times \Lambda_{DR} \otimes \mathbb{C}$ is said to be "absolutely Hodge" iff it lies in the image of the map $\Lambda_{\mathbb{Q}} \to \pi_\ell \Lambda_\ell \times \Lambda_{DR} \otimes \mathbb{C}$ for every $\overline{\sigma}$ and its projection to Λ_{DR} lies in F^0_{Hodge}. In this context, the latter condition is trivial, so we have only to unravel the former. Let e be a basis vector for $\Lambda_{\mathbb{Q}}$ with $(e,e) = (-1)^\gamma$, and let $(e_{\overline{\sigma}\ell}, e_{\overline{\sigma}DR})$ denote its image in $\pi_\ell \Lambda_\ell \times \Lambda_{DR} \otimes \mathbb{C}$. Since the natural maps above are compatible with the intersection forms, $(e_{\overline{\sigma}DR}, e_{\overline{\sigma}DR}) = (-1)^\gamma$, and hence if e_{DR} is a basis for Λ_{DR}, $e_{\overline{\sigma}DR} = ae_{DR}$ for some $a \in \overline{\mathbb{Q}}$. Thus it is clear that $\text{Aut}(\mathbb{C}/\overline{\sigma}(\overline{\mathbb{Q}}))$ acts trivially on $(e_{\overline{\sigma}\ell}, e_{\overline{\sigma}DR})$.

If now $g \in \mathrm{Gal}(\overline{\mathbb{Q}}/\mathbb{Q})$ and $\overline{\sigma}' = \overline{\sigma}\circ g$, $(e_{\overline{\sigma}'\ell}, e_{\overline{\sigma}'\ell}) = (-1)^{\gamma} = (e_{\overline{\sigma}'DR}, e_{\overline{\sigma}'DR})$, and hence we have $e_{\overline{\sigma}'\ell} = \chi_{\ell}(g)e_{\overline{\ell}}$ and $e_{\overline{\sigma}'DR} = \chi_{DR}(g)e_{\overline{\sigma}DR}$, where the χ_{ℓ}'s and χ_{DR} are μ_2-valued characters of $\mathrm{Gal}(\overline{\mathbb{Q}}/\mathbb{Q})$. It is clear that we must prove that they are all equal. Obviously χ_{ℓ} is just the determinant of the standard representation of $\mathrm{Gal}(\overline{\mathbb{Q}}/\mathbb{Q})$ on $H^N_{\acute{e}t}(X_{\mathbb{Q}}, \mathbb{Q}_{\ell})(1/2\, N)$, and χ_{DR} is just the representation associated to the quadratic form on Λ_{DR} multiplied by $(-1)^{\gamma}$, i.e. to the equation $a^2(e_{DR}, e_{DR}) = (-1)^{\gamma} = (e,e)$. Let λ_{DR} be the character associated to the form on Λ_{DR} itself and let $(\frac{-1}{})$ denote the character extending the Legendre symbol in the obvious way; evidently $\chi_{DR} = \lambda_{DR}(\frac{-1}{})^{\gamma}$. Thus, the absolute Hodgeness of e comes down to the equality:

$$\chi_{\ell}\lambda_{DR} = (\frac{-1}{})^{\gamma}.$$

But we saw in the proof of (3.10) that $(\chi_{\ell}\lambda_{DR})(F_p) = (\frac{X(p)}{p})$ for any p not dividing $2M$, and so this equality is equivalent to $(\frac{X(p)}{p}) = (\frac{-1}{p})^{\gamma}$ for all such p. □

Let us now discuss the case of surfaces in somewhat more detail. First of all, if X/k is such that $H^2_{cris}(X/W) \otimes K$ is spanned by algebraic cycles ($\rho = \beta_2$), then the conjecture follows easily from the algebraic Hodge index theorem. (The argument in [18, 7.1] for K3 surfaces applies in general.) Secondly, if X' is obtained from X by blowing up a point, then the conjectures for X and for X' are equivalent, because there is an orthogonal direct sum decomposition $H^2_{cris}(X'/W) = H^2_{cris}(X/W) \oplus W(-1)$ and $W(-1)$ has a basis vector e which is fixed by Φ and satisfies $(e,e) = -1$. Moreover, it follows from (3.6) that the conjecture is invariant under deformation. Next, observe that the conjecture is true for abelian surfaces. For example, one can use (3.5), since $\beta_2 = 6$ is even: choose a three-dimensional subspace K of $H^1_{cris}(X/W)$; then $\Lambda^2 K$

is a totally isotropic three-dimensional subspace of $H^2_{cris}(X/W)$. Evidently $\Lambda^2 K \cap \Lambda^2 \Phi(K)$ has dimension 1 (if $K \neq \Phi(K)$) or 3 (if $K = \Phi(K)$), hence $(\frac{H}{p}) = (\frac{-1}{p})^3 = (\frac{-1}{p})^7$, as predicted by the conjecture. The conjecture is true for Kummer surfaces, hence for any K3 surface which is a deformation (in characteristic p) of a Kummer surface.

(3.14) <u>Theorem</u>: <u>Conjecture</u> (3.11) <u>is true for any nonsingular surface in</u> $\mathbb{P}^3$.

<u>Proof</u>: We begin by verifying the conjecture for Fermat surfaces. Using (3.13), we could appeal to Deligne's very deep result stating that any Hodge tensor on a Fermat surface is absolutely Hodge, but it seems more reasonable to calculate directly. Thus, let X_m denote the surface in $\mathbb{P}^3$ with equation $X_0^m + \ldots + X_3^m = 0$, in characteristic p not dividing m, and let the group $\Gamma = \bigoplus_0^3 \mu_m/\text{diag}$ act on $X_m \subseteq \mathbb{P}^3$ in the usual way. For each character $\chi \in \Gamma_m^* \subseteq \bigoplus_0^3 \mathbb{Z}/m\mathbb{Z}$, let H_χ denote the subspace of the cohomology H of X_m on which Γ_m operates by χ, and recall that the primitive cohomology H_m of X_m admits a decomposition $H_m = \bigoplus H_\chi$, where H_χ has dimension at most one and is nonzero iff no $\chi_i = 0$. In particular, H_m has rank $(m-1)((m-1) + (m-2)^2)$. Since the action of Γ_m preserves the intersection form, we see that H_χ is orthogonal to $H_{\chi'}$ unless $\chi + \chi' = 0$, and since Φ is Frobenius-linear and commutes with the action of Γ_m, $\Phi H_\chi = H_{p\chi}$.

Let us first deal with the case in which m is even, say $m = 2m'$. If χ_0 is the character (m',m',m',m'), I claim that H_{χ_0} has a basis e_0 such that $\Phi(e_0) = pe_0$ and $(e_0,e_{\bar{\theta}}) = -m(m')^2$. For example, X_2 is a quadric, $\rho = \beta_2 = 2$, and $NS(X_2) \otimes \mathbb{Q}$ has a basis (ξ,η) with $\xi^2 = 2$, $(\xi,\eta) = 0$, and $\eta^2 = -2$, where ξ is the class of a hyperplane. It is clear that η induces a basis for the primitive cohomology H_2, on which Γ_2 acts by (1,1,1,1). To analyze the general case, let $f: X_m \to X_2$ be the map sending $(X_0:\ldots X_3)$ to $(X_0^{m'}:\ldots X_3^{m'})$ and let $g: \Gamma_m \to \Gamma_2$ be defined similarly; then $f(\gamma x) = g(\gamma)\, f(x)$. It follows that $f^*H_2 = H_{\chi_0}$, and $f^*(\eta)$ yields the desired basis e_0. Thus, if H'_m is the orthogonal complement of H_{χ_0} in H_m, we evidently have $(\frac{H_m}{p}) = (\frac{H'_m}{p})(\frac{-m}{p})$. Since $H = H_m \oplus H_0$ (orthogonally), where H_0 is the span of a hyperplane class, and since $(\frac{H_0}{p}) = (\frac{m}{p})$, we have finally the equation: $(\frac{X_m}{p}) = (\frac{H'_m}{p})(\frac{-1}{p})$.

The rank $(m-1)((m-1)+(m-2)^2)-1$ of H'_m is divisible by 4, so that we can apply (3.5). Choose a set S' of characters occurring in H'_m such that $S'\amalg -S'$ is the set of all such characters. Then $H_{S'} =: \oplus\{H_\chi : \chi \in S'\}$ is a maximal isotropic subspace of H'_m, and so (3.5) tells us that: $(\frac{H'_m}{p}) = (-1)^{\dim H_{S'}/H_{S'}\cap\Phi H_{S'}} = (-1)^{\mathrm{card}(S'-pS')}$. The same argument used to prove (3.5) shows that this sign is the determinant of any linear map sending each H_χ to $H_{p\chi}$, and hence is just the sign of the permutation "multiplication by p" on the set of all such χ. The conjecture asserts that $(\frac{X_m}{p}) = (\frac{-1}{p})$, i.e. that the sign $(\frac{H'_m}{p})$ of the permutation is +1. Notice that we can reinsert the character χ_0 into our set without changing the sign.

Before we try to calculate this sign, let us look at the case of odd m. Then $(\frac{X_m}{p}) = (\frac{H_m}{p})(\frac{m}{p})$ and H_m has even rank $2r$, with $r \equiv (m-1)/2 \bmod 2$. Using (3.5) in the same way as above to compute $(\frac{H_m}{p})$, we find that $(\frac{H_m}{p}) = (\frac{-1}{p})^{(m-1)/2}\,\mathrm{sgn}(\cdot p)$. The conjecture predicts that $(\frac{X_m}{p}) = +1$ in this case, and so we are reduced to proving:

<u>Claim</u>: If m is even, multiplication by p induces an even permutation on the set of characters occuring in H_m. If m is odd,

the sign of this permutation is $(\frac{m}{p})(\frac{-1}{p})^{(m-1)/2}$

To prove this, note first that if we allow one or two of the χ_i's to vanish, the sign of the permutation on the correspondence enlarged set is the same, since there are an even number of symmetrical ways to do this. If three of the χ_i's vanish they all do, and adding this character also leaves the sign unchanged. Thus we are reduced to computing the sign of p on $\Gamma_m^* \cong \bigoplus_1^3 \mathbb{Z}/m\mathbb{Z}$.

Observe that if A and B are finite sets, the map $\mathrm{Aut}(A) \times \mathrm{Aut}(B) \to \mathrm{Aut}(A \times B) \xrightarrow{\mathrm{sgn}} \mu_2$ takes (σ,τ) to $\mathrm{sgn}(\sigma)^{\mathrm{card}(B)}\mathrm{sgn}(\tau)^{\mathrm{card}(A)}$. This immediately implies that if m is even, $\mathrm{sgn}(\cdot p)$ is $+1$, proving the conjecture in this case. If m is odd, we see that the sign is the same as the sign $\varepsilon_p(m)$ on $\mathbb{Z}/m\mathbb{Z}$.

I claim that ε_p is multiplicative: $\varepsilon_p(m_1m_2) = \varepsilon_p(m_1)\varepsilon_p(m_2)$. If m_1 and m_2 are relatively prime, we have a p-stable product decomposition: $\mathbb{Z}/m_1m_2\mathbb{Z} \cong \mathbb{Z}/m_1\mathbb{Z} \times \mathbb{Z}/m_2\mathbb{Z}$, so this is clear. If $m \neq p$ is an odd prime, we have a disjoint union: $\mathbb{Z}/m^e\mathbb{Z} = \mathbb{Z}/m^e\mathbb{Z}^* \amalg (m)\mathbb{Z}/m^e\mathbb{Z}$, stable by p, and hence $\varepsilon_p(m^e) = \varepsilon_p(m^{e-1})$ times the sign of the permutation p on $(\mathbb{Z}/m^e\mathbb{Z})^*$. If (p_1,p_2) is the image of the class of p under the group isomorphism: $(\mathbb{Z}/m^e\mathbb{Z})^* \cong \mathbb{Z}/m^{e-1} \times (\mathbb{Z}/m\mathbb{Z})^*$, the latter sign is $\mathrm{sgn}(\cdot p_1)^{m-1}\mathrm{sgn}(\cdot p_2)^{m^{e-1}} = \varepsilon_p(m)$. Thus we find that $\varepsilon_p(m^e) = \varepsilon_p(m^{e-1})\varepsilon_p(m)$, proving the claim.

Notice that if m and n are odd, $(mn-1)/2$ has the same parity as $(m-1)/2 + (n-1)/2$, so the function $m \mapsto (\frac{m}{p})(\frac{-1}{p})^{(m-1)/2}$ is also multiplicative in m. Thus, we are reduced to proving that $\varepsilon_p(m) = (\frac{m}{p})(\frac{-1}{p})^{(m-1)/2}$ when m is an odd prime. But then $\varepsilon_p(m)$ is $(\frac{p}{m})$, so our formula reduces to: $(\frac{p}{m}) = (\frac{m}{p})(\frac{-1}{p})^{(m-1)/2}$ -- which is the quadratic reciprocity law!

This proves the conjecture for one nonsingular hypersurface of degree m in characteristic p not dividing m, and hence by (3.6) for all such hypersurfaces. To finish the proof of the theorem, we still have to verify it for one smooth hypersurface of degree m when p divides m. Choose a nonsingular hypersurface of degree m defined over $\mathbf{F}_q$ for some $q = p^d$ with d odd -- clearly this is possible. Let $\mathcal{O}$ be the ring of integers in the cyclotomic field K of $(q-1)^{st}$ roots of unity. This ring admits a unique maximal ideal m lying over p, $\mathcal{O}/m = \mathbf{F}_q$, and $\mathcal{O}$ is unramified over $\mathbf{Z}$ at m. Choose liftings to $\mathcal{O}$ of the coefficients defining our hypersurface. Then for a suitable N, we get a smooth hypersurface X over $\mathcal{O}[1/N]$ which specializes at m to the given one. By (3.10), there is a character $\varepsilon\colon \mathrm{Gal}(\overline{\mathbf{Q}}/K) \to \mu_2$ such that $\varepsilon(F_s) = \left(\frac{X(s)}{p(s)}\right)^{\deg(s)}$ for every closed point s of Spec $\mathcal{O}[1/N]$. We know that if p(s) doesn't divide mN, then in fact $\varepsilon(s) = \left(\frac{-1}{p(s)}\right)^{\gamma\deg(s)}$. It follows that these two characters agree on all of $\mathrm{Gal}(\overline{\mathbf{Q}}/K)$, and since $\deg(m)$ is odd, we have $\left(\frac{X(m)}{p}\right) = \left(\frac{-1}{p}\right)^{\gamma}$ also. □

We close this section with one more result illustrating the close link between Conjecture (3.11) and the theory of absolute Hodge cycles. Let k be an algebraically closed field of characteristic p > 2, let W be its Witt ring, and let X be a smooth proper surface over W with special fiber Y and second Betti number b.

(3.15) <u>Proposition</u>: <u>Assume that the absolute Frobenius endomorphism</u> F_{abs} <u>of</u> Y <u>lifts to an endomorphism</u> F_X (<u>covering the Frobenius endomorphism</u> F_W <u>of</u> W). <u>Assume also that the highest exterior power</u> $\overset{b}{\Lambda} H_{DR}(X/W)(b)$ <u>is spanned by absolutely Hodge cycles (in the sense of Deligne). Then the crystalline discriminant</u>

$(\frac{X}{p}) = (\frac{-1}{p})^{b-1}$.

<u>Proof</u>: Since k is algebraically closed and the pairing on $H^2_{cris}(Y/W)$ is perfect, we can find an element e_{cris} of $\Lambda_{cris} =: \overset{b}{\wedge} H^2_{cris}(Y/W)(b)$ such that $(e_{cris}, e_{cris}) = (-1)^{b-1}$. We have to show that $F^*_{abs}(e_{cris}) = e_{cris}$ (and not $-e_{cris}$, which of course is the only other possibility). Let e_{DR} be the image of e_{cris} via the canonical isomorphism $\Lambda_{cris} \cong \Lambda_{DR}$, where of course $\Lambda_{DR} =: \overset{b}{\wedge} H^2_{DR}(X/W)(b)$. Then we have to show that $F^*_X(e_{DR}) = e_{DR}$.

Now choose an embedding s of W in $\mathbb{C}$, and let Λ_s be $\overset{b}{\wedge} H^2(X_s, \mathbb{Z})(b)$. It follows from the Hodge index theorem that both of the two basis elements $\pm e_s$ for Λ_s satisfy $(e_s, e_s) = (-1)^{b-1}$. Therefore, the canonical map $\Lambda_s \to \Lambda_{DR} \otimes_s \mathbb{C}$ factors through Λ_{DR}, and we can assume that e_s maps to e_{DR}.

We also have a canonical map $\Lambda_s \to \Lambda_\ell$, where $\Lambda_\ell =: \overset{b}{\wedge} H^2(X_{et}, \mathbb{Z}_\ell)(b) \cong \overset{b}{\wedge} H^2(Y_{et}, \mathbb{Z}_\ell)(b)$ and $\ell \neq p$. Denote the image of e_s under this map by e_ℓ. Thus, we can think of e_s as a pair (e_{DR}, e_ℓ) in $\Lambda_{DR} \times \Lambda_\ell$.

Now here is the point. Since F_X is not a W-morphism, it does not induce an endomorphism of the complex analytic space X_s, and in general one can say nothing about $F^*_X(e_{DR}, e_\ell)$. However if we assume that the (Hodge) cycle e_s is absolutely Hodge, then $F^*_X e_s$ is also absolutely Hodge, since this notion is (by definition) independent of the ground field. In particular, $F^*_X(e_{DR}, e_\ell)$ is also rational, and hence in the image of the map $\Lambda_s \to \Lambda_{DR} \times \Lambda_\ell$. Since F^*_X also preserves cup product, we must have $F^*_X(e_{DR}, e_\ell) = \pm(e_{DR}, e_\ell)$.

It is clear that the sign must be plus. Indeed, we can calculate the sign on either factor. (The de Rham factor is what we <u>want</u> to be plus.) But $F^*_X(e_\ell)$ can be computed from the action of F_{abs} on étale cohomology mod p — which of course is trivial! □

4. Absolute Hodge Cycles and Crystalline Cohomology

This section is devoted to a crystalline analogue of Deligne's notion of "absolute Hodge cycles." To explain it, we must introduce some notation. If R is a smooth $\mathbb{Z}$-algebra and X is a smooth proper R-scheme, let $H^i_{DR}(X/R)$ be its ith De Rham cohomology group, a finitely generated R-module endowed with its Hodge filtration $F^{\cdot}$. For convenience we shall always assume that the Hodge cohomology groups $H^q(X,\Omega^p_{X/R})$ are flat R-modules (which in any case becomes true if we localize R a little); then the Hodge to De Rham spectral sequence degenerates at E_1.

If $S = \underline{\text{Spec}}\ R$ and $\sigma \in S(\mathbb{C})$, we obtain by base change a $\mathbb{C}$-scheme X_σ and hence a complex analytic space X^{an}_σ, and we let $H^i_\sigma(X)$ denote $H^i(X^{an},\underline{\mathbb{Q}})$. Recall that there is a canonical isomorphism:

$$\sigma_B : H^i_{DR}(X/R) \otimes \mathbb{C} \to H^i_\sigma(X) \otimes \mathbb{C}.$$

Sometimes we shall also use σ_B to denote the composite:

$$\sigma_B : H^i_{DR}(X/R) \to H^i_{DR}(X/R) \otimes \mathbb{C} \to H^i_\sigma(X) \otimes \mathbb{C}.$$

Now if W is the Witt ring of a perfect field k of characteristic $p > 0$ and if $\sigma \in S(W)$ lies over $\bar{\sigma} \in S(k)$, we obtain by base change a W-scheme X_σ reducing to a k-scheme $X_{\bar{\sigma}}$. Let K be the fraction field of W and let $H^i_{\bar{\sigma}}(X)$ denote $H^i_{cris}(X_{\bar{\sigma}}/W) \otimes K$, which has a natural injective Frobenius-linear endomorphism Φ induced by the absolute Frobenius endomorphism of $X_{\bar{\sigma}}$. Recall that there is a natural isomorphism: $H^i_{DR}(X_\sigma/W) \to H^i_{cris}(X_{\bar{\sigma}}/W)$, and hence we can transport Φ back to $H^i_{DR}(X/R) \otimes_\sigma W \cong H^i_{DR}(X_\sigma/W)$ to obtain a Frobenius-linear endomorphism Φ_σ. (We also use Φ_σ to denote the corresponding action on $H^i_{DR}(X/R) \otimes_\sigma K$.) Or, to emphasize the similarity with the complex case, we let σ_{cris} denote the isomorphism:

$$\sigma_{cris} : H^i_{DR}(X/R) \otimes_\sigma K \xrightarrow{\cong} H^i_{\bar{\sigma}}(X), \quad \text{or the composite}$$

$$H^i_{DR}(X/R) \to H^i_{DR}(X/R) \otimes_\sigma K \xrightarrow{\cong} H^i_{\bar{\sigma}}(X).$$

It will also be important to use tensor products and duals of various cohomology groups. For example, if $i < 0$, $H^{-i}_{DR}(X/R)$ will denote $\mathrm{Hom}(H^i_{DR}(X/R), R)$, with the filtration: $F^{-i}H^{-i} = \mathrm{Ann}(F^{i+1}H^i)$. If $\sigma \in S(W)$, we endow $H^{-i}_{\sigma}(X)$ with the Frobenius-linear endomorphism $\phi \mapsto F_K \circ \phi \circ \Phi_\sigma^{-1}$. (Note that although this map does not leave the lattice $H^{-i}_{cris}(X_{\bar{\sigma}}/W)$ stable, it will do so if we multiply it by p^i, because $p^i H^i_{cris}(X_{\bar{\sigma}}/W) \subseteq \mathrm{Im}\,\Phi_\sigma$ In other words, the "twist" $H^{-i}_{\sigma}(X)(-i) =: H^{-i}_{\sigma}(X) \otimes H^2_{cris}(\mathbb{P}^1)^{\otimes i}$ does correspond to an integral F-crystal.)

More generally, let $X = (X_1, \ldots X_m, i_1, \ldots i_m)$ be a finite collection of R-schemes and indices (positive or negative). Then we can form $H_{DR}(X/R) =: \bigotimes_j H^{i_j}_{DR}(X_j/R)$, endowed with the tensor product filtration. If $\sigma \in S(\mathbb{C})$, we can similarly form $H_\sigma(X) = \bigotimes_j H_\sigma^{i_j}(X_j)$, and there is a canonical isomorphism:

$$\sigma_B : H_{DR}(X/R) \otimes_\sigma \mathbb{C} \to H_\sigma(X) \otimes \mathbb{C} .$$

If $\sigma \in S(W)$, $H_{DR}(X/R) \otimes_\sigma K$ has a Frobenius-linear endomorphism Φ_σ, obtained from the canonical isomorphism:

$$\sigma_{cris} : H_{DR}(X/R) \otimes_\sigma K \to H_{\bar{\sigma}}(X) .$$

(4.1) Terminology: Let ξ be an element of $H_{DR}(X/R)$.

(4.1.1) <u>If</u> $\sigma \in S(\mathbb{C})$, ξ <u>is a "Hodge class at</u> σ" <u>iff</u> $\sigma_B(\xi)$ <u>is a Hodge class in</u> $H_\sigma(X) \otimes \mathbb{C}$, <u>i.e. iff</u> $\sigma_B(\xi) \in (F^0 H_{\bar\sigma}(X) \otimes \mathbb{C}) \cap H_\sigma(X)$.

(4.1.2) <u>If</u> $\sigma \in S(\mathbf{W})$, ξ <u>is a "Tate class at</u> σ" <u>iff</u> $\sigma_{cris}(\xi)$ <u>is a Tate class in</u> $H_{\bar\sigma}(X)$, <u>i.e. iff</u> $\Phi\sigma_{cris}(\xi) = \sigma_{cris}(\xi)$.

(4.1.3) <u>The element</u> ξ <u>is "absolutely Hodge" iff it is a Hodge class at every</u> $\sigma \in S(\mathbb{C})$, <u>and is "absolutely Tate" iff it is a Tate class at every</u> $\sigma \in S(\mathbf{W})$ (<u>for every</u> $\mathbf{W}$).

(4.2) <u>Example</u>: If X is a smooth surface over R with second Betti number β, let $\Lambda_{DR} = H^2_{DR}(X/R)(\beta)$, which can be embedded naturally in $(H^2_{DR}(X/R) \otimes H^{-2}_{DR}(\mathbb{P}^1/R))^{\otimes\beta}$. The isomorphisms σ_β and σ_{cris} are compatible with the analogous embeddings in rational and in crystalline cohomology, and we deduce the existence of isomorphisms: $\sigma_B : \Lambda_{DR} \otimes_\sigma \mathbb{C} \to \Lambda_\sigma \otimes \mathbb{C}$ and $\Lambda_{DR} \otimes_\tau \mathbf{W} \to \Lambda_\tau$, for $\sigma \in S(\mathbb{C})$ and $\tau \in S(\mathbf{W})$. These isomorphisms are compatible with the cup product pairings. Suppose that ξ is an element of Λ_{DR} such that $(\xi,\xi) = (-1)^{\beta-1}$ (and note that such a ξ will always exist after a finite extension of R). Then necessarily $(\sigma_B(\xi), \sigma_B(\xi)) = (-1)^{\beta-1}$ also, which agrees with the self intersection of a generator for the highest exterior power of integral cohomology. It follows that $\sigma_B(\xi)$ must lie in Λ_σ, and hence is a Hodge cycle. Thus, ξ is absolutely Hodge. The statement that ξ be a Tate class at some $\tau \in S(\mathbf{W})$ reduces to Conjecture (3.11) for $X_{\bar\tau}$.

(4.3) <u>Remark</u>: "Hodgeness" of a cohomology class $\xi \in H_{DR}$ at some $\sigma \in S(\mathbb{C})$ says that in terms of the Hodge decomposition $H_\sigma \otimes \mathbb{C} = \oplus H^{pq}$, ξ lies in H^{00} and that its periods, computed with respect to a rational basis of homology, are rational numbers.

It is possible to describe "Tateness" in a similar way, which I would like to take the trouble to explain.

If $\sigma \in S(W)$, then $H_{\overline{\sigma}}$ is an F-isocrystal over the residue field k of W, and hence has a slope decomposition: $H_{\sigma} = \oplus\{H_{\lambda} : \lambda \in \mathbb{Q}\}$ [11]. Recall that if $\overline{k}$ is an algebraic closure of k and $\overline{K}$ is the fraction field of its Witt ring, then $(H_{\lambda} \otimes \overline{K}, \Phi)$ becomes isomorphic to a direct sum of copies of the standard F-isocrystal (S_{λ}, Φ) of slope λ. (If $\lambda = a/b$ in lowest terms, S_{λ} has a basis $x_1 \ldots x_b$ such that $\Phi(x_i) = x_{i+1}$ if $i < b$ and $\Phi(x_b) = p^a x_1$.)

We can define "p-adic periods" in several ways. Perhaps the simplest is the following. Let K_q denote the fraction field of the Witt ring of $\mathbb{F}_{p^b}$, regarded as a subfield of K. Clearly $V_{\lambda} =: \{x \in H_{\overline{\sigma}} \otimes \overline{K} : \Phi^b(x) = p^a x\}$ is a K_q-subspace of $H_{\lambda} \otimes \overline{K}$, and the natural map $V_{\lambda} \otimes_{K_q} \overline{K} \to H_{\lambda} \otimes \overline{K}$ is an isomorphism. Now if $\gamma \in \mathrm{Hom}_{K_q}(V_{\lambda}, K_q)$ and $\omega \in H_{DR}$, we let $\int_{\gamma}\omega$ denote $(\gamma, \sigma_{cris}(\omega))$, an element of $\overline{K}$. It is quite apparent from the definitions that:

(4.3.1) <u>If</u> $\omega \in H_{DR}(X/R)$, <u>then</u> ω <u>is a Tate class at</u> σ <u>iff</u> $\int_{\gamma}\omega = 0$ <u>for every</u> $\gamma \in V_{\lambda}^*$ <u>with</u> $\lambda \neq 0$ <u>and</u> $\int_{\gamma}\omega \in \mathbb{Q}_p$ <u>for every</u> $\gamma \in V_0^*$.

To pursue this digression further, note that $\mathrm{Gal}(\overline{k}/k)$ operates naturally on each $(H_{\lambda} \otimes \overline{K}, \Phi)$ and hence on $V_{\lambda} \subseteq H_{\lambda} \otimes \overline{K}$; this semi-linear action is linear if $\mathbb{F}_q \subseteq k$. (Recall that $q = p^b$ and $\lambda = a/b$.) Thus in this case we have a linear representation of $\mathrm{Gal}(\overline{k}/k)$ in $\mathrm{Gl}(V_{\lambda})$. If k is a finite field containing $\mathbb{F}_q$, this representation determines a conjugacy class in $\mathrm{Gl}(K_q, n_{\lambda})$, viz. the conjugacy class of (geometric) Frobenius, and in fact this conjugacy class determines the representation up to isomorphism. On the other hand, if $|k| = p^d$, then Φ^d is a K-linear endomorphism of H_λ, and it also defines a conjugacy class $\rho(\Phi^d)$ in $Gl(K, n_\lambda)$.

(4.3.2) Claim: The conjugacy class of geometric Frobenius in $Gl(K_q, n_\lambda) \subseteq Gl(K, n_\lambda)$ equals the conjugacy class $\rho(|k|^{-\lambda}\Phi^d)$.

Proof: It suffices to prove this over $\overline{K}$, since conjugacy classes in $Gl(n)$ don't change under field extension. But over $\overline{K}$, there is in fact a canonical isomorphism $V_\lambda \otimes_{K_q} \overline{K} \to H_\lambda \otimes_K \overline{K}$, and we shall see that this isomorphism carries geometric Frobenius $\rho(F) \otimes \text{id}$ to crystalline Frobenius $\Phi^d \otimes |k|^{-\lambda}$, i.e.:

(4.3.3) Lemma: The diagram below commutes.

$$\begin{array}{ccc} H_\lambda \otimes \overline{K} & \xrightarrow{\Phi^d_{cris} \otimes |k|^{-\lambda}} & H_\lambda \otimes \overline{K} \\ \cong \uparrow & & \uparrow \cong \\ V_\lambda \otimes \overline{K} & \xrightarrow{\rho(F) \otimes \text{id}} & V_\lambda \otimes \overline{K} \end{array}$$

Proof: It is convenient to convert the $\overline{K}$-linear maps in the diagram to $F^d_{\overline{K}}$-linear maps by composing on the right with the square of bijections:

$$\begin{array}{ccc} H_\lambda \otimes \overline{K} & \xrightarrow{\text{id} \otimes F^d_{\overline{K}}} & H_\lambda \otimes \overline{K} \\ \cong \uparrow & & \uparrow \cong \\ V_\lambda \otimes \overline{K} & \xrightarrow{\rho(\phi) \otimes F^d_{\overline{K}}} & V_\lambda \otimes \overline{K} \, . \end{array}$$

Here $\phi \in Gal(\bar{k}/k)$ is arithmetic Frobenius, which acts on $\bar{K}$ as $F^d_{\bar{K}}$. Thus, this square commutes by definition of the action of Galois. After composing the original square with this one, we obtain along the top the usual F^d_K-linear extension of $\Phi^d_{cris} \otimes |k|^{-\lambda}$ to $H_\lambda \otimes \bar{K}$. Along the bottom, we obtain the $F^d_{\bar{K}}$-linear extension of the identity endomorphism of V_λ, so commutativity reduces to checking that $\Phi^d(x) = |k|^\lambda x$ for $x \in V_\lambda$. Since $|k| = p^d$ and $q = p^b$, b divides d. By definition, $V_\lambda = \{x \in H_\lambda \otimes \bar{K} : \Phi^b x = p^{ax}\}$, hence $\Phi^d x = (\Phi^b)^{d/b}(x) = (p^a)^{d/b}(x) = p^{\lambda d}x$. □

(4.3.4) <u>Example</u>: Consider the Fermat variety of dimension n and degree m, and recall the well-known decomposition of its cohomology in terms of the action of $\Gamma = \oplus \mu_m/(\text{diag.})$. If we let $K = \mathbf{Q}(\mu_m)$ and work over K, the De Rham cohomology of X splits up into a sum of one-dimensional character spaces $H^n_{DR,\chi}$, [17,§3], and if σ is a place of K and $\mathbf{K}$ is the completion of K at σ, Φ_σ takes $H_{DR,\chi} \otimes \mathbf{K}$ to $H_{DR,p\chi} \otimes \mathbf{K}$. If the residue field $k(\sigma)$ is $\mathbf{F}_q$, with $q = p^d$, then m divides $q - 1$, $q\chi = \chi$, and so $H^n_{DR,\chi} \otimes \mathbf{K}$ is invariant under Φ^d_σ. Moreover, Φ^d_σ is $\mathbf{K}$-linear, and it is known that its eigenvalue on $H^n_{DR,\chi} \otimes \mathbf{K}$ is a "Gauss sum" $g(\chi)$ (c.f. [8,§12] for details and explanation). If ω_χ is a basis for $H_{DR,\chi}$ and γ is a basis for V^*_χ, then the period $\int_\gamma \omega$ belongs to $\bar{\mathbf{K}}$, and its image in $\bar{\mathbf{K}}^*/\mathbf{K}^*$ does not depend on the choices. Let $g_0(\chi)$ be $g(\chi)p^{-\text{ordg}(\chi)}$; then it follows from (4.3.2) that $(\int_\gamma \omega)^{1-F_q} = g_0(\chi)$. This determines the class of $\int_\gamma \omega$ in $\bar{\mathbf{K}}^*/\mathbf{K}^*$ uniquely. I like to think that this justifies a "p-adic period" interpretation of the Gross-Koblitz formulas [8].

Unfortunately it seems to be impossible at present to say anything nontrivial about absolute Tate cycles in the above context. Instead we are forced to consider a weaker notion, in which the ground ring R is replaced by a field k of characteristic zero. Any such field is a filtering direct limit of smooth $\mathbb{Z}$-algebras R, and any finite family X of smooth proper k-schemes will descend to smooth proper R-schemes over some suitable such R. Moreover any element in a tensor product $\xi \in H_{DR}(X/k) = H_{DR}(X_R/R) \otimes k$ will also descend to some (perhaps larger) such R. We will frequently make use of such descents without further explicit mention.

(4.4) Terminology: If k is a field of characteristic zero and $\xi \in H_{DR}(X/k)$, then:
(4.4.1) ξ is "absolutely Hodge" iff for some smooth $\mathbb{Z}$-algebra R contained in k, there is an absolutely Hodge class ξ_R in $H_{DR}(X_R/R)$ inducing ξ.
(4.4.2) ξ is "absolutely Tate" iff for some smooth $R/\mathbb{Z}$ in k, there is an absolutely Tate class $\xi_R \in H_{DR}(X_R/R)$ inducing ξ. We let "$H_{AH}(X/k)$" (resp. "$H_{AT}(X/k)$") denote the set of absolutely Hodge (resp. Tate) classes.

(4.5) Remark: Etale cohomology aside, Deligne's definition of "absolute Hodgeness" of a $\xi \in H_{DR}(X/k)$ is that for every embedding σ of k in $\mathbb{C}$, $\sigma_B(\xi)$ should be a Hodge class in $H_\sigma(X)$. This definition is apparently weaker than ours, but in fact the two are equivalent. Indeed, if ξ satisfies Deligne's definition we can find an $R/\mathbb{Z}$ such that X and ξ descend to R, and we shall see that ξ_R is then automatically absolutely Hodge (over R). First of all we may replace R by $R \otimes \mathbb{Q}$, since the corresponding schemes have the same $\mathbb{C}$-valued points. Then $gr_F.H_{DR}(X/R)$ is projective, and hence if $\sigma: k \subseteq \mathbb{C}$, $H_{DR}(X/R)/F^0 \subseteq (H_{DR}(X/R)/F^0) \otimes \mathbb{C} = (H_\sigma(X) \otimes \mathbb{C})/F^0$, and hence

$\xi_R \in F^0H_{DR}(X/R)$. We also know that for every embedding σ of k in $\mathbb{C}$, i.e. of R in $\mathbb{C}$, $\sigma_B(\xi_R) \in H_\sigma(X) \subseteq H_\sigma(X) \otimes \mathbb{C}$. This says that for every $\mathbb{C}$-valued generic point σ of the smooth $\mathbb{C}$-scheme $S_{\mathbb{C}} =: \mathrm{Spec}(R \otimes \mathbb{C})$, $\sigma_B(\xi) \in H_\sigma(X)$. On the complex analytic space $S_{\mathbb{C}}^{an}$ associated to $S_{\mathbb{C}}$, we have a local system of $\mathbb{Q}$-vector spaces $H_B = Rf_*\underline{\mathbb{Q}}$, where $f : X_\sigma^{an} \to S_{\mathbb{C}}^{an}$ is the structure map. Moreover, there is an isomorphism of coherent sheaves: $H_B \otimes \mathcal{O}_{S_{\mathbb{C}}^{an}} \overset{\mathcal{B}}{\leftarrow} (H_{DR}(X/R) \times \mathbb{C})^{an}$, and ξ_R "is" a global section of this sheaf. There is a canonical bijection $S(\mathbb{C}) = S_{\mathbb{C}}^{an}$, and evaluating $\mathcal{B}$ at some point $\sigma \in S(\mathbb{C})$ we just get back the isomorphism σ_B. Our assumption says that $\xi_R(\sigma)$ belongs to $H_B(\sigma)$ for every generic point σ of S; we want to prove that the same is true at every point. Let U be a small ball in $S_{\mathbb{C}}^{an}$ on which the local system H_B becomes constant; then the coordinates of $\xi|_U$ in terms of a basis for H_B are holomorphic functions on U which are rational at all points corresponding to generic points of S. Since these generic points are known to be dense, these functions are constant, hence rational at every point.□

(4.6) <u>Remark</u>: The above argument is an adaptation of Deligne's proof that an absolute Hodge cycle is horizontal, i.e. annihilated by the Gauss-Manin connection. As a consequence of this fact, one sees easily that if k' is a field containing k and if k is algebraically closed in k', then any absolutely Hodge class in $H_{DR}(X_{k'}/k')$ lies in $H_{DR}(X_k/k)$. We show below that a similar result holds for absolutely Tate cycles.

(4.7) <u>Proposition</u>: <u>An absolutely Tate class is horizontal under the action of the Gauss-Manin connection</u>.

Proof: To prove this, we may assume that ξ descends to an absolute Tate cycle ξ in $H_{DR}(X/R)$, where X/R and $R/\mathbb{Z}$ are smooth. Since the Gauss-Manin connection is integrable, it is enough to show that the restriction of ξ to every relative curve is horizontal, so we may assume that R has relative dimension one. Also, we may localize R around an arbitrary closed point $\overline{\tau}$. Then the residue field $K(\overline{\tau})$ at $\overline{\tau}$ is a finite field $\mathbb{F}_q$, and we may replace R by $R \otimes \mathbb{W}_q$ and choose a $\mathbb{W}_q$-valued point τ of R extending $\overline{\tau}$. The module with connection $(H_{DR}(X/R),\nabla)$ induces a crystal H on $\mathrm{Cris}(S_0/\mathbb{W}_q)$ ($S_0 = \underline{\mathrm{Spec}}(R \otimes \mathbb{F}_q)$, canonically isomorphic to the crystalline cohomology H_{cris} of the family X_0/S_0. This crystal is in fact an F-isocrystal, i.e. we have a map $\Phi : F_{S_0}^*(H_{cris}) \otimes \mathbb{Q} \to H_{cris} \otimes \mathbb{Q}$, which, after we multiply by the nth power of p, induces a map $\Phi : F_{S_0}^*(H_{cris}) \to H_{cris}$. The scheme $S = \underline{\mathrm{Spec}}\, R$ is a PD-thickening of S_0, and ξ can be viewed as a global section of $H_{DR} \cong H_S$ (the value of the crystal H on this particular PD-thickening).

Now choose a uniformizing parameter t at the point $\overline{\tau}$, so that $t(\overline{\tau}) = 0$ and $R/\mathbb{W}_q[t]$ is étale. Then there is a unique lifting F_S of the absolute Frobenius endomorhism of S_0 to S such that $F_S^*(t) = t^p$, and corresponding to F_S we have a horizontal map: $\Phi_S : F_S^*(H,\nabla) \to (H,\nabla)$. If $\sigma \in S(\mathbb{W}_{q'})$ is any Teichmuller point (i.e, a $\mathbb{W}_{q'}$-valued point such that $F_{\mathbb{W}_{q'}}^* \circ \sigma^* = \sigma^* \circ F_S^*$), then the F-isocrystal $(H_{\overline{\sigma}},\Phi_{\overline{\sigma}})$ is obtained by evaluating Φ_S at σ. Thus, our assumption on ξ implies that $\sigma^*(\Phi_S(1 \otimes \xi) - p^n\xi) = 0$ for all Teichmuller points σ.

It is clear that this implies that $\Phi_S(1 \otimes \xi) = p^n\xi$. Indeed, recall that any $\mathbb{F}_{q'}$-valued point $\overline{\sigma}$ of S can be lifted to a (unique) Teichmuller point σ, so that $\delta =: \Phi_S(1 \otimes \xi) - p^n\xi$ is killed by every σ. This implies that $\delta|_{S_0}$ vanishes at every closed point of S_0 and hence identically, so $\delta = p\delta'$ for some δ'. Repeating the argument, we find that δ is divisible by every power of p, hence vanishes.

Now we can easily prove that ξ is horizontal. It suffices to show, by induction on i, that $\nabla\xi \in (t^i)\Omega^1_S \otimes H$ for every i. Since $F^*_S(dt) = pt^{p-1}dt$, F^*_S maps Ω^1_S into $(t)\Omega^1_S$. Now $p^n\nabla\xi = \nabla p^n\xi = \nabla\Phi_S(1 \otimes \xi) = (F^*_S \otimes \Phi_S)(\nabla\xi)$. Thus if $\nabla\xi$ belongs to $(t^i)\Omega^1_S \otimes H$, $p^n\nabla\xi$ belongs to $(t^{i+1})\Omega^1_S \otimes H$, and hence so does $\nabla\xi$.□

(4.8) Corollary: If X/k is as above and k'/k is a field extension:

(4.8.1) If $\xi \in H_{DR}(X/k)$ becomes absolutely Tate in $H_{DR}(X_{k'}/k')$, it is already absolutely Tate in $H_{DR}(X/k)$.

(4.8.2) If k is algebraically closed, any absolutely Tate class of $H_{DR}(X_{k'}/k')$ lies in $H_{DR}(X/k)$.

(4.8.3) If k'/k is Galois with group G, then the set of absolute Tate classes in $H_{DR}(X_{k'}/k') \cong H_{DR}(X/k) \otimes k'$ is stable under the natural action of G, and its G-invariants are precisely the absolute Tate classes of $H_{DR}(X/k)$.

Proof: The first statement is practically obvious. Choose $R' \subseteq k'$ and $R \subseteq k$ so that $R \subseteq R'$, ξ is defined over R, and ξ is absolutely Tate in $H_{DR}(X_{R'}/R')$. Applying Chevalley's theorem and the theorem on generic smoothness, we see that we can localize R and R' and assume that the map of spectra $S' \to S$ is smooth and surjective. Then if $\sigma \in S(W_q)$, there exists a W'_q/W_q and a $\sigma' \in S'(W_{q'})$ lying over σ, and hence a canonical isomorphism $H_{\sigma'}(X_{R'}) \cong H_\sigma(X_R) \otimes W_{q'}$. Since this contains $H_\sigma(X_R)$, it is clear that $\sigma(\xi)$ is a Tate class.

Statement (4.8 2) follows from (4.7), since any element of $H_{DR}(X'/k')$ fixed by the Gauss-Manin connection is annihilated by any derivation of k'/k and hence lies in $H_{DR}(X/k)$. Statement (4.8.3) follows from the fact that absolute Tateness is preserved by base change and from (4.8.1). □

It is clear that the set $H_{AH}(X/k)$ of absolutely Hodge cycles forms a finite dimensional $\mathbb{Q}$-vector space, and that it is really "just a $\mathbb{Q}$-subspace" of $H_{DR}(X/k)$, i.e. that the map $H_{AH}(X/k) \otimes_{\mathbb{Q}} k \to H_{DR}(X/k)$ is injective (tensor with $\mathbb{C}$). The same is true for absolutely Tate cycles, although this result is less trivial to prove.

(4.9) Proposition: The natural map: $H_{AT}(X/k) \otimes_{\mathbb{Q}} k \xrightarrow{\nu} H_{DR}(X/k)$ is injective. In particular, $H_{AT}(X/k)$ is a finite dimensional $\mathbb{Q}$-vector space.

Proof: We may assume without loss of generality that k is algebraically closed. Endow $H_{AT}(X/k) \otimes k$ with the "constant" connection $id \otimes d_{k/\mathbb{Q}}$ and $H_{DR}(X/k)$ with the Gauss-Manin connection ∇. Then it follows from (4.7) that the map θ is compatible with the connections, and hence its kernel K is a horizontal subspace. As is well-known, this implies that K is defined over the algebraic closure of $\mathbb{Q}$ in k. Therefore it suffices to prove the injectivity of each restriction of ν to each $H_{AT}(X/k) \otimes_{\mathbb{Q}} k'$, for every finite Galois extension k' of $\mathbb{Q}$ contained in k. It even suffices to look at the restrictions of ν to each $H'_{AT}(X/k) \otimes k'$, where $H'_{AT}(X/k)$ is a finite dimensional $\mathbb{Q}$-subspace of $H_{AT}(X/k)$.

Choose a smooth k'-algebra R over which the elements of $H'_{AT}(X/k)$ are defined and absolutely Tate, and let m be a maximal ideal of R. Then $k'' =: R/m$ is a finite extension of $\mathbb{Q}$, and we obtain by specialization a set of k"-schemes $X_{k''}$, plus an isomorphism: $H_{DR}(X_R/R) \otimes_R R/m \cong H_{DR}(X_{k''}/k'')$. It is clear that under this isomorphism, $H'_{AT}(X/k)$ maps to $H'_{AT}(X_{k''}/k'')$. Moreover, it follows from (4.7) that this map is injective, and (hence) so is the map $H'_{AT}(X/k) \otimes k'' \to H_{AT}(X_{k''}/k'') \otimes k''$. We have a commutative diagram:

$$\begin{array}{ccc} H'_{AT}(X/k) \otimes k' & \xrightarrow{\nu'} & H_{DR}(X_R/R) \\ \downarrow & & \downarrow \\ H_{AT}(X_{k''}/k'') \otimes k'' & \xrightarrow{\nu''} & H_{DR}(X_{k''}/k'') \end{array}$$

Since $H_{DR}(X_R/R) \subseteq H_{DR}(X_k/k)$, it suffices to show that ν' is injective, and from the diagram we see that it is enough to prove that ν'' is injective. In other words, we are reduced to the case of schemes defined over a numberfield.

Changing notation, we assume X is defined over a finite Galois extension k of $\mathbb{Q}$, with ring of integers O. Choose N so that if $R =: O[1/N]$, X and H_{AT} exist over R. To prove that $H_{AT}(X/k) \otimes k \to H_{DT}(X/k)$ is injective, it suffices to prove that its kernel K is defined over $\mathbb{Q}$, i.e. that $K = (K \cap H_{AT}(X/k)) \otimes k$, since $H_{AT}(X/k)$ is by definition contained in $H_{DR}(X/k)$. We shall show that if $g \in Gal(k/\mathbb{Q})$, K is invariant under $id \otimes g$. Chebataroff assures us that there exists an unramified prime p of R such that g is the Frobenius element ϕ_p of $Gal(k/\mathbb{Q})$ and in particular leaves the completion k_p of k at p invariant. Clearly it suffices to show that $K \otimes k_p$ is g-invariant. Of course, the residue field $k(p)$ is a finite field $\mathbb{F}_q$, $k_p \cong W_q$, and we can identify $k \to k_p$ with a W_q-valued point σ of S=: Spec R. Moreover, g acts on $k \cong W_q$ as the (absolute) Frobenius endomorphism of W_q. By the definition of absolute Tate cycles, the map $H_{AT}(X/R) \otimes k_p \to H_{DR}(X/R) \otimes_\sigma K_q$ therefore takes $(1 \otimes \phi_p)$ to Φ_σ, and hence its kernel must indeed be invariant under $1 \otimes \phi_p$. □

If X is defined over k and $\bar{k}$ is an algebraic closure of k, we see from (4.8.3) and (4.9) that $H_{AT}(X_{\bar{k}}/\bar{k})$ is a finite dimensional $\mathbb{Q}$-vector space on which $Gal(\bar{k}/k)$ operates continuously. Suppose k is a numberfield, and choose a finite Galois extension k' of k through which this action factors. Then the technique above together with (4.3.3) gives the following formula for the action of a geometric Frobenius element $F_{p'} \in Gal(k'/k)$ at a suitable prime p' of k', lying over p of k.

(4.10) Corollary: For almost all unramified primes p' of k', the action of geometric Frobenius $F_{p'}$ on $H_{AT}(X_{\bar{k}}/\bar{k}) \subseteq H_{DR}(X/k) \otimes k'_{p'}$ is compatible with the action of $\Phi_p^{\deg(p)} \otimes$ id, where Φ_p is the action of absolute Frobenius on crystalline cohomology. □

The obvious difficulty with the notion of absolute Tate cycles is to produce some. It is clear that cohomology classes of algebraic cycles will be absolutely Tate, and it is also easy to see that the Kunneth components of the diagonal are absolutely Tate, although they are not known to be algebraic. The value of the notion lies in the fact that it is easier to construct absolute Tate cycles than algebraic cycles, as we shall see in our attempt to analyze the following hopes.

(4.11) Hopes: Let k be a field of characteristic zero and let X be a finite family of smooth k-schemes and indices.

(4.11.1) (Deligne) If $\xi \in H_{DR}(X/k)$ is a Hodge class at some $\sigma : k \to \mathbb{C}$, then it is absolutely Hodge.

(4.11.2) Any absolutely Hodge class is absolutely Tate.

(4.11.3) Any absolutely Tate class is absolutely Hodge.

Deligne's main theorem on absolute Hodge cycles asserts that (with his stronger definition of "absolute Hodge" involving also étale cohomology) if the X_i's are abelian varieties, then (4.11.1) is true. His proof uses two principles of construction which can easily be made to work for absolutely Tate cycles.

(4.12) Construction: Let k be an algebraically closed field of characteristic zero, Y a smooth affine connected k-scheme, and X a finite family of smooth proper Y-schemes and indices. Let $\xi \in H_{DR}(X/Y)$ be an element such that:

(a) $\nabla\xi = 0$, where ∇ is the Gauss-Manin connection.

(b) For some $y_0 \in Y(k)$, $y_0^*(\xi) \in H_{DR}(X(y_0)/k)$ is absolutely Tate. Then $y^*(\xi)$ is absolutely Tate for every $y \in Y(k)$.

Proof: As usual, we may choose a smooth $\mathbb{Z}$-algebra $R \subseteq k$ such that everything descends to $S =: \text{Spec } R$. Notice that since the geometric generic fiber of Y_S/S is connected, the same is true of all geometric fibers in some nonempty open set of S. (This follows, for instance, from [4], although it is of course much more elementary.) Localizing, we may assume that all the fibers are geometrically connected, and also that y_0 and y are given by two sections of Y_S/S, and that $y_0^*(\xi)$ is absolutely Tate on S. I claim that then $y^*(\xi)$ is absolutely Tate on S as well.. Indeed, let σ be a W_q-valued point of S; then Y_σ is a smooth W_q-scheme, and ξ is a horizontal section of the module with integrable connection $H_{DR}(X_S/Y_S) \otimes W_q$. Choose a lifting F of absolute Frobenius to the p-adic completion $\hat{Y}_\sigma$ of Y, and let $\Phi_F : F^*H \to H$ be the corresponding horizontal map. Since ξ is horizontal, $\Phi_F(1 \otimes \xi)$ is also horizontal and furthermore is independent of the choice of F [7, 1.1.3.4]. I claim that in fact $\Phi_F(1 \otimes \xi) = \xi$. Since $Y_{\bar{\sigma}}$ is irreducible, $\hat{Y}_\sigma$ is integral, and we can check this after restricting to any nonempty open set. We may, for example, assume that Y is étale over Spec $W[t_1 \ldots t_n]$, where the t_i's are uniformizing parameters along the section y_0. But then there exists a lifting F' of Frobenius such that y_0 is a Teichmuller point, and hence $\Phi_{F'}(1 \otimes \xi) = \xi$ at y_0. Since both these elements are horizontal, it follows that they are equal everywhere, and hence the same is true for an arbitrary lifting F. Applying the same reasoning at the section y, we see that ξ is also a Tate class at y. □

The second construction method is slightly more subtle and requires a bit of preliminary explanation. Let X be a family of smooth proper k-schemes and indices, as above, say $X = \{X_1 \ldots . X_m, i_1 \ldots i_m\}$. Fix an embedding σ of k in $\mathbb{C}$, and suppose $\{\omega_1 \ldots \omega_n\}$ is a finite set of absolutely Hodge-Tate classes living in various tensor functors $T^i_{DR}(X/k)$, say $T^i_{DR}(X/k) = H^{i_1}_{DR}(X_1/k)^{\otimes n_1(i)} \otimes H^{i_2}_{DR}(X_2/k)^{\otimes n_2(i)} \otimes \ldots .$ As usual, the complex embedding σ gives us similarly constructed $\mathbb{Q}$-vector spaces $T^i_\sigma(X)$ and isomorphisms: σ_B: $T^i_{DR}(X/k) \otimes \mathbb{C} \to T^i_\sigma(X) \otimes \mathbb{C}$. Let G be the algebraic group contained in $\prod_j \mathrm{Aut}_{\mathbb{Q}}(H^{i_j}_\sigma(X_j))$ stabilizing the $\sigma_B(\omega_i)$'s. It is clear that this is an algebraic group over $\mathbb{Q}$; if A is any $\mathbb{Q}$-algebra, $\underline{G}(A) = \{g \in \prod_j \mathrm{Aut}_A(H\sigma^{i_j}(X_j) \otimes A : g(\sigma_B(\omega_i)) = \sigma_B(\omega_i)$ for all $i\}$.

(4.13) <u>Construction</u>: <u>With the above notation, suppose</u> $T_{DR}(X/k)$ <u>is another tensor space and</u> $\omega_\sigma \in T_\sigma(X/k)$ <u>is fixed by</u> G. <u>Then if</u> k <u>is algebraically closed, there is an absolutely Hodge-Tate class</u> $\omega \in T_{DR}(X/k)$ <u>such that</u> $\sigma_B(\omega) = \omega_\sigma$.

<u>Proof</u>: Assume that all the X_j's and ω_i's are defined over a smooth $\mathbb{Z}$-algebra R. If A is any $\mathbb{Q} \otimes$ R-algebra, let <u>P</u>(A) denote the set of all families $\{\delta_j\}$ of isomorphisms: $\delta_j : H_{DR}^{i_j}(X_j/R) \otimes_R A \xrightarrow{\cong} H_\sigma^{i_j}(X_j) \otimes_{\mathbb{Q}} A$ such that $\delta_j(\omega_i \otimes 1) = \sigma_B(\omega_i) \otimes 1$ for all i. It is clear that <u>P</u> is represented by an R $\otimes$ $\mathbb{Q}$-scheme P of finite type. Moreover, the canonical maps σ_B define a $\mathbb{C}$-valued point of P, so P is not empty. By the Nullstellensatz, there is a k-valued point δ of P; after changing R, we may even assume that $\delta \in P(R \otimes \mathbb{Q})$. Notice that <u>P</u> is a torsor under the obvious action of G, so our assumption on ω_σ implies that $\omega =: \delta^{-1}(\omega_\sigma \otimes 1)$ is independent of the choice of δ.

Deligne explains that ω is absolutely Hodge; here is why it is absolutely Tate. After localizing R, we may assume that ω is defined over R and that all the ω_i's are absolutely Tate over R. We have to prove that if τ is a W_q-valued point of <u>Spec</u>(R), then the image $\omega \otimes 1$ of ω in $T_{DR}(X/R) \otimes K_q$ is fixed by Φ_τ. To do this, let us consider the Frobenius-linear endomorphism $\psi_\sigma =: \mathrm{id} \otimes F_{K_q}^*$ of $H_\sigma^{i_j}(X_j) \otimes K_q$, which acts by functoriality on the various tensor spaces, fixing the rational classes and in particular the $\sigma_B(\omega_i \otimes 1)$'s and $\omega_\sigma \otimes 1$. Our chosen δ pulls back to a set of isomorphisms $\tau^*(\delta_j): H_{DR}^{i_j}(X_j/R) \otimes K_q \to H_\sigma^{i_j}(X_j) \otimes K_q$, taking the $\omega_i \otimes 1$'s to the $\sigma_B(\omega_i \otimes 1)$'s and $\omega \otimes 1$ to $\omega_\sigma \otimes 1$. Let $\delta': H_{DR}^{i_j}(X_j/R) \otimes K_q \to H_\sigma^{i_j}(X_j) \otimes K_q$ be $\psi_\sigma \circ \tau^*(\delta) \circ \Phi_\tau^{-1}$. It is clear that δ' is K_q-linear and takes each $\omega_i \otimes 1$ to $\sigma_B(\omega_i \otimes 1)$

since each ω_i is absolutely Tate and rational. Thus, δ' is another K_q-valued point of $\underline{P}$, and hence $\delta' = g.\tau^*(\delta)$ for some $g \in G(K_q)$. This implies that $\delta'(\omega \otimes 1) = \tau^*(\delta)(\omega \otimes 1) = \omega_\sigma \otimes 1$, i.e. $\psi_\sigma \tau^*(\delta)\Phi_\tau^{-1}(\omega \otimes 1) = \omega_\sigma \otimes 1$. Since ω_σ is rational, it is fixed by ψ_σ^{-1}, so $\tau^*(\delta)\Phi_\tau^{-1}(\omega \otimes 1) = \omega_\sigma = \tau^*(\delta)(\omega \otimes 1)$, and hence $\Phi_\tau(\omega \otimes 1) = \omega \otimes 1$. □

Deligne has shown [6] that repeated applications of the construction principles (4.12) and (4.13) obtain every Hodge cycle in $H_B(X)$ if the X_i's are abelian varieties, Fermat hypersurfaces, or K3 surfaces over $\mathbb{C}$. In our case, we can conclude:

(4.14) Theorem: Hopes (4.11.1) and (4.11.2) are true for abelian varieties, Fermat hypersurfaces, K3-surfaces, and projective spaces.

Let us now take a brief look at Hope (4.11.3). As evidence, we can offer one general result and one specific case.

(4.15) Proposition: If $\xi \in H_{DR}(X/k)$ is absolutely Tate, then $\xi \in F^0 H_{DR}(X/k)$. Furthermore, there exists an $R/\mathbb{Z}$ such that for every closed point $s \in \mathrm{Spec}(R)$, $\xi(s) \in F^0_{con} H_{DR}(X(s)/k(s))$. That is, $\xi \in C^0_{DR}(X/k)$ in the notation of (2.15).

Proof: This is based on the remarkable theorem of Mazur [1, §8, 15], which asserts that the mod p Hodge filtration and the conjugate filtration of a variety over a perfect field k of characteristic p are equal to the mod p Hodge and conjugate filtrations of the corresponding F-crystal. In particular, if Y/W is a smooth proper W-scheme with torsion-free Hodge groups and if Y_0/k is its reduction mod p, then:

$$F^jH^i_{DR}(Y_0/k) = \mathrm{Im}\{\xi \in H^i_{cris}(Y_0/W) : \Phi(\xi) \in p^jH^i_{cris}(Y_0/W)\} ,$$

and

$$F^j_{con}H^i(Y_0/k) = \mathrm{Im}\ \{p^{-j}\Phi(\xi) : \Phi(\xi) \in p^jH^i_{cris}(Y_0/W)\} .$$

It is clear that the Hodge and conjugate filtrations of an F-crystal constructed as a tensor product of two F-crystals are just the tensor product filtrations of the corresponding Hodge and conjugate filtrations. Moreover, if H is an F-crystal, then although its dual H^* is not defined over W in general, a suitable twist $H^*(-n)$ will be, and hence will have Hodge and conjugate filtrations. We then can define the Hodge and conjugate filtrations on $H^* \otimes k$ by shifting back--$F^jH^* \otimes k =: F^{j+n}(H^*(-n) \otimes k)$--and it is easy to see that the result is independent of the choice of n. Thus, we can extend Mazur's result to tensor functors of the form we are considering. But it then is clear from the definitions that if $\xi \in H_{DR}(X/R)$ is a Tate cycle at σ, then (mod p), $\xi(\bar{\sigma}) \in F^0_{con}H_{DR}(X(\bar{\sigma})/k(\bar{\sigma}))$ and $\in F^0H_{DR}(X(\bar{\sigma})/k(\bar{\sigma}))$. This shows that $\xi \in C^0H_{DR}(X/k)$ and since we may assume that $H_{DR}(X/R)/F^0$ is torsion-free (and finitely generated), that $\xi \in F^0H_{DR}(X/R) \subseteq F^0H_{DR}(X/k)$. □

(4.16) <u>Theorem</u>: <u>Hope</u> (4.11.3) <u>is true if the</u> X_i's <u>are all abelian varieties of CM type, Fermat hypersurfaces, or projective spaces</u>.

<u>Proof</u>: We know from (4.14) (Hope (4.11.2)) that $T_{AH}(X/k) \subseteq T_{AT}(X/k)$, so to prove they are equal, it suffices to prove that they have the same dimension over $\mathbb{Q}$. We may assume that k is algebraically closed, since both satisfy Galois descent,

and we may even assume that $k = \mathbb{C}$, by (4.8.2) (which Deligne has also proved for absolutely Hodge cycles). Since the maps: $H_{AH}(X/\mathbb{C}) \otimes \mathbb{C} \to H_{DR}(X/\mathbb{C})$ and $H_{AT}(X/\mathbb{C}) \otimes \mathbb{C} \to H_{DR}(X/C)$ are injective, it suffices to prove that they have the same image. Deligne's result that (4.11.1) holds tells us that the image of $H_{AH}(X/\mathbb{C}) \otimes \mathbb{C}$ is $N^0H_{DR}(X/\mathbb{C})$, the complex span of the Hodge cycles. On the other hand, (4.15) tells us that the image of $T_{AT}(X/\mathbb{C}) \otimes \mathbb{C}$ is contained in $C^0_{DR}(X/\mathbb{C})$. But a completely straightforward generalization of (2.15) shows that if the X_i's are as above, then $C^0_{DR}(X/\mathbb{C}) = N^0_{DR}(X/\mathbb{C})$. □

References

1. Berthelot, P., and Ogus, A. Notes on Crystalline Cohomology, Princeton University Press, Princeton, N.J. (1978)

2. Bloch, S., and Ogus, A. "Gersten's conjecture and the homology of schemes" Ann Sci. E.N.S., 4e serie, t. 7(1974) 181-202

3. Deligne, P. "Relèvement des surfaces K3 en caracteristique nulle" to appear

4. Deligne, P. "Théorèmes de finitude en cohomologie l-adique" SGA 4-1/2, Lecture Notes in Math. No. 569, Springer, N.Y. (1977)

5. Deligne, P. "La conjecture de Weil pour les surfaces K3" Inv. Math. 15 (1972) 206-226

6. Deligne, P. "Hodge cycles on abelian varieties" this volume

7. Deligne, P., and Illusie, L. "Cristaux ordinaires et coordonnées canoniques" to appear

8. Gross, B. and Koblitz, N. "Gauss sums and the p-adic Γ-function" Ann. of Math. 109 (1979) 569-581

9. Katz, N. "Algebraic solutions of differential equations" Inv. Math. 18 (1972) 1-118

10. Katz, N. "Travaux de Dwork" Sem. Bourbaki No. 409 (1971-72), Lecture Notes in Math. No. 317, Springer, N.Y. 1973

11. Katz, H. "Slope filtration of F-crystals" Asterisque No. 64 (1979) 113-164

12. Koblitz, N. A Short Course on Some Current Research in p-adic Analysis to appear

13. Koblitz, N. and Ogus, A. "Algebraicity of some products of values of the Γ-function" (appendix to "Valeurs de fonctions L et periodes d'integrales" in Automorphic Forms, Respresentations, and L-functions, Prod. of Symposia in Pure Math. Vol. XXXIII part 2, A.M.S. pp. 343-346

14. Kuga, M. and Satake, I. "Abelian varieties attached to polarized K3 surfaces" Math Ann. 169 (1967) 239-242

15. Mazur, B. "Frobenius and the Hodge filtration (estimates)" Ann of Math. 98 (1975) 58-95

16. Ogus, A. "F-crystals and Griffiths transversality" Proceedings of the International Symposium on Algebraic Geometry, Kyoto, 1977, Kinokuniya Book-Store, Tokyo (1978) 15-44

17. Ogus, A. "Griffiths transversality in crystalline cohomology" Ann. of Math. 108 (1978) 395-419

18. Ogus, A. "Supersingular K3 crystals" Asterisque no. 64, 3-86

19. Pjatekii-Sapiro, I. and Safarevic, I. "A Torelli theorem for algebraic surfaces of type K3" Math. U.S.S.R. Izvestija vol. 5 (1971) No. 3.

20. Serre, J.P. Abelian l-adic represenations and elliptic curves Benjamin, N.Y. (1968)

21. Weil, A. Variétés Kaehleriennes Hermann, Paris (1958).

ADDENDUM 1989

The theory developed in the first article of this volume is applied in [4] to give explicit relations between the periods of abelian varieties. In [10] it is applied to the study of the periods of the motives attached to Hecke characters.

The gap in the theory of Tannakian categories noted on p 160 (namely, the lack of a proof that two fibre functors are locally isomorphic for the fpqc topology) is filled in [5] (under the necessary assumption that $k = \mathrm{End}^{\otimes}(1)$).

An extension of the Taniyama group is constructed in [1], and is applied to prove a relation between the critical values of certain Hecke L-functions and values of the classical gamma function.

The explicit rule of J. Tate mentioned on p263 is contained in an unpublished manuscript of Tate, whose contents can be found in §1 – §3 of Chapter VII of [6]. The last section of the same chapter (based on a letter from Deligne to Tate) gives a slightly different approach to the proof of the main theorem of article IV of this volume.

Langlands's conjecture (p311) is proved in complete generality in ([2], [3]) and in [7]. The theorem of Kazhdan on which these works are based has been given a simpler proof in [9]. The survey article [8] reviews some of the material in this volume, and explains how the Taniyama group (together with the period torsor) controls the rationality properties of holomorphic automorphic forms and other objects.

Blasius (unpublished) has proved results related to the conjectures in article VI.

[1] Anderson, G., Cyclotomy and an extension of the Taniyama group, Comp. Math., 57 (1986), 153–217.

[2] Borovoi, M., Langlands's conjecture concerning conjugation of Shimura varieties, Sel. Math. Sov. 3 (1983/4), 3–39.

[3] Borovoi, M., On the group of points of a semisimple group over a totally real field, in Problems in Group Theory and Homological Algebra, Yaroslavl, 1987, 142–149.

[4] Deligne, P., Cycles de Hodge absolus et périodes des intégrals des variétiés abéliennes, Soc. Math. de France, Mémoire no. 2, 1980, 23–33.

[5] Deligne, P., Catégories tannakiennes, in Grothendieck Festschrift, Progress in Math., Birkhäuser, (to appear, 1989).

[6] Lang, S., Complex Multiplication, Springer, Heidelberg, 1983.

[7] Milne, J., The action of an automorphism of **C** on a Shimura variety and its special points, Progress in Math., 35, Birkhäuser, 1983, 239–265.

[8] Milne, J., Canonical models of (mixed) Shimura varieties and automorphic vector bundles, in Automorphic Forms, Shimura Varieties, and L-functions, (Editors Laurent Clozel and J.S. Milne), Academic Press, Boston, 1989, Vol I, 283–414.

[9] Nori, M., and Raghunathan, M., On the conjugation of Shimura varieties (preprint 1989).

[10] Schappacher, N,. On the Periods of Hecke Characters, Lecture Notes in Math., Springer, Heidelberg, 1988.

ERRATUM 1989

CONTENTS: The title of the third article is "Langlands's construction of the Taniyama group".

General Introduction: The authors considered it so well-known that Grothendieck was the originator of the theory of motives and the theory of Tannakian categories that they neglected to mention it; perhaps they should have.

p8$_2$: motivic Galois group

p15$_4$: This is not quite so transparent as the "and so" suggests.

p21^8: $0 \longrightarrow \mathcal{O}_X^\times \longrightarrow 0 \longrightarrow \dots$

p27$_{11}$: ... and remain true, if ...

p28$_2$: $H^i(X)(d)$

p42^3: from

p42^{12}: The complex conjugate $\overline{\mu(\lambda)}$ of $\mu(\lambda)$ satisfies $\overline{\mu(\lambda)}\cdot v^{pq} = \bar{\lambda}^{-q}\cdot v^{pq}$.

p43$_9$: It is more natural to let ν act as ν.

p45^6: complex conjugation on $H_\sigma(\mathbb{C})$ corresponds to $\sigma\circ$(complex conjugation) on $H(\mathbb{C})$.

p56$_9$: and an

p61^9: to $\psi = \mathrm{Tr}_{E/Q}(f\varphi)$.

p75^2: There is no need to refer to Borel-Springer for the proof, since it is given in the remainder of the paragraph.

p80$_3$: When all $a_i = 0$, the dimension of $H^n(V,\mathbb{C})_{\mathbf{a}}$ is 1 only if n is even; otherwise it is zero.

p85$_6$: Replace F_q^{n+1} with F_q^{n+2}.

p85$_5$: Replace P^n with P^{n+1}.

p89^3: $\Sigma a_i \equiv 0 \pmod d$.

p98^{15}: Springer.

p101^6: Replace 149 with 147.

p104^3: $(X,Y) \mapsto X\otimes Y$.

p119^1: $(\underline{C},\otimes)$

p124$_9$: indeterminate

p147$_7$: form

p148^{10}: representable

p154$_7$: if and only if

p157^5: $\underline{\mathrm{Aut}}^\otimes(\omega)$

p168$_4$: $\xrightarrow{1\otimes a^{-1}}$

p198^4: $H^{2r-s}(X)$

p199^{10}: $\xrightarrow{\mathrm{id}\otimes\ *}$

p216$_8$: [2.0.10]

p218$_8$: Kuga-Satake

p231^{11}: For any L Galois over $\mathbb{Q}$,

p232^7: $\lambda(\iota\sigma) + \lambda(\sigma)$

p232^9: $\Lambda^L \subset \Lambda^F$ where $F = L\cap\mathbb{Q}^{cm}$

p232^{11}: $\Lambda^L \supset \Lambda^F$

p232$_1$: The diagram should be:

$$\begin{array}{ccccccc} & & F^\times/F_0^\times & \xrightarrow{\approx} & S^F/hw(\mathbb{Q}^\times) & & \\ & & \uparrow & & \uparrow & & \\ 1\to \mathrm{Ker} & \longrightarrow & F^\times & \longrightarrow & S^F & & \to 1 \\ \uparrow\approx & & \uparrow & & \uparrow hw & & \\ 1\to \mathrm{Ker} & \longrightarrow & F_0^\times & \xrightarrow{\text{norm}} & \mathbb{Q}^\times & & \to 1 \end{array}$$

p259^1: Delete the second b from the first diagram.

p264$_{14}$: $z^{-p}\bar{z}^{-q}$

p271^2: ${}^K S^\circ$

p286$_4$: $\phi^0(\tau;\mu',\mu)\circ\phi^0_{\tau,\mu} = \phi^0_{\tau,\mu'}$.

p331^1: Delete "Shimura Varieties V.7"

p343^4: being in $G^{ad}(\mathbb{R})^+$.

p381^{14}: $\mathrm{disco}(H_d)$

General Remarks

Lecture Notes are printed by photo-offset from the master-copy delivered in camera-ready form by the authors. For this purpose Springer-Verlag provides technical instructions for the preparation of manuscripts.

Careful preparation of manuscripts will help keep production time short and ensure a satisfactory appearance of the finished book. The actual production of a Lecture Notes volume normally takes approximately 8 weeks.

Authors receive 50 free copies of their book. No royalty is paid on Lecture Notes volumes.

Authors are entitled to purchase further copies of their book and other Springer mathematics books for their personal use, at a discount of 33,3 % directly from Springer-Verlag.

Commitment to publish is made by letter of intent rather than by signing a formal contract. Springer-Verlag secures the copyright for each volume.

Addresses:

Professor A. Dold
Mathematisches Institut
Universität Heidelberg
Im Neuenheimer Feld 288
D-69120 Heidelberg
Federal Republic of Germany

Professor B. Eckmann
Mathematik, ETH-Zentrum
CH-8092 Zürich, Switzerland

Springer-Verlag, Mathematics Editorial
Tiergartenstr. 17
D-69121 Heidelberg
Federal Republic of Germany
Tel.: *49 (6221) 487-410